FUNDAMENTALS OF RENEWABLE ENERGY

About the Authors

Prof. N. S. Rathore is Vice Chancellor, Maharana Pratap University of Agriculture and Technology, Udaipur. He did M.Tech with specialization in Energy Studies from I.I.T., Delhi and Ph.D. from Rajasthan Agricultural University, Bikaner. He has served the ICAR as Deputy Director General (Engg.), Deputy Director General (Education). He is also founder Vice Chancellor of Sri Karan Narendra Agriculture University, Jobner- Jaipur, Rajasthan. He has contributed more than 200 technical papers and 36 books on various aspects of Renewable Energy, Environment and Agricultural Engineering. He has undertaken 50 research projects and organized 20 International trainings, Seminars/winter, Workshops and Conferences and 12 ISTE sponsored Summer School. There are 100 reports, proceedings and popular articles to his credit. He was honored by the Indian Society of Agricultural Engineers and other government and private organization.

Dr. N.L. Panwar is an Assistant Professor of Renewable Energy Engineering in the Faculty of Engineering at MPUAT, Udaipur, Rajasthan (India). He has been awarded PhD from the Centre for Energy Studies, Indian Institute of Technology Delhi, India. His areas of interest cover energy and energy analyses of thermal systems and energy efficiency & management. He has contributed more than 100 papers in international journals and several books on renewable energy aspect. He has been a recipient of ‘Prakritik Urja Puraskar’ from Ministry of New and Renewable Energy, Government of India for his outstanding book on Alternative Energy Resources. He has also been awarded by “Shrimati Vijay-Usha Sodha Research Award" from Indian Institute of Technology Delhi in 2014, and Rajasthan Energy Conservation Award 2018 from Government of Rajasthan.

FUNDAMENTALS OF RENEWABLE ENERGY

N.S. Rathore
Vice Chancellor
Maharana Pratap University of Agriculture and Technology
Udaipur (Rajasthan) India

Formerly
Vice Chancellor
Sri Karan Narendra Agriculture University, Jobner- Jaipur, Rajasthan
Deputy Director General (Education)
Deputy Director General (Agricultural Engineering)
Indian Council of Agricultural Research
(New Delhi) India

N.L. Panwar
Department of Renewable Energy Engineering
Maharana Pratap University of Agriculture and Technology
Udaipur (Rajasthan) India

NEW INDIA PUBLISHING AGENCY
New Delhi-110 034

NEW INDIA PUBLISHING AGENCY
101, Vikas Surya Plaza, CU Block, LSC Market
Pitam Pura, New Delhi – 110 034, India
Email: info@nipabooks.com
Web: www.nipabooks.com

For customer assistance, please contact
Phone: + 91-11-27 34 17 17 Fax: + 91-11- 27 34 16 16
E-Mail: feedbacks@nipabooks.com

ISBN: 978-93-89907-41-4

Composed and Designed by NIPA.

Preface

Energy is acting as wheels for the national economic development and its progress. Rapid industrialization, infrastructural growth, population increase and our expanding economy have stressed the need to acquire additional sources of energy. However, conventional energy reserves are finite in shape and will only be available for a limited period and with steadily increasing price. The fossil fuels use is also linked to environmental problems particularly global warming and climate change. The challenges of the present energy scene offer us a window of opportunity in the form of renewable energy sources to reduce dependency on fossil fuels by expanding and diversifying our energy supply mix and shifting the development part towards sustainability as well as environmental and social responsibility. To achieve this objective, the knowledge of basics and fundamentals of renewable energy technology is essential. In this context, this book presents a systematic collection of all aspects of renewable energy resources & devices.

This book is a ready reference material for students from a wide variety of engineering and science backgrounds wish to study the engineering aspects of renewable energy engineering. Number of examples has been included in the text for better understanding of matter.

The contents of book include seven major aspects of non- conventional energy sources such as solar energy, solar thermal energy technology and devices, solar photovoltaic technology, biomass production, utilization and management, gasification technology, wind energy and other renewable energy sources such as geothermal, ocean energy etc.

This book has been divided into seventeen chapters which cover wide spectrum of renewable energy engineering. Most of the information presented in this book reflects a basis to acquire the understanding of the renewable energy sources. In this book, all fundamentals, design, present state of art of technology and future prospective are illustrated through graphs, figures, tables, flowchart, equations etc. to make subject clear and useful. This can be used as text book for students in undergraduate and post graduate degree, and persons working in the field of renewable energy.

Authors

Contents

Nomenclature

A_{ab}	Absorber base area (m^2)
A_c	Area of the tunnel dryer cover material (m^2)
A_f	Floor area (m^2)
A_{gl}	Area of lower glass cover (m^2)
A_{gu}	Area of upper glass cover (m^2)
A_{in}	Cross section area of the air inlet to tunnel dryer (m^2)
A_{out}	Cross section area of the air outlet from tunnel dryer (m^2)
A_p	Area of the drying product (m^2)
A_v	Vessel area (m^2)
A_{vb}	Area of vessel base (m^2)
A_{vc}	Area of vessel cover (m^2)
A_{vf}	Wetted fluid area of vessel (m^2)
C_d	Discharge coefficient
C_p	Air specific heat (J kg^{-1} K^{-1})
C_{pa}	Specific heat of air (J kg^{-1} K^{-1})
C_{pc}	Specific heat of cover material of tunnel dryer (J kg^{-1} K^{-1})
C_{pf}	Specific heat of tunnel dryer floor (J kg^{-1} K^{-1})
C_{pl}	Specific heat of liquid (J kg^{-1} K^{-1})
C_{pp}	Specific heat of drying product (J kg^{-1} K^{-1})
C_{pv}	Specific heat of water vapour (J kg^{-1} K^{-1})
D_p	Thickness of the drying product (m)
F_p	Fraction of solar radiation falling on the drying product (decimal)
H_{in}	Humidity ratio of air entering tunnel dryer
H_{out}	Humidity ratio of the air leaving from tunnel
I_s	Incident solar radiation (Wm^{-2})
L_p	Latent heat of vaporization of moisture from drying product (J kg^{-1})
M_e	Equilibrium moisture content of drying product (% db)
M_o	Initial moisture content of drying product(% db)
M_p	Moisture content of dry product (% db)
P_{vp}	Vapour pressure inside the air (kPa)
P_{ws}	Air saturated vapour pressure inside air (kPa)
S_o	Solar flux to cooker (W m^{-2})
T_∞	Reference temperature (K)
T_a	Air temperature in the dryer (K)
T_{ab}	Temperature of absorber plate (K)
T_{air}	Temperature of cooker air (K)
T_{am}	Ambient temperature (K)

T_c	Cover temperature of tunnel dryer (K)
T_{cf}	Temperature of cooker fluid (K)
T_f	Floor temperature of tunnel dryer (K)
T_{gl}	Temperature of lower glass cover (K)
T_{gu}	Temperature of upper glass cover (K)
T_{out}	Chimney outlet temperature (K)
T_p	Temperature of drying product (K)
T_{sky}	Sky temperature, K
T_v	Temperature of vessel (K)
U_c	Overall heat loss coefficient from tunnel dryer cover to ambient air ($Wm^{-2}\ K^{-1}$)
V_{in}	Inlet airflow rate to tunnel dryer (m^3s^{-1})
V_{out}	Outlet airflow rate from tunnel dryer (m^3s^{-1})
k_a	Thermal conductivity of air ($Wm^{-1}\ K^{-1}$)
k_c	Thermal conductivity of insulation material ($W\ m^{-1}\ K^{-1}$)
k_f	Thermal conductivity of tunnel dryer floor material ($W\ m^{-1}\ K^{-1}$)
m_a	Mass of air inside the tunnel dryer (kg)
m_c	Mass of tunnel dryer cover (kg)
m_f	Mass of concrete floor of tunnel dryer (kg)
m_p	Mass of drying product (kg)
v_{in}	Inlet air speed ($m\ s^{-1}$)
v_{out}	Outlet air speed ($m\ s^{-1}$)
h_1	Enthalpy at state point (kJ)
h_o	Enthalpy at reference state (kJ)
s_1	Entropy at state point ($kJ\ K^{-1}$)
s_o	Entropy at reference state ($kJ\ K^{-1}$)
D	Average distance between the floor and the cover of tunnel dryer (m)
H	Humidity ratio of drying air inside the tunnel dryer ($kg\ kg^{-1}$)
M	Moisture content of drying product (% db)
Nu	Nusselt number
Rh	Relative humidity (%)
Re	Reynolds number
V	Volume of tunnel dryer (m^3)
W	Width of tunnel dryer floor (m)
g	Gravitational acceleration ($m\ s^{-1}$)
t	Time (hours)
$h_{c,ab-a}$	Convective heat transfer between absorber plate and cooker air ($Wm^{-2}\ K^{-1}$)
$h_{c,a-gl}$	Convective heat transfer between cooker air and lower glass cover ($Wm^{-2}\ K^{-1}$)
$h_{c,a-gc}$	Convective heat transfer between air and lower glass cover ($W\ m^{-2}\ K^{-1}$)
$h_{c,c-a}$	Convective heat transfer between tunnel dryer cover and drying air ($Wm^{-2}\ K^{-1}$)
$h_{c,f-a}$	Convective heat transfer between dryer floor cover and drying air ($W\ m^{-2}\ K^{-1}$)

$h_{c,gl-gu}$ Convective heat transfer between lower and upper glass cover ($W\ m^{-2}\ K^{-1}$)

$h_{c,gu-an}$ Convective heat transfer between upper glass cover and ambient ($W\ m^{-2}\ K^{-1}$)

$h_{c,p-a}$ Convective heat transfer between drying product and the drying air ($Wm^{-2}K^{-1}$)

$h_{c,v-a}$ Convective heat transfer between vessel cover and cooker air ($W\ m^{-2}\ K^{-1}$)

$h_{c,v-f}$ Convective heat transfer between vessel and cooker fluid ($W\ m^{-2}\ K^{-1}$)

$h_{r,ab-gl}$ Radiative heat transfer between absorber base and lower glass cover ($Wm^{-2}K^{-1}$)

$h_{r,c-s}$ radiative heat transfer between the tunnel dryer cover and the sky ($W\ m^{-1}\ K^{-1}$)

$h_{r,gl-gu}$ Radiative heat transfer between lower and upper glass cover ($W\ m^{-2}\ K^{-1}$)

$h_{r,gu-sky}$ Radiative heat transfer between upper glass cover and the sky ($W\ m^{-2}\ K^{-1}$)

$h_{r,p-c}$ Radiative heat transfer between drying product and tunnel dryer cover ($Wm^{-2}K^{-1}$)

$h_{r,v-gl}$ Radiative heat transfer between vessel and lower glass cover ($W\ m^{-2}\ K^{-1}$)

Greek letters

Ξi Exergy input

Ξo Exergy output

α_{ab} Absorptance of cooker absorber plate

α_c Absorptance of tunnel dryer cover material

α_f Absorptance of tunnel dryer floor (decimal)

α_{gl} Absorptance lower glass cover

α_{gu} Absorptance upper glass cover

α_p Absorptance of drying product

δ_c Thickness of tunnel dryer cover (m)

$\square_c$ Emissivity of tunnel dryer cover material

$\square_p$ Emissivity of the drying product

ρ_a Density of air ($kg\ m^{-3}$)

ρ_c Density of tunnel dryer cover material ($kg\ m^{-3}$)

ρ_d Density of the dry product ($kg\ m^{-3}$)

ρ_p Density of the drying product ($kg\ m^{-3}$)

τ_c Transmittance of tunnel dryer cover material

τ_{gl} Transmittance lower glass cover

τ_{gu} Transmittance upper glass cover

α Absorptance of shading nets

λ Thermodynamic psychrometric constant ($kPa\ K^{-1}$)

λ Latent heat of vaporization ($kJ\ kg^{-1}$)

υ Viscosity of air ($m^2\ s^{-1}$)

σ Stefan Boltzmann's constant ($W\ m^{-2}\ K^{-4}$)

Ψ Exergy efficiency

Units of Conversion

Length SI unit: Meter (m) 1 kilometer = 1000. meters = 0.62137 mile 1 meter = 100. centimeters 1 centimeter = 10. millimeters 1 nanometer = 1.00×10^{-9} meters 1 picometer = 1.00×10^{-12} meters 1 inch = 2.54 centimeters (exact) 1 Ångstrom = 1.00×10^{-10} meters **Energy** SI unit: Joule (J) 1 joule = 1 kg * m^2/s^2 = 0.23901 calorie = 1C x 1V 1 calorie = 4.184 joules **Temperature** SI unit: Kelvin (K) K = 273.15 °C K = °C + 273.15 °C ? °C = (5 °C/9 °F)(°F - 32 °F) ? °F = (9 °F/5 °C)°C + 32 °F	**Volume** SI unit: Cubic meter (m^3) 1 liter (L) = $1.00 \times 10^{-3} m^3$ = 1000. cm^3 = 1.056710 quarts 1 gallon = 4.00 quarts **Pressure** SI unit: Pascal (Pa) 1 pascal = 1 N/m^2 = 1 kg/m * s^2 1 atmosphere = 101.325 kilopascals = 760. mmHg = 760 torr = 14.70 lb/in^2 = 1.01325 bar 1 bar = 10^5 Pa (exact) **Mass** SI unit: Kilogram (kg) 1 kilogram = 1000. grams 1 gram = 1000. milligrams 1 pound = 453.59237 grams = 16 ounces 1 ton = 2000. pounds
Prefixes	
T = tera = 10^{12} G = giga = 10^9 M = mega = 10^6 k = kilo = 10^3	c = centi = 10^{-2} m = milli = 10^{-3} μ = micro = 10^{-6} n = nano = 10^{-9} p = pico = 10^{-12} f = femto = 10^{-15}

1

Introduction

1.1 General

Renewable technologies are considered to be clean sources of energy. The optimal use of these resources minimize environmental impacts, produce minimum secondary wastes, and are sustainable based on current and future economic and social needs. Sustainable development requires methods and tools to measure and compare human activities' and its environmental impacts for various products. Renewable energy sources (RES), which include biomass, hydropower, geothermal, solar, wind, and marine energies, supply 14% of the world's total energy demand. Renewable resources refer to primary, domestic, and clean or inexhaustible energy resources. Large-scale hydropower energy supplies 20 % of the global electricity need. Wind power in coastal and other windy regions is a promising source of energy.

Renewable energy resources can make a decisive contribution to the economic, social and sustainable development of rural regions in developing countries, yet the consumption of fossil fuels is dramatically increasing along with improvements in the quality of life, the industrialization of developing nations, and the increase of the world's population. It has long been recognized that excessive consumption of fossil fuel leads not only to a diminishing fossil fuel reserves more quickly, but also has a significant adverse impact on the environment. Such impacts result in increased health risks and the threat of global climate change. Changes to improve environmental conditions are becoming more politically acceptable globally, especially in developed countries. Society is slowly moving towards seeking more sustainable production methods, minimizing waste, reducing air pollution from vehicles, generating distributed energy, conserving of native forests, and reducing greenhouse gas emissions. Increasing consumption of fossil fuel to meet out current energy demands, however, has sounded alarms over regarding potential energy crisis. This has generated a resurgence of interest in promoting renewable alternatives to meet the developing world's growing energy needs. Excessive use of fossil fuels has caused global warming from carbon dioxide emissions, therefore, promoting renewable forms of clean energy is eagerly required. To monitor the level of these greenhouse gas emissions, an agreement that has fulfills the objectives of the Kyoto Protocol was create with overall pollution prevention targets.

Access to Energy is a crucial enabling condition for achieving sustainable development. Prudent energy policies and research can play an important role in steering both industrialized and developing countries onto more sustainable energy development paths. Specifically, they can strengthen the three pillars of sustainable development: the economy, by boosting productivity; social welfare, by improving living standards and enhancing safety and security, and the environment, by reducing indoor and outdoor pollution and remediating environmental degradation.

1.2 Energy and it's Requirement

Energy in any form is necessary to meet many of our requirements. It's varying definitions shows the importance of energy in our life. Everything happens in Universe is essentially a energy process.

1.2.1 Energy

Energy is the capacity of doing work, in fact it is a means for performing activities which are directly related to human being.

1.2.2 Fuel

Available sources of energy are sometimes referred to as fuel. The term originally came into use to describe energy sources and to show the phenomenon of burnt oil, coal, wood, oil etc.

1.2.3 Principle Form of Energy

There are six forms of principle energy:

1. Heat
2. Mechanical
3. Electrical
4. Electromagnetic wave (light, Radio Waves)
5. Chemical
6. Nuclear

1.2.4 Energy Sources

Principally, there are two types of energy sources on the earth

1. Conventional Energy Sources
2. Non Conventional Energy Sources

There are five ultimate sources of useful energy for mankind

1. The sun.
2. The motion & gravitational potential of sun, moon & earth.
3. Geothermal energy from cooling, chemical reaction, & radioactive decay on the earth.
4. Nuclear reaction on the earth.
5. Chemical reaction from mineral sources.

The conventional energy sources are derived from sources (1) Fossil fuels, (2) Hot rock, (3) Nuclear and (4) Chemical reaction.

Whereas non-conventional energy sources is derived from sources (1) Sun, (2) Motion & gravitational forces and (3) Geothermal energy or molten magma

1.2.5 Definitions

(a) Conventional Energy: It is energy obtained from static storage of energy sources (fossil fuel, hot rock, nuclear and chemical) which remain static bound in position until or unless it is released by human interaction.

Examples are coal, oil & natural gas. Hence initially the energy remains in static nature which otherwise reacted by human activities and converted in kinetic form and there by energy is used. These sources in present context are known as finite sources of energy or non renewable. Example are petroleum product, coal etc.

(b) Renewable Energy Sources: Renewable energy resources will play an important role in our planet's future. Major renewable energy sources and their usage forms are presented in Table 1.1. The share of RESs is expected to increase significantly (30–80% by 2100). The energy resources have been split into three categories: fossil fuels, renewable resources and nuclear resources. Renewable energy sources are those resources which can be used to produce energy again and again, e.g. solar energy, wind energy, biomass energy, geothermal energy, etc. They are often called "alternative" sources of energy. Renewable energy sources that meet domestic energy requirements have the potential to provide energy services with zero or almost zero emissions of both air pollutants and greenhouse gases. Renewable energy system development will make it possible to fulfil the most crucial current tasks, including: improving energy supply reliability and organic fuel economy; solving problems of local energy and water supply; increasing the standard of living and level of employment of the local population; ensuring sustainable development of remote regions in the desert and mountain zones; and fulfilling obligations of countries with regard to international agreements relating to environmental protection. Development and implementation of renewable energy project in rural areas can create job opportunities and thus minimize migration toward urban areas. Harvesting renewable energy in a decentralized manner is one of the options to meet rural and small-scale energy needs in a reliable, affordable and environmentally sustainable way.

Table 1.1: Main Renewable Energy Sources and their Usage Form

Energy Sources	Energy conversion and usage option
Hydropower	Power generation
Modern biomass	Heat and power generation, pyrolysis, gasification, digestion
Geothermal	Urban heating, power generation, hydrothermal, hot dry rock
Solar	Solar home system, solar dryers, solar cookers
Direct solar	Photovoltaic, thermal power generation, water heater
Wind	Power generation, wind generator, windmills, water pumps
Wave	Numerous designs
Tidal	Barrage, tidal stream

Examples are Solar Energy, Wind Energy, Geothermal Energy and Bio energy. These sources also known as renewable energy. Comparison of these sources is given in Table 1.2 and the definition is illustrated in Fig 1.1. The global renewable energy scenario forecasted for 2040 is presented in Table 1.3

Table 1.2: Comparison of Conventional & Renewable Energy Sources

S. No.	Parameters	Conventional Energy Supplies	Non Conventional Energy Supplies
1.	Source	Static Stores in Earth	Natural Sources
2.	Examples	Coal, Oil, Gas	Wind, Solar, Biomass, Tidal
3.	Nature of occupancy	Static Store of energy	A current of energy
4.	Normal State	Remain in bound position	Continuously or Repetitively released
5.	Initial Intensity	Released @ 100 kW/sqm & more	Low intensity, dispersed 300 W/sqm or less
6.	Life time of supply	Finite	Infinite
7.	Cost at Sources	Increasingly expensive	Free
8.	Conversion process/ Utilization device/ Transmission & Distribution media	Established / Commercialized	Under R&D stages, certain items are established
9.	Cost equipment	Moderate	Quite high in present context
10.	Variation & Control	Steady, best controlled by adjusting source with feedback control	Fluctuating, best controlled by change of load using feed forward control
11.	Location for use	General use	Site & location specific
12.	Scale of Production	Large scale	Small scale
13.	Skills or Production	Strongly available	Under R&D stage
14.	Skills for utilization	Strongly available	Under R & D stage
15.	Context of utilization	Urban, Centralized	Rural, Decentralized
16.	Dependence	System dependent on outside inputs	Selfsufficient encouraged
17.	Safety	Most dangerous when faulty	Usually safe when out of action
18.	Pollution & Environmental Damages	Environmental pollution common	Usually little harmful to the environmental especially at moderate scale
19.	Ecology Damage	Permanent damage common from Mining	Hazards from excessive wood burning/soil erosion from excessive biofuel use.
20.	Esthetics	Small structure may produce little aesthetic difficulty	Local perturbations may be serious.
21.	Economics	Availability is costly, but harnessing is within economic limit	Availability free, but harnessing is quite costly.
22.	Over benefit in present context	Finite source, may not remain in future	Infinite, locally available. No danger on source availability

Fig. 1.1: Conventional & Renewable Energy Flow Pattern

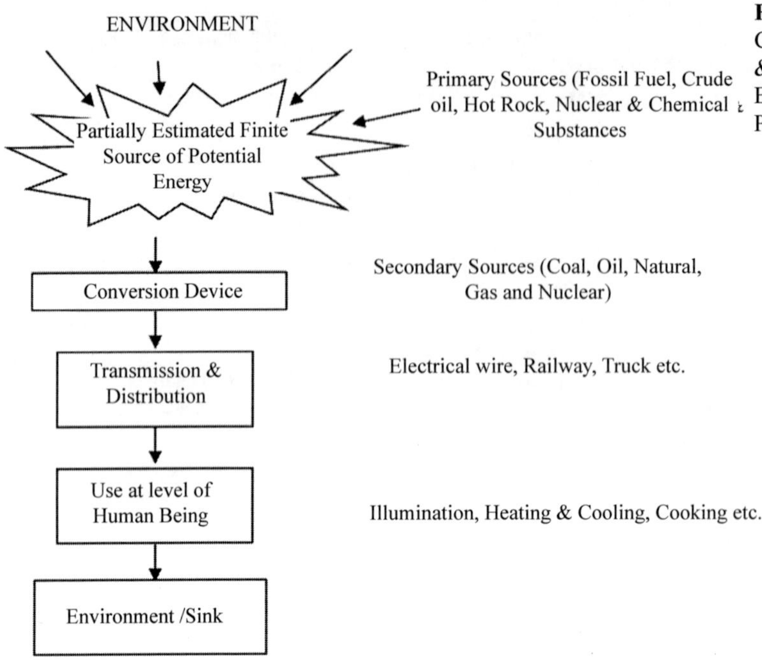

(a) Conventional Energy Sources

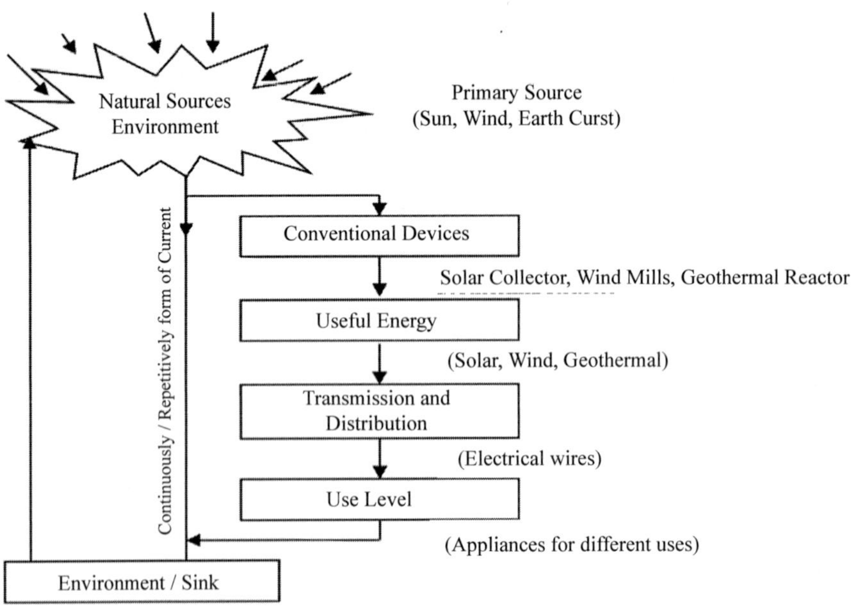

(b) Renewable Energy Sources

Table 1.3: Global Renewable Energy Scenario by 2040

	2001	2010	2020	2030	2040
Total consumption (Mtoe)	10,038	10,549	11,425	12,325	13,310
Biomass	1080	1313	1791	2483	3271
Large hydro	22.7	266	309	341	358
Geothermal	43.2	86	186	333	493
Small hydro	9.5	19	49	106	189
Wind	4.7	44	266	542	688
Solar thermal	4.1	15	66	244	480
Photovoltaic	0.1	0.4	3	16	68
Solar thermal electricity	0.1	0.4	3	16	68
Marine (tidal/wave/ocean)	0.05	0.1	0.4	3	20
Total RES	1164.45	1745.5	2694.4	4289	6351
Renewable energy source contribution (%)	11.60	16.6	23.6	34.7	47.7

1.3 Reforms in the Energy Sector

Reforms in the energy sector were initiated to supplement the Government's efforts in the development of the sector and to make it more efficient. The Government has been endeavouring to provide a policy environment that encourages free and fair competition in each element of the energy value chain and attracts capital from all sources-public and private, domestic and foreign. Encouraging such capital formation is crucial for India's to meet its energy needs. Significant progress has been made in establishing independent and transport regulatory authorities in the power sector to facilitate the rationalization of electricity tariff as well as to encourage competition while protecting the interests of all stakeholders.

The thrust of the reforms has been to deregulate the prices of commercial energy resources (which, until recently, were entirely administered), increase competition through institutional, legislative and regulatory reforms and reduce subsidies.

1.3.1 Energy priorities for developing countries

Achieving development goals in rural areas where poverty conditions are currently the greatest will require greater access and improved energy services as a means to reach education, health, water and other goals. In particular, expanding energy services is a means to generate increased employment and income generating opportunities, and is therefore a pre-requisite to increased value adding activities in rural areas. For the policy-maker, helping create a sustainable energy pathway will require broad societal consensus around the strategic choices of economic, environmental and social development.

Depending on their current state of development and priorities, countries will likely pursue different paths towards a variety of sustainable development options.

1.3.2 Access and affordability

Access to commercial energy will mean that energy must be available at prices which are both affordable (low enough for access by the poorest people) and sustainable (prices which cover the real costs of energy production, transmission and distribution to support the financial ability of companies to maintain and develop energy services and which internalize the social/environmental costs associated with these activities, if practicable). The best way to ensure that a growing number of people will be able to afford commercial energy is to accelerate economic growth and assist those with the lowest income to become wealthier.

1.3.3 Energy and the environment

Environment is one of many factors that societal decision-makers account for when considering energy choices. For example, energy production and consumption can raise environmental challenges, including land use, global climate change, water and urban air quality.

1.3.4 Energy efficiency

Energy efficiency is essentially using less energy to provide the same service. In this sense, energy efficiency can also be thought of as a supply resource – often considered an important, cost effective near- to mid-term supply option. Investments in energy efficiency can provide *additional* economic value by preserving the resource base and (especially combined with pollution prevention technologies) mitigating environmental problems.

Improvements in energy efficiency can produce direct environmental benefits in a number of ways, not only reducing pollution but also delaying the need to develop new fuel resources. In addition, energy efficiency improvements can considerably reduce the cost of pollution abatement. Industrial, commercial, and consumer equipment today can be as much as 80% more efficient than equipment installed just twenty years ago. For example, improved efficiency in any power plant can produce significant reductions in CO_2 emissions. Typically, a 1%-point gain in efficiency reduces CO_2 output by 2%.

Efficiency can also be increased at the point of consumption. For example, compact fluorescent lamps are available which use 75% less electricity than conventional light bulbs, and last much longer, while providing light of a similar quality. Although cost-effective over the lifetime of the bulb, compact lamps are not used everywhere due to the greater initial cost.

This is an example of one of the major barriers to greater energy efficiency - capital outlay. The ability to use the full range of market-based energy and energy technology resources, along with cleaner technologies and fuel systems, will help drive the innovation needed to optimize business and societal activities within the framework of sustainable development.

1.4 Climate Change Scenario

Climate change is one of the primary concerns for humanity in the 21st century. It may affect health through a range of pathways, for example, as a result of increased frequency and intensity of heat waves, reduce cold-related deaths, increased floods and droughts, changes in the distribution of vector-borne diseases and effects on the risk of disasters and malnutrition. The overall balance of effects on health is likely to be negative and populations in low income countries are likely to be particularly vulnerable to the adverse effects. The experience of the 2003 heat wave in Europe showed that high-income countries may also be adversely affected. The potentially most important environmental problem relating to energy is global climate change (global warming or the greenhouse effect). The increasing concentration of greenhouse gases such as CO_2, CH_4 , CFCs, halons, N_2O, ozone, and peroxyacetyl nitrate in the atmosphere is acting to trap heat radiated from the Earth's surface and is raising Earth's surface temperature.

Many scientific studies reveal that overall CO_2 levels have increased 31% in the past 200 years, and 20 Gt of carbon has been added to the environment since 1800 due solely to deforestation, while the concentration of methane gas which is responsible for ozone layer depletion has more than doubled since then. The global mean surface temperature has increased by 0.4–0.8 °C in the last century above the baseline of 14 °C. Increasing global temperatures have ultimately increased global mean sea levels at an average annual rate of 1–2mm over the last century. Arctic sea ice thinned by 40% and decreased in extent by 10–15% in summer since the 1950s. Industry contributes directly and indirectly (through electricity consumption) about 37% of global greenhouse gas emissions, of which over 80% is from energy use. Total energy- related emissions, which were 9.9 Gt CO_2 in 2004, have grown by 65% since 1971. There is ample scope to minimize emission of greenhouse gases if efficient utilization of renewable energy sources in actual energy meeting routes is promoted.

1.5 Role of Energy in Economic Development

Humanity has searched a variety of energy sources in order to get greater comfort, enhanced security from want and the satisfaction of wants through increasingly elaborate artifacts. History reveals three energy areas: wood, coal and petroleum fuel are being predominately used.

The use of wood as a fuel for space heating was even used during prehistoric times, but it employment as a fuel for manufacturing or industrial purpose probably dates back less than 50,000 years. Coal began to be employed on a large scale in Europe only after the elimination of forest cover from areas of intensive fuel use. In fact industrialized countries used wood as major energy sources only upto the eighteenth century. During the nineteenth century, coal replaced wood, wood and coal, coupled with the development of steam generation technology, enabled a massive substitution of human labour by inanimate energy.

When crude oil was first marketed on large scale in the 1860's, it did not immediately begin to crowd coal out of the market. However, the lubricants derived from petroleum were unique suitable to the needs of high-speed machinery and many advances in mechanical engineering would have been impossible without the improved lubricants derived from crude oil. Eventually it began to complete more effectively with coal when its value as a boiler fuel was recognized and its costs became competitive with those of coal. However, it was not until the development of internal combustion engine that crude displaced coal, in the first part of the twentieth century. Natural gas began to complete with crude as a source of nationwide energy supplies in the 1930's when the technology of long-distance low-cost gas pipeline was developed. For a number of years gas consumption grew more rapidly than that of coal and gas displaced oil in same market because of its superior convenience, cleanliness and economy.

Infact, Wood, coal and petroleum energy played a vital role in economic development. As an economy develops, it demands locally available more energy quantity wise and quality wise in more convenient forms. In general, the more rapid the rate of economic growth, the more rapid the replacement of solid fuels by fluid fuels.

In earlier years petroleum product made considerable contributions to the economic growth in direct as well as indirect ways. Later on, with the invention of electric power and the internal combustion engine, the entire population got enhanced mobility. In the farm, the IC engine made it possible to turn out more foodstuffs. Further, electrification in turn spurred the development of new industrial process which greatly enhanced labour productivity. Similarly, such advances are more evident in the home, with the advent of the electric refrigerator, microwave oven, stove, dishwasher, radio and many other home appliances.

1.6 Role of Renewable Energy Sources

Per capita consumption of electricity in India continues to be extremely low, only slightly above 1181 kWh per annum. Only onethird of the total electricity supply is consumed in the rural areas, despite twothirds of the population living in these areas.

In April 2018 the Government of India announced that India had achieved its goal of providing electricity to every village in India (600 000 villages). A year later, in April 2019, the Government of India announced that 211.88 million rural households were provided with electricity, close to 100%.

Over the last 25 years, the technical, operational and economic viability of grid connected and offgird electricity production from renewable energy has been progressively demonstrated with increasing success. Furthermore, several renewable energy technologies are now showing signs of being able to stand on their own in the foreseeable future, without subsidies. The recent surge of interest in non- conventional is on account of these developments, as also their declining costs, while fossil fuels are showing rising cost trends. In addition, the steadily growing awareness of the importance of environmental protection is a major factor favoring non- conventional.

Renewable energy sources can create a significant impact in the generation of electricity. Several technologies for grid connected power generation such as wind power, small hydropower and biomass-based power have now matured. Renewable non-conventional electric technologies also offer possibilities of distributed generation, at or near actual load centers, thereby saving on costly establishment and maintenance of transmission and distribution networks. A major niche area for renewable energy technologies, which should be of interest to utilities, is a range of demandside applications, particularly in our urban centers and industry, such as air and water heating, industrial process heating, solar building designs, as well as energy recovery from urban, municipal and industrial wastes. It is unlikely though that a single renewable non-conventional energy technology will predominate. Variations in the availability of the energy resource, its economics & range and speed of applicability, will favour some renewable over others for different enduses.

In our rural areas, energy for cooking, lighting, water pumping, agro and rural industry and other productive activities can be effectively provided through locally available renewable energy sources. In remote areas, where transmission of grid power has been found to be totally uneconomical, offgird electrification can be undertaken through non-conventional renewable energy systems such as solar photovoltaics. Where appropriate electricity supply requirements can also be met by hybrid systems integrating two or more sources, in conjunction with storage.

Renewable energy options also make economic sense, as several technologies are now competitive, and markets are growing. Besides opportunities in production of equipment, renewable offer vast employment opportunities in the service sector, in local management, and in maintenance and repair. Moreover, the income and employment benefits are more evenly distributed, in accordance with the spread of resources.

India is endowed with abundant sunlight, wind, water and biomass as renewable sources of energy and over the last two decades vigorous efforts have been made to tap these renewable energy resources to utilize them for variety of end use applications i.e., cooking, water heating, drying, water pumping, lighting and power generation for meeting the decentralized energy requirements in villages, urban areas, schools, hospitals, etc. India today has the world's largest programmes for renewable energy. It is now recognized that nonconventional energy source can provide the basis for sustainable energy development on account of their inexhaustible nature and environmentfriendly features. During the last two decades, several non-conventional energy technologies have been developed and deployed in villages and cities. The world is increasingly looking at the renewable energy sector as a viable option to bridge the energy gap and also to address environmental issues. Table 1.4 summarizes the status of non- conventional energy potential and achievement its India as on February 29, 2020.

Table 1.4: Renewable Energy at a Glance in India (Grid Interactive Renewable Power)

Sector	MW (as on 29/02/2020)	% of Total
Central sector	93476.93	25.26
State sector	173039.30	46.76
Private sector	103531.74	27.98
Total	370047.97	
Fuel		
Total Thermal	230809.57	62.37
Coal	205344.50	55.49
Gas	24955.36	6.74
Diesel	509.71	0.14
Hydro	45699.22	12.35
Nuclear	6780.00	1.82
Renewable Energy Sources (RES)	86759.19	23.45

Break up of RES all India as on 29/02/2020 is given below (in MW):

Small Hydro Power	Wind Power	Bio-Power		Solar Power	Total Capacity
		Biomass Power/ Cogeneration	Waste to Energy		
4683.16	37669.25	9861.31	139.80	34405.67	86759.19

It is widely acknowledged that there are many hidden and intangible subsidies for conventional energy and power that renewable does not get, and also there are many tangible and intangible benefits that are not captured in conventional economic accounting. These include:

1. Social and economic development: Renewable energy sources, such as biomass and biogas can provide economic development and employment opportunities, particularly in rural areas; Can mitigate rural poverty, and reduce urban migration.

2. Land restoration: Biomass plantations prevent soil erosion, provide a better habitat for wildlife, and support rural development.

3. Reduced air pollution: Renewable sources, such as methanol or H_2 for fuel cell vehicles cause reduced air pollution.

4. Abatement of global warming: Renewable sources of energy make little contribution to global warming. In biomass utilization, the CO_2 emission is more or less balanced by the CO_2 absorption by the biomass during its growth.

5. Fuel supply diversity: Renewable offer a wide range of choices, enhancing energy security.

6. Operational facility: Renewable energy sources are having quality such as Adaptability, Efficiency, Inclusivity. Opportunity and Universalism.

2

Thermodynamic Basis of Energy and Exergy Analysis

2.1 Introduction

The name thermodynamics stems from the Greek words therme (heat) and dynamis (power), which is most descriptive of the early efforts to convert heat into power. It may be defined as the science that deals with the relations among heat, work and properties of a system. Today, thermodynamics is broadly interpreted to include all aspects of energy and energy transformations, including power generation, refrigeration, and relationships among the properties of matter. It is viewed as the science of energy, and thermal engineering is concerned with making the best use of available energy resources.

Thermodynamics basically entails four laws or axioms known as Zeroth, First, Second and Third law of thermodynamics.

- The zeroth law deals with thermal equilibrium and establishes a concept of temperature.
- The first law pertains to the conservation of energy and introduces the concept of internal energy.
- The second law indicates the limit of converting heat into work and introduces the concept of entropy.
- The third law defines the absolute zero of entropy.

The concept of energy was introduced by Newton in the field of mechanics during generalizing the hypothesis of kinetic and potential energies. Energy may be defined as the capacity to do work and it is a scalar quantity. It cannot be seen or observed. Its presence can, however, be felt by the properties of the system. There occurs a change in one or more properties of the system in case of energy transfer. The absolute value of energy of a system is difficult to measure, whereas the energy change is rather easy to calculate. Energy follows the conservation law, one of the most fundamental laws of nature, which simply states that during an interaction, energy can change its form from one form to another but the total amount of energy remains constant, that is, energy cannot be created or destroyed only form can be changed.

The energy of a system can take the form: stored energy, i.e., energy contained within the system boundaries, e.g., potential energy, kinetic energy and internal energy; and energy in transit, that is, the energy which crosses the system boundaries, e.g., heat work and electrical energy. The energy can also be classified in macroscopic and microscopic forms. Macroscopic forms of energy: are those in which an overall system possesses energy with respect to a reference frame, e.g., kinetic and potential energies. The macroscopic energy of a system is related to motion and the influence of external effects such as gravity, magnetism, electricity and surface tension. Microscopic forms of energy: are those related to the molecular structure of a system and the degree of molecular activity, and are independent of outside reference frames. The sum of all the microscopic forms of energy of a system is its internal energy.

2.2 The First Law of Thermodynamics

The first law of thermodynamics is commonly called the law of conservation of energy. In mechanics, the study of conservation of energy emphasizes changes in kinetic and potential energy and their relationship to work. A more general form of conservation of energy includes the effects of heat transfer and internal energy changes as well. This more general form is usually called the first law of thermodynamics. The first law of thermodynamics was stated for a cycle: whenever a system undergoes a cycle, the net heat transfer is equal to the net work done. This can be expressed in equation form

$$\sum Q = \sum W$$

$$\oint \delta Q = \oint dw$$

by where, the symbol $\oint$ implies integration around a complete cycle.

2.2.1 The First Law Applied to a Non-flow Process

For a closed system undergoing an infinitesimally slow process, during which the only allowed interactions with its environment are those involving heat, and work, and change in internal energy (*dU*), the first law can be expressed quantitatively as follows:

$$dQ - dW = dU \tag{2.1}$$

2.2.2 The First Law Applied to a Steady-Flow Process

For a process during which a fluid flows through a control volume steadily, that is, the fluid properties can change from point to point within the control volume, but at any point, they remain constant during the entire process, the application of the first law yields the following steady flow energy equation

$$\dot{Q}_{in} + \dot{W}_{in} + \sum_{in} \dot{m}\left(h_1 + \frac{V_1^2}{2} + gz_1\right) = \dot{Q}_{out} + \dot{W}_{out} + \sum_{out} \dot{m}\left(h_2 + \frac{V_2^2}{2} + gz_2\right)$$

$$\dot{Q} - \dot{W} = \sum_{out} \dot{m}\left(h_2 + \frac{V_2^2}{2} + gz_2\right) - \sum_{in} \dot{m}\left(h_1 + \frac{V_1^2}{2} + gz_1\right)$$

$$\dot{Q} - \dot{W} = \dot{m}\left[h_2 - h_1 + \frac{V_2^2 - V_1^2}{2} + g(z_2 - z_1)\right]$$

$$q - w = h_2 - h_1 + \frac{V_2^2 - V_1^2}{2} + g(z_2 - z_1) \quad (2.2)$$

Where $q = \dot{Q}/\dot{m}$ and $w = \dot{W}/\dot{m}$ the heat transfer and work done per unit mass of the working fluid, respectively. When the fluid experiences negligible changes in its kinetic and potential energies (i.e.,), the energy balance equation is reduced further to

$$q - w = h_2 - h_1 \quad (2.3)$$

2.2.3 The First Law Applied to an Unsteady-flow Processes

In engineering practice, the variable flow process applications are as common as the steady flow processes. These processes involve changes within the control volume with time, and are called unsteady-flow, or transient flow, processes. The general energy balance equation for any process is as follows:

$$\begin{pmatrix}\text{Net energy transfer by} \\ \text{heat, work, and mass}\end{pmatrix} = \begin{pmatrix}\text{Change in internal,} \\ \text{kinetic, potential, etc. energies}\end{pmatrix} \quad (2.4)$$

The energy balance for an unsteady state uniform flow process can be written as:

$$\left[Q_{in} + W_{in} + \sum_{in} m\,\theta\right] - \left[Q_{out} + W_{out} + \sum_{out} m\,\theta\right] = (m_2 e_2 - m_1 e_1)_{system}$$

Where, $= h + \frac{V^2}{2} + gz$, and $e = u + KE + PE$

$$Q - W = \sum_{out} mh - \sum_{in} mh + (m_2 u_2 - m_1 u_1)_{system} \quad (2.5)$$

where $Q = Q_{in} - Q_{out}$ the net heat is input and $W = W_{out} - W_{in}$ is the network output. It may be noted that if no mass enters or leaves the control volume during a process ($m_i = m_e = 0$, and $m_1 = m_2 = m$), this equation reduces to the energy balance relation for closed systems. It may also be noted that an unsteady-flow system may involve boundary work as well as electrical and shaft work

2.3 The Second Law of Thermodynamics

The first law of thermodynamics deals with the conservation and conversion of energy. This law, however, fails to state the conditions under which energy conversion are possible. This law places no restriction on the direction of a process. Merely satisfying the first law does not ensure that the process can actually occur. This inadequacy of the first law to identify whether a process can take place is remedied by introducing another general principle, the second law of thermodynamics. The first law is concerned with the quantity of energy and the transformations of energy from one form to another with no regard to its quality. Preserving the quality (capacity to cause change) of energy is a major concern to engineers, and the second law provides the necessary means to determine the quality as well as the degree of degradation of energy during a process.

The implications of the Second Law are manifold. The condition of the increase of entropy can be used to predict processes, chemical reactions, transformations between various energy forms, or directions of heat transfer can and cannot occur. From the condition that a state of equilibrium of an isolated, two part system corresponds to a maximum of entropy of the system, it can be shown that the conditions of thermal, mechanical and chemical equilibrium correspond respectively to equality of temperature, pressure and chemical potential. In addition, the Second Law governs the limits to energy conversion between different energy forms, leading to concept of energy quality.

2.3.1 The Principle of Increase of Entropy

Consider a cycle comprising of two processes: process 1-2, which is arbitrary (reversible or irreversible), and process 2-1, which is internally reversible, as shown in Fig. 2.1. From the Clausius inequality,

$$\oint \frac{\delta Q}{T} \leq 0$$

$$\int_1^2 \frac{\delta Q}{T} + \int_2^1 \left(\frac{\delta Q}{T}\right)_{rev} \leq 0 \qquad (2.6)$$

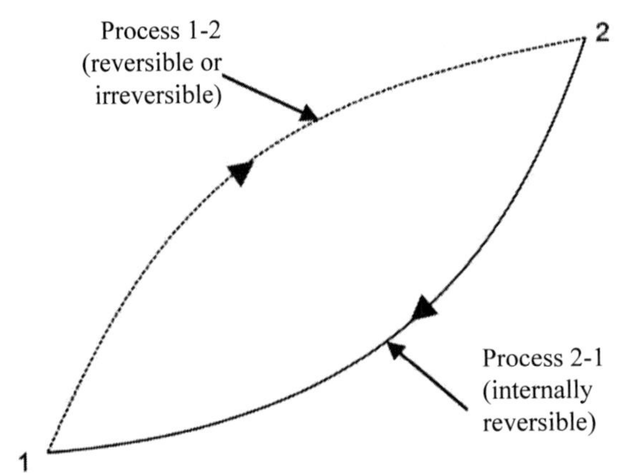

Fig. 2.1: A Cycle Composed of a Reversible and an Irreversible Process

Where $\frac{dQ}{T} = ds$ is defiened as change in entropy for a reversible process and thus the second integral in the equation (6) is recognized as the entropy change $S_1 - S_2$. Therefore

$$\int_1^2 \frac{\delta Q}{T} + S_1 - S_2 \leq 0 \tag{2.7}$$

This can be rearranged as

$$S_2 - S_1 \geq \int_1^2 \frac{\delta Q}{T}\ ;\ dS \geq \int_1^2 \frac{\delta Q}{T}$$

The term $\Delta S = S_2 - S_1$ represents the change in entropy of the system. For a reversible process, it becomes equal to $\int_1^2 \frac{dQ}{T}$, which represents the entropy transfer with heat.

In any process entropy is generated due to irreversibilities. The entropy generated during a process is called entropy generation and can be expressed by S_{gen}

$$\Delta S_{sys} = S_2 - S_1 = \int_1^2 \frac{\delta Q}{T} + S_{gen} \tag{2.8}$$

The entropy generation S_{gen} is always a positive quantity or zero. Its value depends on the process, and thus it is not a property of the system.

$$S_{gen} = \Delta S_{total} = \Delta S_{sys} + \Delta S_{surr} \geq 0 \tag{2.9}$$

In a larger irreversible process, more entropy is being generated. No entropy is generated during reversible processes ($S_{gen} = 0$). The increase of entropy principle does not imply that the entropy of a system cannot decrease. The entropy change of a system can be negative during a process, but entropy generation cannot be negetive. The increase of entropy principle can be summarized as follows:

$$S_{gen} \begin{cases} > 0\ \textit{Irreversible process} \\ = 0\ \textit{Reversible process} \\ < 0\ \textit{Impossible process} \end{cases}$$

Basically, entropy is a non-conservable property therefore, there is no such law of the conservation of entropy. Entropy can be conserved only during the idealized reversible processes and increases during all actual processes.

2.4 Thermodynamic Basis of Exergy Analysis

The first law analysis is based on the law of conservation of energy, which is valid whether a process is physically possible or not. The performance evaluation based on first law efficiency is inadequate, and more meaningful evaluation must include the second law analysis. The second law analysis is based on law of degradation of available energy i.e. quality of energy which is more realistic, rational and true measure of assessing the deviation of actual system from ideal system.

The first law gives no distinction between heat and work, no provision for quantifying the quality of heat, no accounting for the work lost in a process and no information about the optimal conversion of energy. The second law of thermodynamics applied in the form of exergy balances for components and processes can locate and quantify the irreversibilities which cause loss of work and inefficiency in thermal systems. Identifying the main sites of exergy loss shows the direction for potential improvements. The principles and methodologies of exergetic analysis are well established. Many of the analysis might be more effective if a greater prominence is given to the exergetic analysis. The second law analysis provides the tool for clear distinction between energy losses to the environment and internal irreversibilities in the process.

Exergy is the maximum work potential which can be obtained from a given form of energy. The exergy losses have a significant effect on environmental impact and can be used as a criterion for assessing the depletion of natural resources. Any thermodynamic assessment of the system is incomplete, unless the exergy concept becomes a part of the analysis. This concept has been successfully applied to various types of thermal systems including thermal power plants.

The aim of exergetic analysis is to detect and to evaluate quantitatively the effect of irreversible phenomena which increase the thermodynamic imperfection of the considered process. The optimal design criteria for thermal system can be achieved by maximizing the exergy output / exergy efficiency of the system or by reducing the irreversibility of the system. Thermodynamically there exists a direct relationship between the irreversibility of the process and the amount of useful work dissipated in the process.

2.4.1 Different Components of Exergy

Exergy is a generic term for a group of concepts that define the maximum work potential of a system, a stream of matter or a heat interaction; the state of the (conceptual) environment being used as the datum state. In an open flow system, there are three types of energy transfers across the control surface namely work-transfer, heat-transfer, and energy associated with mass transfer or flow. The work transfer is equivalent to exergy in every respect as exergy is maximum work, which can be obtained from that form of energy. The exergy of heat transfer Q from the control surface at temperature T_k is determined from maximum rate of conversion of thermal energy to work W_{max}. The W_{max} is given by

$$W_{max} = Q(1 - \frac{T_o}{T_k}) \quad (2.10)$$

Exergy of steady flow stream of matter is the sum of kinetic, potential, physical and chemical exergy. The kinetic and potential energy are again equivalent to exergy. The physical and chemical components of exergy depend on the states of matter and the environmental state.

2.4.1.1 Physical exergy

Physical exergy is equal to the maximum amount of work obtainable when the stream of substance is brought from its initial state define (P_1 and T_1) to the environment state defined by and by physical process involving only thermal interaction with the environment as shown in Fig 2.2.

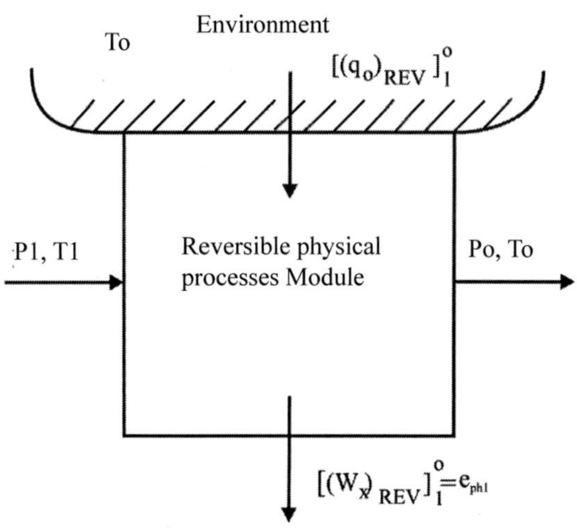

Fig.2.2: Reversible Module for Determining the Physical Exergy of a Steady Stream of Matter

To understand the physical exergy of process module as shown in Fig 2.2, we are neglecting kinetic and potential energy. The only interaction associated with the module is reversible heat transfer with the environment which per unit mass is:

$$\left[(q_o)_{REV}\right]_1^o = T_o(s_o - s_1) \quad (2.11)$$

The steady flow energy equation for the module, per unit mass, is:

$$\left[(q_o)_{REV}\right]_1^o - \left[(W_X)_{REV}\right]_1^o = (h_o - h_1) \quad (2.12)$$

As per the definition, the reversible work delivered by the module is equal to the specific physical exergy of the stream. Thus:

$$\varepsilon_{ph1} = (h_1 - T_o s_1) - (h_o - T_o s_o) \quad (2.13)$$

Physical exergy of perfect gas

The specific physical exergy of a perfect gas can be expressed as:

$$\varepsilon_{ph} = c_p\left(T - T_o\right) - T_o\left(c_p \ln\frac{T}{T_o} - R\ln\frac{P}{P_o}\right) \tag{2.14}$$

2.4.1.2 Chemical exergy

Chemical exergy is equal to the maximum amount of work obtainable when the substance under consideration is brought from the initial state to the dead state by processes involving heat transfer and exchange of substance only with the environment. The concept of reversibility, that all flow all direction and interaction can be reversed, gives the alternative definition: chemical exergy is equal to the minimum amount of work necessary synthesis, and to deliver in the environmental state, the substance under consideration from environmental substances by means of processes involving heat transfer and exchange of substance only with the environment.

2.4.2 Exergy balance for a control volume

Energy analysis is based only on the first law of thermodynamics, which is related to the conservation of energy and the conservation of mass whereas exergetic analysis is a method that uses the conservation of mass and conservation of energy principles together with the second law of thermodynamics for the analysis for actual, design and improvement of energy systems. Exergetic analysis is a useful method for complementing rather to replace energy analysis. Unlike the mass and energy the exergy is not conserved.

The exergy balance for steady flow process of an open system is given by

$$\sum\left(1 - \frac{T_o}{T_k}\right)\dot{Q}_k + \sum_{in}\dot{m}\Psi_i = \dot{E}x_W + \sum_{out}\dot{m}\Psi_o + \dot{I} \tag{2.15}$$

where Ψ_i and are respectively the exergy associated with mass inflow and outflow and can be expresses as follows:

$$Ø_i = h_i - T_oS_i + \frac{V_i^2}{2} + gz_i; Ø_o = h_o - T_oS_o + \frac{V_o^2}{2} + gz_o$$

$\sum\left(1 - \frac{T_o}{T_k}\right)\dot{Q}_k$ is the exergy associated with heat transfer, is exergy associated with work transfer and I is irreversibility of process. The irreversibility is also given by Gouy-Stodola theorem as

$$\dot{I} = T_o\dot{S}_{gen} \tag{2.16}$$

Exergetic analysis yields useful results because it deals with irreversibility minimization or maximum exergy delivery. The irreversibility may be due to heat transfer through finite temperature difference, mixing of fluids and mechanical friction. Irreversible energy transfer is a process that leads to exergy loss or entropy generation. Exergetic analysis is an effective means, to pinpoint losses due to irreversibility in a real situation and evaluate various thermodynamic losses in terms of various entropy generations. Exergetic analysis of a complex system can be performed by analyzing the components of the system separately.

2.5 Energetic and Exergetic Efficiencies of a Thermal System

The energetic or first law efficiency of a system or system component which is based on first law of thermodynamics is defined as the ratio of energy output to the energy input of system or system component i.e.

$$\eta_{\mathrm{I}} = \frac{\text{Desired output energy}}{\text{Input energy supplied}} \tag{2.17}$$

The exergetic or second law efficiency is defined as

$$\eta_{\mathrm{II}} = \frac{\text{Desired output}}{\text{Maximum possible output}} = \frac{\text{Exergy output}}{\text{Exergy input}} \tag{2.18}$$

2.6 Exergetic Analysis of Solar Radiations

The maximum useful work from solar radiation sets the upper limit of performance for solar energy conversion devices. The exergy of thermal radiation and, in particular the exergy of solar radiation have been the subject of a number of fundamental studies. Petela (1964) derived the formulae related to radiation exergy for different radiation categories. Different relations have been suggested in order to evaluate the exergy of radiation in a period of around 20 years. It can be concluded from these relations that thermal radiation from the sun is relatively rich in exergy.

The exergy of radiation differs from exergy of heat by irreversibility occurring during conversion of radiation into heat (i.e., absorption) or reverse conversion (i.e., emission). Absorption and emission phenomena are thermodynamically irreversible, while transmission of radiation is reversible. Petela (1964) derived the formulae related to radiation exergy for different radiation categories. Gribik and Osterle (1984) reported four different expressions for the maximum thermodynamically permitted efficiency of solar devices for converting undiluted black body radiation from the sun into useful work. Bejan (1987) presented a unified theory to explain the exergy of solar radiation and raised some doubts in Petela's theory. Petela (2003) explained and clarified the points raised by Bejan (1987) and accordingly, the exergy of radiation emitted by a black body, which is at temperature, is calculated by multiplying the energy of radiation by exergetic efficiency term , as given by

$$\eta_s = \left[1 - \frac{4}{3}\left(\frac{T_o}{T_s}\right) + \frac{1}{3}\left(\frac{T_o}{T_s}\right)^4\right] \tag{2.19}$$

The exergetic efficiency term for solar beam radiation is calculated at black body temperature (≈5600 K) of the sun (by assuming the sun as a black body radiation source). For simplicity of the analysis, the effective temperature of diffuse solar radiation may also be taken equal to that of sun. As the temperature of radiating surface for the diffused solar radiation is unknown, the measured data on the spectrum of diffused radiation at that instant of time is required for the determination of exergy of solar diffused radiation.

3

Solar Radiation

3.1 The Sun

The Sun is a star. It is a huge, spinning, glowing sphere of hot gas. It appears so much larger and brighter than the other stars because we are so close to it. The Sun is the center of our Solar System and contains most of the mass in the Solar System. All of the planets in our Solar System, including Earth, orbit around the Sun. In sun the energy generation is a multi-step process of thermo-nuclear fusion reactions. In the thermo nuclear fusion hydrogen (four protons) combines to form helium (one helium nucleus). The mass of the helium nucleus is less than that of four hydrogen protons and mass having been lost in the reaction is converted to energy. The reaction is represented as

$$4\,{}^{1}_{1}\mathrm{H} \rightarrow {}^{4}_{2}\mathrm{He} + 26.7\ \mathrm{MeV}$$

3.2 The Sun-Earth Relationship

The diameter of sun is 1.39 X 10^9 m. The diameter of earth is 1.27 X 107 m. The average distance between the sun and earth is 1.496 X10^{11} m, which varies by ± 1.7% as the path of earth's revolution around sun is elliptical. The sun subtends an angle of 32' with earth. Fig. 3.1 shows schematically the geometry of the sun-earth relationship.

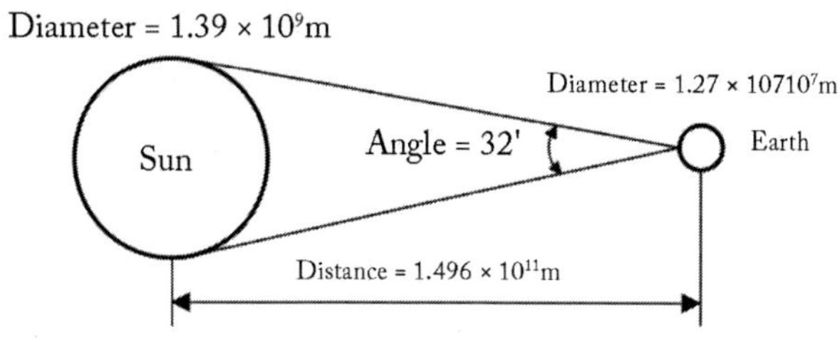

Fig. 3.1: Sun-Earth Relationship

3.3 The Structure of the Sun

The Sun is a huge ball of gas. Unlike Earth, it doesn't have a solid surface where we could stand even, if it were cool enough. And, just like a golf ball, the Sun is made up of layers: a core, a surface, and surrounding atmospheric layers, each of which have their own layers as illustrated in Fig. 3.2.

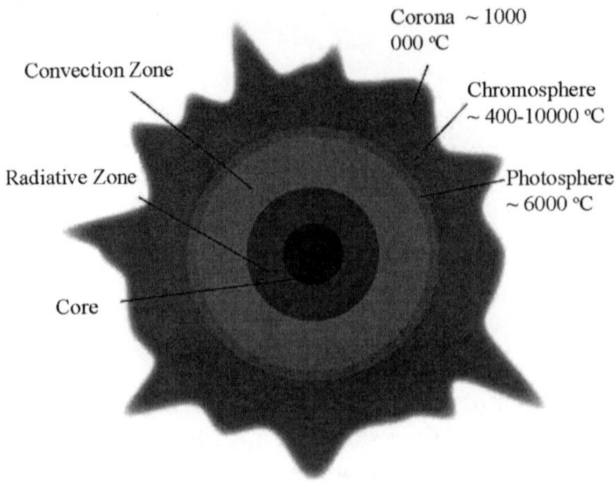

Fig. 3.2: Structure of the Sun

a. **Core:** The temperature at the very center of the Sun is about 15 million degrees Celsius. The temperature cools down through the radiative and convective layers that make up the Sun's core.

b. **Photosphere** - The photosphere is the deepest layer of the Sun that we can observe directly. It reaches from the surface visible at the center of the solar disk to about 400 km above that. The temperature in the photosphere varies between about 6200 °C at the bottom and 3700 °C at the top Most of the photosphere is covered by granulation.

c. **Chromosphere** - The chromosphere is a layer in the Sun between about 400 km and 2100 km above the solar surface i.e., the photosphere. The temperature in the chromosphere varies between about 3700 °C to 7700 °C.

d. **Transition Region** - The transition region is a very narrow (100 km) layer between the chromosphere and the corona where the temperature rises abruptly from about, 7700 to 500,000 °C.

e. **Corona** - The corona is the outermost layer of the Sun, starting at about 2100 km above the solar surface (the photosphere). The temperature in the corona is 500,000 °C or more, up to a few million degree Celsius. The corona cannot be seen with the naked eye except during a total solar eclipse, or with the use of a coronagraph. The corona does not have an upper limit

3.4 Solar Constant (G_{sc})

The solar constant is a measure of flux density, is the amount of incoming solar electromagnetic radiation per unit area that would be incident on a plane perpendicular to the rays, at a distance of one astronomical unit (AU) (roughly the mean distance from the Sun to Earth). The solar constant includes all types of solar radiation, not just the visible light. Its average value was thought to be approximately 1366 W/m^2, varying slightly with solar activity, but recent recalibrations of the relevant satellite observations indicate a value closer to 1361 W/m^2 is more realistic. The solar irradiance G_{on} out of the atmosphere on normal surface of the sun and earth undergoes certain changes because there is minor change in sun – earth distance. The value of G_{on} changes within the range of ±3%. G_{on} on the nth day of the year can be determined through the following

$$G_{on} = G_{sc}\left[1.0 + 0.033\cos\frac{360^\circ n}{365}\right]$$

3.5 Solar Radiation Outside the Earth's Atmosphere (Extra Terrestrial Radiation)

The sun emits electromagnetic radiation and its spectrum is much closed to that of black body with a temperature about 5800 K. Nuclear fusion is taking place in the sun it produces gamma rays. The internal absorption and thermalization convert high energy photon into lower energy level photons before reaching to the earth surface. The sun also emits X-rays, ultraviolet, visible light, infrared, and radio waves. The spectrum of solar radiation striking on earth's atmosphere ranging from 100 nm to 1mm (1,000,000 nm) and can be divided into five regions in increasing order of wavelengths: (i) Ultraviolent C (UVC) in the range of 100 to 280nm and invisible to human eye as it has higher frequency than violet light. Due to absorption by the atmosphere very little reaches Earth's surface, (ii) Ultraviolent B (UVB) in the range of 280 to 315 nm. It absorbed by atmosphere along with UVC and it is essential for vitamin D synthesis in the skin and fur of mammals, (iii) Ultraviolent A (UVA) in the range of 315 to 400 nm. It is responsible for damaging DNA via indirect routes (formation of free radicals and reactive oxygen species), and can cause cancer, (iv) Visible range in the range 380 to 700 nm. This range is visible to the naked eye. It is also the strongest output range of the Sun's total irradiance spectrum, and (v) Infrared range in the range of 700 nm to 1,000,000 nm. Infra means below. It comprises an important part of the electromagnetic radiation that reaches as shown in Fig. 3.3.

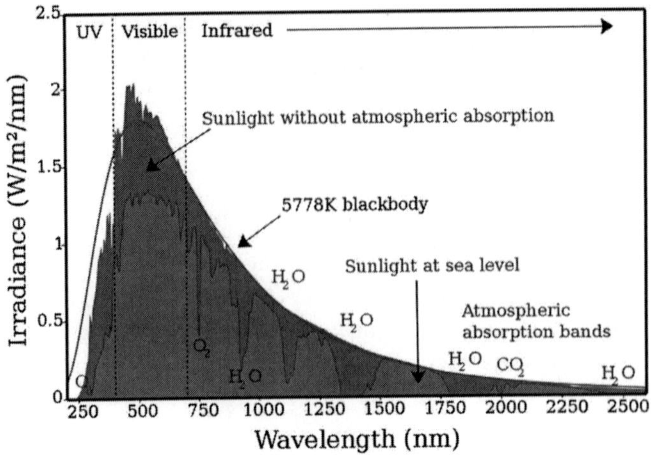

Fig. 3.3: Solar Irradiance Spectrum above Atmosphere and at Surface

3.6 Solar Radiation at the Earth's Surface (Terrestrial Radiation)

The solar radiation received at the surface of the earth is entirely different than the radiation received outside the atmosphere due to the absorption and scattering in the atmospheric. A fraction of the radiation reaching the earth's surface is reflected back into the atmosphere and subjected to atmospheric phenomenon and the remainder is received by the earth surface.

The solar radiation reaching earth's surface can be classified into two components known as direct radiation and diffuse radiation.

(a) **Direct radiation**: That portion of the incident solar radiation which comes directly from sun without reflection and change of direction from atmospheric components is called as direct radiation or beam radiation. If beam radiation measured at surface of earth at a given location with a surface perpendicular to the sun is called Direct Normal Irradiance (DNI).

(b) **Diffuse radiation**: It is the radiation at the Earth's surface from light scattered by the atmosphere. It is measured on a horizontal surface with radiation coming from all points in the sky excluding radiation coming from the sun disk. There would be almost no DHI in the absence of atmosphere.

(c) **Total radiation:** It is the summation of direct radiation and diffuse radiation. It is also known as global solar radiation

3.7 Determination of Solar Time

Information on the availability of solar radiation is necessary for the design and estimation of the performance of any solar energy device. While designing a solar system. One needs to know monthly average daily global and diffuse radiation on a horizontal surface, solar insolation at hourly intervals for inclined surface and knowledge of hourly solar insolation on an inclined surface.

The apparent solar time is known as true solar time. the value of time in solar energy computations are expressed in apparent or true solar time. time as measured by the apparent diurnal motion of the sun is called solar apparent solar time or solar time. the length of apparent solar day i. e. interval between two successive passages of the sun through the meridian, is not constant. Local civil time may deviate from true solar time as much as 4.5°.

The difference between local solar time and local civil time is called the equation of time thus

LST = LCT + Eq.of time

Local civil time (LCT) can be derived from the Indian standard time, with the help of the following equation –

$$LCT = \text{Standard time} \pm (L_{st} - L_{local}) \times 4$$

and Solar time

$$LST = \text{Standard time} + E \pm (L_{st} - L_{local}) \times 4$$

(+ve sign for west and -ve sign for east)

Where

E = The equation of time in minutes

L_{st} = The standard meridian for the local time zone and

L_{local} = The longitude of the location in equation in degree east or west

For India, negative sign is taken and hence

$$LST = \text{Indian standard time} + E \pm (L_{st} - L_{local}) \times 4$$

3.8 Solar Radiation Geometry

The location of the sun can be specified in terms of two angles, the solar zenith angle θ_z and solar azimuth angle Y_s. The solar zenith angle is measured from the vertical axis with origin at the observer's position. The azimuth angle is measured in the horizontal plane between a due south line and the projection of the site to sun line on the horizontal plane. Slope β is the angle between the plane surface in question and the horizontal. These angle i.e. zenith angle; azimuth angle and slope are shown in Fig. 3.4.

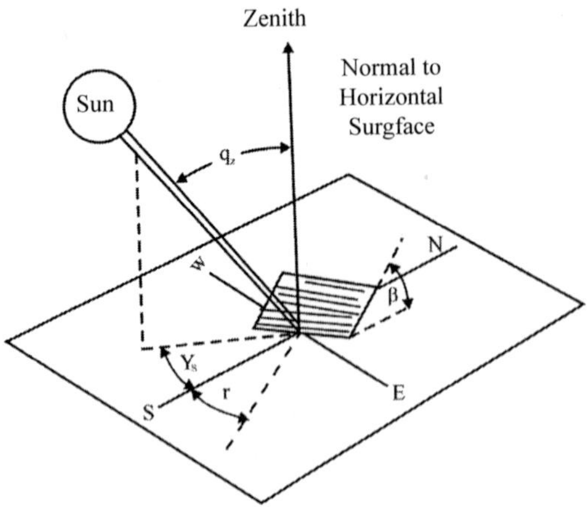

Fig. 3.4: Solar Radiation Geometry

(a) **Sun at Zenith**: The position of sun directly above observer's head

(b) **Solar Zenith angle(θz):** the angle made by the vertical line to the zenith (i.e. the point directly overhead) and ling of sight to the sun i.e. angular distance of sun from zenith is known as zenith angle.

$\cos\theta_z = \cos\delta \cos\varnothing \cos\omega + \sin\delta \sin\varnothing$

(c) **Altitude angle(α_s):** It is the vertical angle between the projection of the sun's rays on the horizontal plane and direction of sun's rays i.e. angular distance of sun from horizontal is known as altitude angle.

$\alpha_s = 90^o - \theta z$

(d) **Solar Azimuth angle (γ_s):** This is the angle between the projection line of sun's rays on the horizontal ground to the center of sun and the south direction. It can be calculated using following equation:

$$\gamma_s = \text{arc cos}\left(\frac{\sin\delta\,\cos\phi - \cos\delta\,\cos\omega\,\sin\phi}{\cos\alpha_s}\right) - 180^o$$

$\gamma_s = -\gamma_{s,}$ if $\sin\omega > 0$

(e) **Air Mass (AM):** It is basically the path length which sunlight takes through the atmosphere normalized to the shortest possible path length(when sun is directly overhead i.e., at zenith). The Air Mass quantifies the reduction of solar power as it absorbed by air and dust. It can be defined as:

$$\text{AM} = \frac{1}{\cos\theta_z} = \sec\theta_z$$

(f) **Geographic Poles of the earth:** The Geographic poles are the place where the axis of rotation intersects its surface. The geographic poles of the Earth, i.e., North Pole and South are located where the line of longitude (Meridian) converges.

(g) **Magnetic Poles of the Earth:** The earth behaves like a strong magnetic bar with its north pole at the geographical south direction and South Pole at geographical north direction. The directions of Earth's geographical poles and magnetic poles are actually opposite to each other. It is because of this reason the compass needle always points towards the magnetic south pole or geographical north pole when kept in the northern hemisphere and towards the magnetic north pole or geographical south pole when kept in the southern hemisphere.

(h) The **equator** is 0° latitude. This imaginary line, which runs through parts of South America, Africa, and Asia, is officially the halfway point between the North Pole and the South Pole.

(i) **Prime meridian:** It is basically 0 degrees longitude. This imaginary line runs through Greenwich, England and is at 0° longitude. By using the equator and prime meridian, we can divide the world into four hemispheres, north, south, east, and west. For instance, the United States is in the Western Hemisphere (because it is west of the prime meridian) and also in the Northern Hemisphere (because it is north of the equator).

(j) **Latitude and Longitude:** Latitude lines run east and west and measure north and south; longitude lines run north and south and measure east and west. The lines measure distances in degrees.

(k) **The Hour angle (ω):** The hour angle (h or ha) of a point on the earth's surface is the angle through which the earth would turn to bring the meridian of the point directly under the sun. The earth is rotating, so this angular displacement represents time. At midday solar hour angle is zero, has a negative value in the morning and a positive value in the afternoon. The variation speed of the solar hour angle is 15 degrees per hour.

(l) **Declination (δ):** The rotation of the earth axis is always inclined at an angle of 23.45° from the ecliptic axis, which is normal to the ecliptic plane. The angular position of the sun at solar noon with respect to the plane of the earth's equator is termed as declination. The declination can be determined from the equation. In other words, we can say the angle between a line joining center of the sun to the center of the earth and the projection of this line upon the earth' equatorial plane. Declinations north of the equator are positive, and that south are negative. The declination ranges from 0° at the spring equinox to $+23.45^o$ at the summer solstice, 0° at the fall equinox, and -23.45^o at the winter solstice as shown in Fig. 3.5. Declination for different months on n^{th} day are presented in Table 3.1.

$$\delta = 23.45 \sin\left[360 \frac{284+n}{365}\right] \quad (1)$$

Table 3.1: Average Days for the months and Value of n by month

Month	n for i^{th} Day of month	For the Average Day of the month		
		Date	n, Day of Year	δ, Declination
January	i	17	17	-20.9
February	31+i	16	47	-13.0
March	59+i	16	75	-2.4
April	90+i	15	105	9.4
May	120+i	15	135	18.8
June	151+i	11	162	23.1
July	181+i	17	198	21.2
August	212+i	16	228	13.5
September	243+i	15	258	2.2
October	273+i	15	288	-9.6
November	304+i	14	318	-18.9
December	334+i	10	344	-23.0

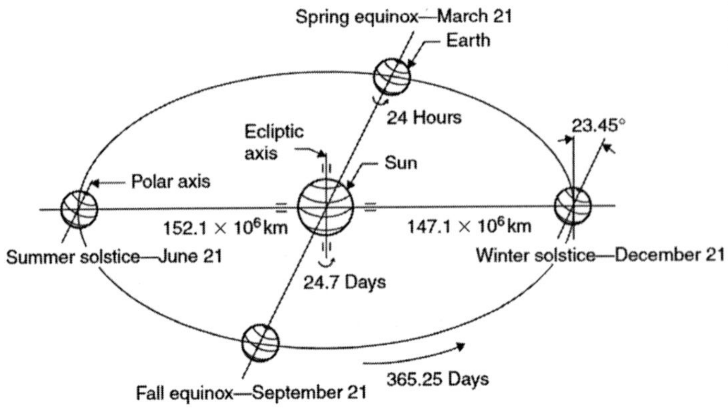

Fig. 3.5: Annual Motion of the Earth about the Sun

(m) **Slope (β):** The angle between horizontal and the plane is known as slope of the plane.

(n) **Incident angle (θ):** The angle measured between the beam of the rays and normal to the plane is known as incident angle

(o) **Surface Azimuth angle (γ):** The angle made by projection normal to the surface with south is known as surface azimuth angle. East Positive and west negative.

(p) From the solar geometry the relationship between angle of incidence of beam radiation on surface θ and other angles are as follows:

cosθ = sinδ sinϕ cosβ – sinδ cosϕ sinβ cosγ + cosδ cosϕ cosβ cosω + cosδ sinϕ sinβ cosγ cosω + cosδ sinβ sinγ sinω (2)

For vertical surface facing the south γ = 0, β = 90 E and incident angle is given by

cosθ = cosδ sinϕ cosω – sinδ cosϕ (3)

For horizontal surface β = 0, the incidence angle is given by

cosθ = cosδ sinϕ cosω + sinδ cosϕ (4)

The sun is said to rise or set when the surface incidence angle is 90E or the altitude angle is zero. The sunset hour angle ω_δ is:

$$\cos\omega_\delta = -\frac{\sin\phi \sin\delta}{\cos\phi \cos\delta} = -\tan\phi \tan\delta \quad (5)$$

(q) **Sunshine Duration (N):** It represents the difference between the daily sunrise and sunset times and is determined by local latitude (φ) and daily solar declination angle. It can be estimated using following formulae:

$N = 2 \cos^{-1} (-\tan Ø \tan\delta)/15$ (6)

The equation (1) is plotted in Fig. 3.6 and its accuracy of prediction is adequate for engineering purpose.

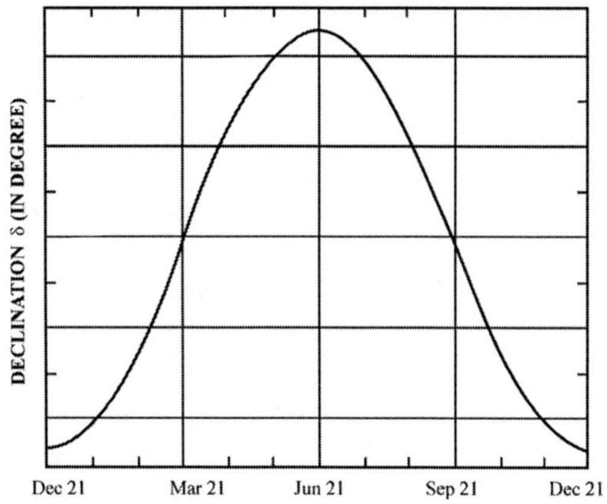

Fig. 3.6: Variation of Declination Over the Year

Example 3.1: Calculate the solar incident angle for a collector that faces due south and has a slop of 24.5° in a city of January 26 at solar noon. The latitude of city is 24.5° N

Solution – For January 26 n = 26 $\gamma_s = 0$

From equation (1), we have

$$\delta = 23.45 \sin\left[\frac{360}{365}(284+26)\right] = -19.03$$

For $\varphi = 24.5°$, $\beta = 24.5°$, $\delta = -19.03°$ & $\omega = 0$

From equation (2), we have

$\cos\theta = \sin(-19.03°)\sin 24.5° \cos 24.5° - \cos(-19.03°) \cos 24.5° \sin 24.5° + \cos(-19.03°)\cos 24.5° \cos 24.5° \cos 0° + \cos(-19.03°)\sin 24.5° \sin 24.5° \cos 0°$

$= \cos(-19.03°) (\cos^2 24.5° + \sin^2 24.5°) \cos 0^0 = 0.95$

Hence the solar incidence angle is

$\theta = 19.03^0$

3.9 Extraterrestrial Radiation on Horizontal Surface

At any point in time, solar radiation on an extraterrestrial horizontal plane is given as,

$$I_t = I_{sc}\,(1+.03)\cos\left(\frac{360\text{ n}}{365}\right)\times(\sin\phi\,\sin\delta + \cos\phi\,\cos\delta\,\cos\omega) \qquad (7)$$

The hourly extraterrestrial radiation on a horizontal surface is obtained by integrating equation (7) for a period of one hour defined by hour angles ω_1 and ω_2 (where ω_2 is the larger).

$$I_o = \frac{12\times3600}{\pi}\times I_{sc}\left(1+0.033\cos\frac{360\text{ n}}{365}\right)$$

$$\times\left[\frac{2\pi(\omega_2-\omega_1)}{360}\sin\phi\,\sin\delta + \cos\phi\,\cos\delta(\sin\omega_2 - \sin\omega_1)\right] \qquad (8)$$

The hourly radiation can also be approximated by the following equation at ω for the midpoint of the hour

$$I_o = 3600\times I_{sc}(1+0.033\cos\frac{360\text{ n}}{365})\times(\sin\phi\,\sin\delta+\cos\phi\,\cos\delta\,\cos\omega) \qquad (9)$$

The daily extraterrestrial radiation on a horizontal surface is obtained by integrating equation (7) over the period from sunrise to sunset. The integration yields.

$$H_o = \frac{24\times3600}{\pi}\times(\cos\phi\,\cos\delta\,\sin\omega_s + \frac{2\pi\omega_s}{360}\sin\phi\,\sin\delta) \qquad (10)$$

Example- 3.2: Compute the radiation on a horizontal surface in the absence of the atmosphere at latitude 24.5° N on January 26 between the intervals of one hour from 8:30 to 9:30

Solution- From example, $\delta = -19.03°$

From equation (8), we have

$$I_o = \frac{12\times3600}{\pi}\times1.353\times10^{-3}\left[1+0.033\,cos\frac{360\times26}{365}\right]$$
$$\times\frac{\{2\pi(-375)-(52.5)\}}{360}sin24.5^o\sin\left(-19.03^o\right)$$
$$+cos24.5^o\cos\left(-19.03^o\right)\left[\sin(-37.5^o)-\sin(-52.5^o)\right]=2.36\frac{MJ}{m^2}$$

From the approximate equation (9), we have

$$I_o = 3600\times1.353\times10^{-3}\left[1+0.033\,cos\frac{360\times26}{365}\right]\times sin24.5^o\sin\left(-19.03^o\right)$$
$$+cos24.5^o\cos\left(-19.03^o\right)\cos(-45^o)=2.37\frac{MJ}{m^2}$$

Similarly, the hourly extraterrestrial radiations on the horizontal plane are computed for other intervals and the result are shown in following table

Time	I_o (MJ/m²)	
	From Eq.(8)	Approx Eq (9)
8:30 - 9:30	2.36	2.37
9:30 - 10:30	3.05	3.06
10:30 - 11:30	3.48	3.49
11:30 – 12:30	3.62	3.64
12:30 – 1:30	3.48	3.49
1:30 – 2:30	3.05	3.06
2:30 – 3:30	2.36	2.37
3:30 – 4:30	1.47	1.48

3.10 Terrestrial Solar Radiation

The sun is a sphere of intensely hot gaseous matter, the heat being generated by various kinds of fusion reactions. In addition to total energy in solar spectrum, it is useful to know the spectral distribution of the extraterrestrial radiation. It should be noted that 99% of the sun radiation energy is obtained upto a wavelength of 4 mm.

Further, solar radiation received on the earth's surface is considered for utilization in solar energy devices & these are different from extraterrestrial radiation for reasons of scattering, absorption etc., in the earth's atmosphere, As solar radiation passes through the atmosphere, the short wave ultraviolet rays are absorbed by the ozone and the longwave infrared rays by CO_2 and moisture. This makes the bandwidth narrow. Terrestrial solar radiation lies within a range of 0.29 - 2.5 µm. Also, water vapor, dust etc, scatter a portion of the radiation and convert it into diffuse radiation. Hence terrestrial radiation consisted of beam & diffuse radiations.

3.11 Estimation of Daily Solar Radiation

Informations on measured solar radiation data are not available for most of the area as and also measurement of solar radiation on hourly basis are rarely carried out in the meteorological stations. However, through the data from nearby locations of similar climate, it is possible to use empirical relationship to estimate radiation from hours of sunshine or cloudiness.

The Angstron equation relates the solar radiation to the number of sunshine hours as:

$$\frac{\overline{H_o}}{H_c} = a + b\frac{\overline{S}}{S_o} \tag{11}$$

where

$\overline{H}_o$ = Monthly average daily radiation on a horizontal surface.

H_c = Average clear sky daily radiation

a,b = constants

$\overline{S}$ = Monthly average of sunshine hours per day

S_o = Monthly average of maximum possible sunshine hours.

others have modified equation in the following form.

$$\frac{\overline{H}}{H_c} = a + b\frac{\overline{S}}{S_o}$$

where $\overline{H}$ = Monthly average daily radiation on an extraterrestrial horizontal surface.

Example 3.3: Estimate the average daily global radiation on a horizontal surface during the month of January at a city, if the average sunshine hour per day is 7.3. Use the following correlation for computation of the average daily global radiation on a horizontal surface:

$$\frac{\overline{H}}{H_o} = 0.18 + 0.39\frac{\overline{S}}{S_o}$$

Solution –

From equation (1) for n = 17 as the average day

$\delta = -20.92^o$

from equation (5) we have

$\cos \omega_\delta = -\tan 24.5^o \tan(-20.92^o) = 0.174$

Hence $\omega_\delta = 79.97^o$

From equation (6), the day length is

$$\bar{S}_o = \frac{2}{15}\omega_\delta = \frac{2}{15}\times 77.91^o$$

From equation (10), we have

$$\bar{H}_o = \frac{24\times1.35\times3600}{\pi\times1000}\left(1+0.033\ \cos\frac{360\times17}{365}\right)\times\frac{2\pi\times80.96}{360}\sin24.5^o\ \sin\left(-20.92^o\right)$$

$$+\cos24.5^o\ \cos(-20.92^o)\ \sin79.97^o$$

$$= 24.20\,\frac{MJ}{m^2}$$

From the equation

$$\frac{\bar{H}}{H_o} = 0.18 + 0.39\frac{\bar{S}}{S_o}$$

We have

$$\bar{H} = \left[0.18+0.39\frac{7.3}{10.66}\right]\times 24.20$$

$$= 10.82\,\frac{MJ}{m^2}$$

3.12 Relationship of Hourly Diffuse and Global Radiation

Estimation of hourly radiation on a tilted surface is required in the simulation and optimization of different solar systems. This necessitates splitting the hourly global insolation into direct and diffuse component for computing total radiation on surfaces of other orientation.

It has been observed that hourly relationship between diffuse ratio (I_d/I_b) and clearness index ($K_T= I_h/I_o$) is location dependent and can not be generalized. Muneer et. al. (1984) developed the following relationship between hourly diffuse and global radiations for New Delhi –

$$\frac{I_d}{I_h} = 0.95;\ \text{when } K_T < 0.175 \tag{12}$$

where

I_d = Hourly diffuse radiation

I_n = Hourly total radiation

K_T = Clearness index

and I_o= Hourly radiation on an extraterrestrial horizontal surface also

$$\frac{I_d}{I_h} = 0.9698 + 0.4353K_T - 3.4499K_T^2 + 2.1888K_T^3$$

where $0.175 > K_T > 0.775$

$$\frac{I_d}{I_h} = 0.26;\ \text{when } K_T > 0.775 \tag{13}$$

3.13 Relationship of Daily Diffuse and Global Radiation

Several researchers have developed many correlations of daily diffuse fraction with daily clearness index. Muneer et al., developed a correlation between daily diffuse and global radiation $(K_T = \frac{H}{H_o})$ for India. The correlation developed by Muneer et al., (1984) is

$$\frac{H_d}{H} = 0.98, \; K_T < 0.2$$

$$\frac{H_d}{H} = 1.024 + 0.47K_T - 3.622K_T^3; 0.2 \leq K_T \leq 0.77$$

$$\frac{H_d}{H} = 0.16, \; K_T < 0.77 \qquad (14)$$

Where

H_d = Daily total diffuse radiation

H = Daily total radiation

3.14 Relationship of Monthly Average Daily Values of Diffuse and Global Radiation

Studies of monthly average values of solar radiation show that the monthly average fraction which is diffuse, $\frac{\bar{H}_d}{H_o}$ is a fraction of $\bar{K}_T\left(\frac{\bar{H}}{H_o}\right)$, where $\bar{H}_d$, $\bar{H}$ and $\bar{H}_o$ are monthly average daily value of diffuse, total and extraterrestrial radiations on a horizontal surface Hawas and Muner (1984) developed the following linear correlation for India.

$$\frac{\bar{H}_d}{\bar{H}} = 1.35 - 1.60\bar{K}_T \qquad (15)$$

where

$\bar{H}_d$ = Monthly average diffuse radiation

$\bar{H}$ = Monthly average daily radiation on a horizontal surface.

The relationship developed by Hawas and Muneer (1984) differ markedly from all other proposed values and predict much higher diffuse radiation than that predicted from other relationship for value of $\bar{K}_T < 0.6$

Collares-Pereira and Rabl (1979) found some effect of changing season upon the relationship between monthly diffuse and total radiations. But Hawas and Muneer (1984) did not find any noticeable effect of location or seasonal variations.

3.15 Relationship of Hourly & Daily Values of Solar Radiation

This can be developed by starting with the daily solar radiation and then estimation of hourly radiation from daily values.

Lin and Jordan (1960) have developed long term hourly and daily insolation values on a horizontal surface. This is widely used to estimate the diffuse and total irradiance at any time of the day. The expression derived by Lin and Jordan for the ratio of hourly diffuse radiation ($r_d = I_d/H_d$) as a function of the hour angle ω and the sunset hour angle ω_s is given as

$$r_d = \frac{\pi}{24} \frac{\cos\omega - \cos\omega_\delta}{\sin\omega_\delta - \frac{\pi}{180}\omega_\delta \cos\omega_\delta} \tag{16}$$

where ω_δ is in degrees.

Percira and Rabl (1979), in their analysis of the data of 5 stations in the USA found close agreement with Lin and Jordan Correlations for both diffuse and total radiation and developed an analytical expression for the ratio of hourly to daily global radiation ($Y_T = I_n/H$) as

$$\gamma_d = \frac{\pi}{24}(a + b\cos\omega) \frac{\cos\omega - \cos\omega_\delta}{\sin\omega_\delta - \frac{2\pi\omega_\delta}{360}\cos\omega_\delta} \tag{17}$$

where

$a = 0.4090 + 0.5019 \sin(\omega_\delta - 60^o)$

$b = 0.6609 - 0.0.4767 \sin(\omega_\delta - 60^o)$

Example 3.4: The total radiation on January 26 was 14.8 MJ/m^2. Estimate the hourly radiation at an interval of one hour from 9 A.M. to 4 P.M. Also compute the diffuse radiation using the correlation proposed by muneer et al.

Solution: From example 2, $\omega_\delta = 80.96^o$

For the 9 A.M. ω = - 45

The coefficient a & b of the equation (17) are

$a = 0.4090 + 0.5019 \sin(80.96^o - 60^o) = 0.5884$

$b = 0.6609 - 0.0.4767 \sin(80.96^o - 60^o) = 0.4904$

From equation (17), we have

$$I_n = \frac{\pi}{24}\left[0.5884 + 0.4904\cos(-45^o)\right] \times \frac{\cos(-45^o) - \cos 80.96^o}{\sin 80.96^o - \frac{2\pi \times 80.96^o}{360} \times \cos 80.96^o}$$

$$= 1.30 \frac{\text{MJ}}{\text{m}^2}$$

From example (2) at 9AM, we have

$I_o = 2.37$, hence $K_T = \frac{1.30}{2.37} = 0.55$

From equation (13), we have

$I_d = 0.9698 + 0.4353 \times 0.55\text{-}3.4499 \times (0.55)\ 2 + 2.1888 \times (0.55)^2$

$$= 069 \frac{MJ}{m^2}$$

Hence $I_b = I_h - I_d$

$$= 1.30 - 0.69 = 0.61 \frac{\text{MJ}}{\text{m}^2}$$

Similarly, the hourly global, beam and diffuse radiations are computed and the results are tabulated in the following table.

Time	I_h (MJ/m²)	I_b (MJ/m²)	I_d (MJ/m²)
9 AM	1.30	0.61	0.69
10 AM	1.82	0.96	0.85
11 AM	2.17	1.24	0.94
12 AM	2.30	1.34	0.96
1 PM	2.17	1.24	0.94
2 PM	1.82	0.96	0.85
3 PM	1.30	0.61	0.69
4 PM	0.75	0.28	0.44

3.16 Estimation of Total Radiation on Fixed Sloped Surfaces

The solar flux incident upon an inclined surface consists of three components. The beam solar flux, the diffuse solar flux from the sky and the diffuse solar flux from the reflection of the global solar radiation by the ground. The diffuse components (ρ) of the sky and ground solar radiation do not depend on the orientation of the inclined surface, but the inclined surface does not receive the diffuse radiations from the entire sky or ground .

The hourly solar radiation on a tilted surface is given by Duffie and Beckwan (1980).

$$I_T = I_b R_b + I_d \left(\frac{1 + \cos\beta}{2} \right) + (I_b + I_d)\rho \left(\frac{1 - \cos\beta}{2} \right) \tag{18}$$

where I_b is the beam radiation on the tilted surface and the geometric ratio, R_b is given by Duffie and Beckman (1980).

$$R_b = \frac{\sin\delta \sin(\phi - \beta) + \cos\delta \cos\omega \cos(\phi - \beta)}{\sin\delta \sin\phi + \cos\delta \cos\phi \cos\omega} \tag{19}$$

Example 3.5: The total radiation on January is 14.8 MJ/m^2. Compute the hourly solar heat flux on a south facing collector with 24.2^o slop during the period of 9 AM to 4 PM at an interval of 1 hour. Use the correlation of Muneer et al. for computation the correlation of Lin & Jordan & Peeira and Rabl for computation of hourly diffuse and global radiation

Solution – From equation (10) for n =26, H_o = 25.28 MJ/m^2

$$H = 14.8\frac{MJ}{m^2} \text{ and } K_T = \frac{14.80}{25.28} = 0.59$$

From equation (14), we have

H_d=[1.024+0.47×0.59-3.622×(0.59)^2+2×(0.59)^2]×14.8

$$= 6.79\frac{MJ}{m^2}$$

From equation (16), we have for ω = -45

$$I_d = r_d H_d = 6.79 \times \frac{}{24} \frac{\cos(-45^o) - \cos 80.90^o}{\sin 80.90^o \quad \frac{}{360} \cos 80.90^o}$$

$$= 0.67\frac{MJ}{m^2}$$

From example (4), I_h = 1.30 MJ/m^2

$$I_b = I_h - I_d = 1.30 - 0.67 = 0.63\frac{MJ}{m^2}$$

From equation (19), we have

$$R_b = \frac{\sin(-19.03^o)\sin 0^o + \cos(-19.03^o)\cos(-45^o)\cos 0^o}{\sin(-19.03^o)\sin 24.5^o + \cos(-19.03^o)\cos 24.4^o \times \cos(-45^o)}$$

=1.41

For ρ = 0.2, from equation (18)

$$I_T = 0.63 \times 1.14 + 0.67\frac{1+\cos 24.5^o}{2} + (0.63 + 0.67) \times 0.2\frac{1-\cos 24.5^o}{2}$$

$$= 1.54\frac{MJ}{m^2}$$

Similarly I_h, I_d, I_b, and I_T are computed for 9 AM to 4 AM. The computed results are shown in a tabular form.

Time	I_h	I_d	I_b	Rb	I_T
9 AM	1.30	0.63	0.67	1.41	1.54
10 AM	1.82	0.82	0.99	1.34	2.12
11 AM	2.17	0.94	1.23	1.31	2.54
12 AM	2.30	0.98	1.32	1.30	2.68
1 PM	2.17	0.94	1.23	1.31	2.54
2 PM	1.82	0.82	0.99	1.34	2.14
3 PM	1.30	0.64	0.66	1.41	1.56
4 PM	0.72	0.40	0.32	1.60	0.91

Example 3.6: The intensity of sunlight impinging on earth is about 900 W/m^2. What is the power emitted by the sun in the spectrum of frequencies? (Earth-Sun distance = 1.5×10^8 km, Earth radius = 6.4×10^3 km)

Solution:

Given

Intensity of solar radiation (I) on earth = 900 W/m^2

Distance between the sun and the earth = 1.5×10^8 km $=1.5 \times 10^{11}$ m

The energy created by the sun is distributed isotropically, thus power (W) emitted by the sun is:

W = Surface area of the sphere at the earth's orbit around the sun (S)×Solar Intensity(I)

$=4\pi r^2 \times I$

$= 4\pi \times \left(1.5\times10^{11}\,\text{m}\right)^2 \times 900\frac{\text{W}}{\text{m}^2}$

$=2.54 \times 10^{26}$ W

3.17 Measurement of Solar Radiation

It is necessary to collect or measure solar radiation over a sufficient period for designing solar energy system. There are basically two types of solar radiation measuring instruments commonly being used namely, pyranometer and pyrheliometer.

(A) Pyranometer

Pyranometer is an instrument to measure the total radiation (direct and diffuse) in term of energy per unit time per unit area, on a horizontal surface within its hemispherical field of view. The working of this instrument is based on the temperature difference between black surface (which absorb most of the solar radiation) and white surface (which generally reflects most of solar radiation). This difference can be detected through the use of thermopiles, temperature measuring devices consisting of thermocouple connected in series, which give mill volt signals that could readily be detected, recorded, and integrated over (Fig 3.7). The instrument is usually calibrated in a horizontal position. If the pyranometer is used at some angles it may produce some error. Some manufacturers, however, have measured such effects and could thus provide corrections at various tilt angles.

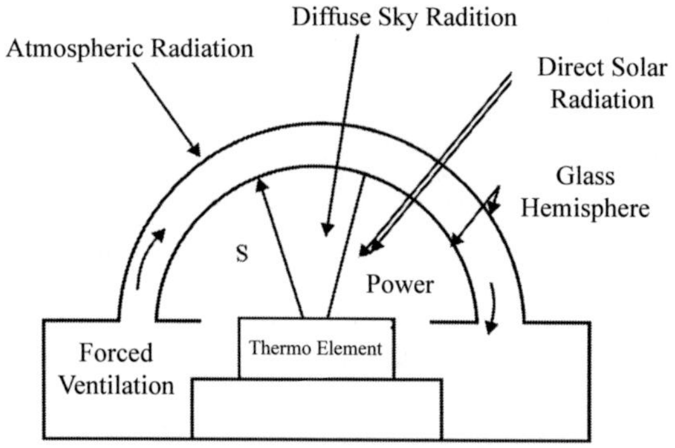

Fig. 3.7: Pyranometer for the Measurement of Global Solar Radiation

Eppley pyranometers, one of the most widely known instruments of the kind, are widely available in the market. It comes in various kinds and depending on user needs, he may prefer one type from the other. Two type of the Eppley pyranometers are available (1) The Eppley Precision Spectral Pyranometer and (2) The Eppley Black White Startype Pyranometers

(i) **The Eppley Precision Spectral Pyranometer (PSP)** It uses a circular wirewound multi junction thermopile developed by Eppley which is coated with a permanent Parsons Eppley black lacquer. The sensing element is mounted under two removable precisionground concentric hemispheres. It has a chromed brass stand with a white enameled grand disk, which provides a hole through which the spirit level may be observed. The Eppley PSP's response time is about one second.

(ii) **The Eppley Black and White Pyranometer** It uses three black (Eppley Parsons permanent black) and three white (barium sulphate) segments. The thermopile is a newly developed radial wirewound plate type with thermistor temperature compensation. The arc is a transparent precisionground optical glass. The response time is usually three to four seconds. Orientation has no effect upon the instruments output, since its segments are symmetrical.

The term solarimeter is used instead of pyranometer. If shaded from the beam radiation by a shade ring, measures diffuse radiation.

(iii) **Photovoltaic Pyranometer** Pyranometer working on photovoltaic solar cell detectors are known as photovoltaic pyranometer. Photovoltaic pyranometer may be used such that silicon cells have short circuit currents at normal radiation level in a linear function with the incident radiation intensity. The inherent spectral sensitivity of silicon cells make them more responsive to the near infrared portion of the solar spectrum than to the visible ultraviolet, and this characteristic must be recognized when they are used as solar pyranometer. On a daylong basis, however, it has been shown that their integrated output agrees within a few percentages with the output of thermopile pyranometer operating under the same conditions.

They have the disadvantages that the spectral response is not linear, so instrument calibration is a function of the spectral distribution of the incident radiation.

In addition, the use of a low resistance with negative temperature characteristics and is thermal but not in electrical contact with the cells virtually eliminates variation of cell output with changes in ambient temperature within the range encountered in actual practice.

(B) Pyrheliometer

It is an instrument for measuring beam radiation. To measure the direct solar radiation, the receiving surface must be normal to the direct solar rays. Diffuse sky radiation is blocked from the sensor surface by mounting the solar sensor at the base of the tube, which total is pointed directly at the sun during the day (Fig.3.8).

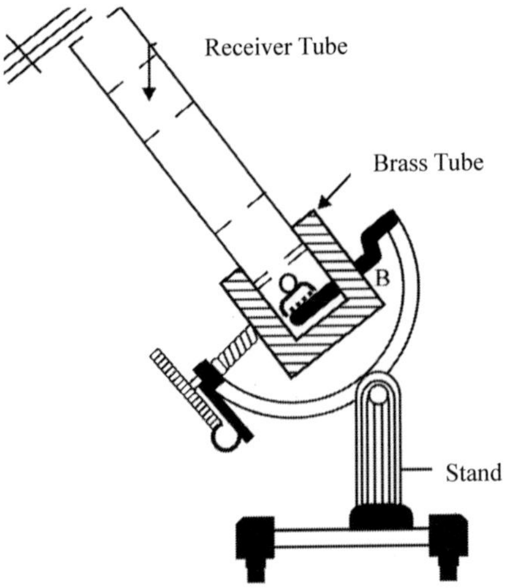

Fig. 3.8: Pyranometer for the Measurement of Global Solar Radiation

The most widely used Pyrheliometer, the Eppley Normal Incidence Pyrheliometer (NIP), uses a multijunction thermopile mounted at the bottom of a brass tube, which is chromeplated externally and the tube to its length is 1 to 10, subtending an angle of 5.73E. The heat sink of the thermopile also contains a thermistor, which compensates for the temperature dependence of the Bismuth Silver wire and keep the response constant within +1 percent over the range of 20 to +40EC. The front flange of the instrument supports a rotatable disk with four apertures, which can contain any three selected filters, while the fourth aperture is left open to permit the full spectral intensity to be measured. The sensitivity of the instrument is taken to be 4.7 mill volts per calorie/cmmin. The response time is one second.

A normal incidence pyrheliometer must obviously "follow thesun" and so the front flange of the NIP has a small hole through which the sun's rays may pass to strike a clearly defined target on the rear flange when the alignment is correct. When the mounting is correctly adjusted for the local latitude and the declination corresponding to the date, the mounting will rotate precisely 15 degrees per hour to enable the instruments to track the apparent motion of the sun across the sky. Other pyrheliometers are also available and may be utilized depending on the system or user preference.

(C) Albedometer

Albedometer measure the amount of short-wave radiation reflected by the ground immediately in front of the collector. This may also be significant since the reflectivity of the ground could increase the amount of short-wave radiation received by the collector.

(D) Pyrgeometer

Pyrgeometer measure the effective long wave outgoing radiation necessary to assess the amount of energy, which a solar collector can lose by radiation during the night. Usually the net atmospheric radiation on a horizontal black surface facing upward at the ambient air temperature is measured with the help of a instrument known as pyranometer. The net atmospheric radiation is the measurement of both solar and terrestrial radiation.

(E) Sunshine Recorders

Sunshine recorders are used to record the duration of bright sunshine during the day. Through this, calculation of global solar radiation can be made by the correlation of the radiation to the amounts of sunshine or clouds. Examples of these are the Campbell Stocks Sunshine recorders. It consists of a glass sphere, (10 cm in diameter) with an axis mounted on a section of a spherical bowl parallel to that of the earth. Here the sphere acts as a lens and it focused sun's image

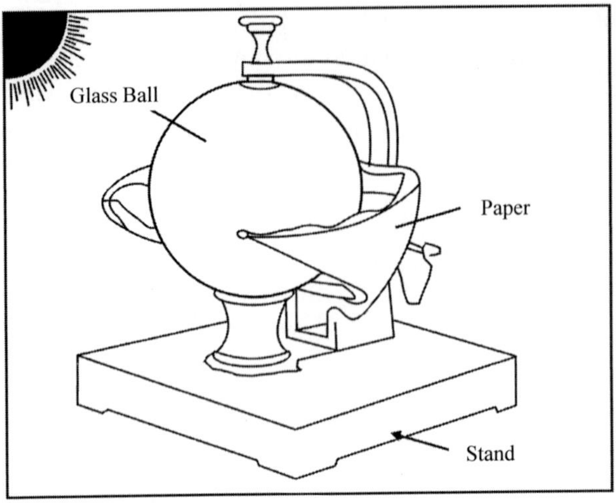

Fig. 3.9: Sunshine Recorder

along a specially prepared paper bearing a time scale. A burning impression on the paper is recorded whenever the solar radiation intensity increases above 200 w/m^2 (Fig 3.9).

4

Heat Transfer Principles

4.1 Introduction

Heat is energy which flows due to difference in temperature. Whereas energy is stored heat. Therefore, when the heat energy transfer is completed. It is stored in the form of potential or kinetic energy, in general as internal energy. Heat as energy is measured in terms of observed changes in other forms of energy & other physical properties. A heat transfer analysis is required for estimating the size, the efficiency and cost of the equipment necessary to transfer a specified amount of heat in a given time. It is well known fact that the size of a solar collector or a heat exchanger depends not so much on the amount of heat to be transmitted but rather on the rate at which heat is to be transferred under given conditions.

4.2 Modes of Heat Transfer

Heat is transferred from one body to another in three possible ways, conduction, convection, and radiation.

(A) Heat Conduction

Conduction is heat transfer by means of molecular agitation within a material without any motion of the material as a whole. If one end of a metal rod is at a higher temperature, then energy will be transferred down the rod toward the colder end because the higher speed particles will collide with the slower ones with a net transfer of energy to the slower ones (Fig. 4.1). For heat transfer between two plane surfaces, such as heat loss through the wall of a house, the rate of conduction heat transfer is:

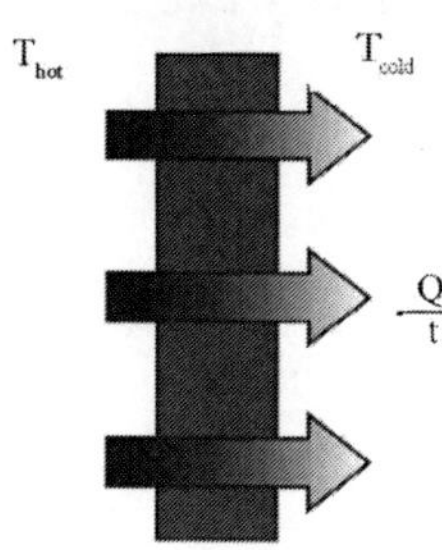

Fig. 4.1: Conduction Heat Flow

$$\frac{Q}{t} = \frac{kA(T_{hot} - T_{cold})}{d}$$

Where:-

Q = Heat Transfer in time t

k = Thermal Conductivity of Barrier

A = Area

d = Thickness of Barrier

(B) Convection

Convection is heat transfer by mass motion of a fluid such as air or water when the heated fluid is caused to move away from the source of heat, carrying energy with it. Convection above a hot surface occurs because hot air expands, and becomes less dense, and rises. Hot water is likewise less dense than cold water and rises, causing convection currents which transport energy (Fig. 4.2)

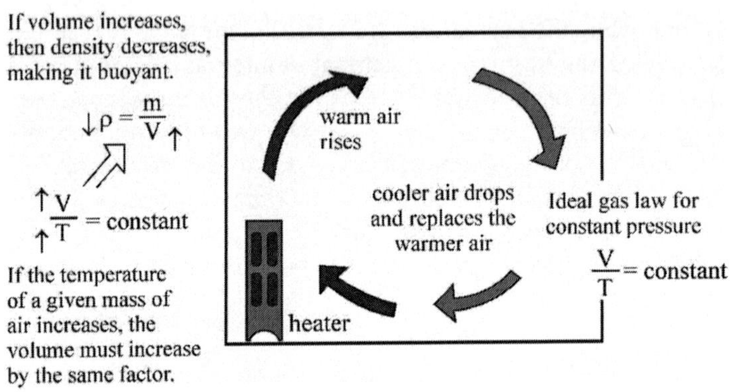

Fig. 4.2: Convective Heat Flow

(C) Radiation

Radiation of heat is the transfer of heat energy in the form of waves in the infrared region of the electromagnetic spectrum. This process can take place in a vacuum, and is in fact the way in which heat from the sun reaches the earth across 150 million km of empty space. All objects above absolute zero radiate heat energy to a greater or lesser extent.

4.3 Steady State Coefficient of Thermal Conductivity and Diffusivity

Steady State: Consider any cross section of a bar heated at one end. Heat received by this section is used in the following three parts:

(i) In increasing its temperature,

(ii) A part is conducted to the next section,

(iii) A part is radiated.

When the temperature of various points of the bar is changing, the state is said as variable state. After some time, a state is reached when the temperature of each cross section becomes steady. This state is known as steady state. In this state, any heat received by any cross section is partly conducted to the next section and partly radiated i.e. no heat is absorbed by the cross section.

Coefficient of thermal conductivity Consider a parallelfaced slab of a material of crosssectional area A and thickness d. Let its faces be maintained at temperature T_1 and T_2 ($T_1 >> T_2$) as shown in Fig. 4.3

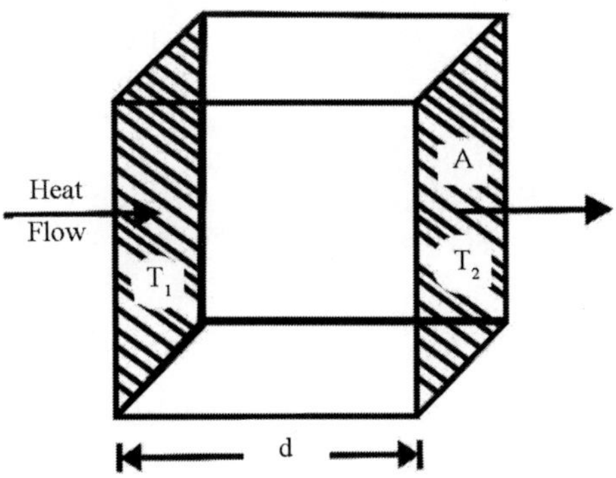

Fig. 4.3: Heat Flow Through Conduction

It has been observed that the quantity of heat Q which passes by conduction, in steady state is

(i) Directly proportional to the area of cross section A,

(ii) Directly proportional to the temperature difference,

(iii) Directly proportional to time t, and

(iv) Inversely proportional to thickness d.

Combining all these, we have

$$Q \ \alpha \ \frac{A(T_1 - T_2)t}{d}$$

or

$$Q = \frac{kA(T_1 - T_2)t}{d}$$

where the constant k is known as Coefficient of Thermal Conductivity.

The quantity $\frac{T_1 - T_2}{d}$ is known as Temperature Gradient. Sometimes, this is denoted by dT/dx

$$Q = kA\frac{dT}{dx}$$

The Coefficient of Thermal Conductivity is equal to the quantity of heat flowing per second through unit area of cross section of the material of unit thickness under unit temperature gradient when steady state is reached.

The unit of k $= \dfrac{W}{m\ ^{\circ}C}$

Thermometric Conductivity or Diffusivity: The diffusivity is defined as the ratio of coefficient of thermal conductivity to the thermal capacity per unit volume of the material.

Thermal capacity per unit volume = d × s

where d is the density of material and s is its specific heat.

Diffusivity D $= \dfrac{k}{d \times s}$

Thermal Resistance The thermal resistance of a body is a measure of its opposition to the flow of heat through it. It is defined as

$$\text{Thermal resistance} = \frac{\text{Temperature difference at the two ends}}{\text{Rate of flow of heat through it}}$$

It can easily be shown that –

$$\text{Thermal resistance} = \frac{\text{Lenght or thickness of the material}}{\text{Thermal conductivity} \times \text{Area}}$$

$$\text{Thermal resistance} = \frac{1}{kA}$$

4.4 Heat Flow through a Composite Slab

Consider a composite slab made of two different materials A and B of thickness d_1 and d_2 as shown in Fig 4.4 with thermal conductivity k_1 and k_2. Let T_1 and T_2 be the temperature of the end faces and T the temperature at common surface.

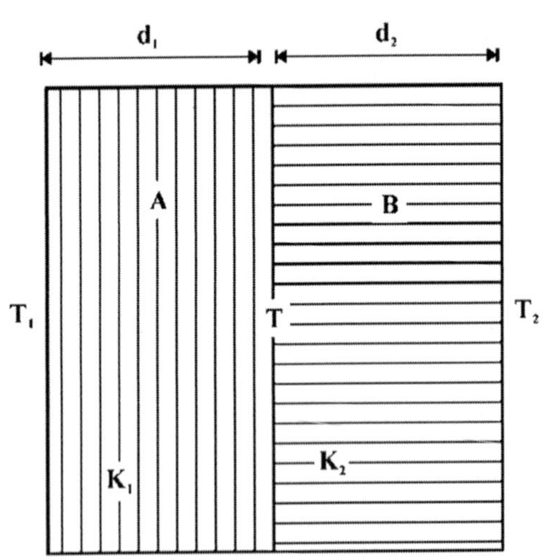

Fig. 4.4: Heat Flow Through Composite Slab

The rate of flow of heat through A

$$\frac{Q}{t}=\frac{k_1A(T_1-T)}{d_1} \qquad (1)$$

In the steady state, the rate of flow of heat thorough B is the same i.e.

$$\frac{Q}{t}=\frac{k_2A(T_1-T_2)}{d_2} \qquad (2)$$

From equations (1) and (2), we have

$$\frac{k_1A(T_1-T)}{d_1}=\frac{k_2A(T_1-T_2)}{d_2}$$

Solving we get

$$T=\frac{\frac{k_1T_1}{d_1}+\frac{k_2T_2}{d_2}}{\frac{k_1}{d_1}+\frac{k_2}{d_2}} \qquad (3)$$

From equation (3) substituting the value of T in equation (1) we get

$$\frac{Q}{t}=\frac{k_1A}{d_1}\left[T_1-\left\{\frac{\frac{k_1T_1}{d_1}+\frac{k_2T_2}{d_2}}{\frac{k_1}{d_1}+\frac{k_2}{d_2}}\right\}\right]$$

$$=\frac{k_1A}{d_1}\left[T_1-\left\{\frac{k_1T_1d_2+k_2T_2d_1}{k_1d_2+k_2d_1}\right\}\right]$$

$$=\frac{k_1A}{d_1}\left[\frac{k_2d_1(T_1-T_2)}{k_1d_2+k_2d_1}\right]$$

$$=\frac{k_1A}{d_1}\left[\frac{k_2d_1(T_1-T_2)}{k_1k_2\left(\frac{d_2}{k_2}+\frac{d_1}{k_1}\right)}\right]$$

$$=\frac{A(T_1-T_2)}{\left(\frac{d_1}{k_1}+\frac{d_2}{k_2}\right)}$$

In general

$$\frac{Q}{t}=\frac{A\left(T_1-T_2\right)}{\sum\frac{d}{k}}$$

4.5 Radial Flow of Heat through a Cylindrical Tube

Consider the case of a cylindrical tube of length inner radius r_1 and outer radius r_2 as shown in Fig. 4.5. Let the tube be heated along its axis by placing a radially from the inner side towards the outer side across the walls of the tube. In the steady, let T_1 and T_2 be the constant temperature of the inner and outer surfaces respectively.

Consider a thin cylindrical shell of thickness dr at a distance r from the axis. Let. T and T+dT be the temperatures of the inner and outer surfaces of this cylindrical shell respectively. The rate of heat flowing through this shell is given by

$$Q = (-)k2\pi r\frac{dT}{dr} \quad (1)$$

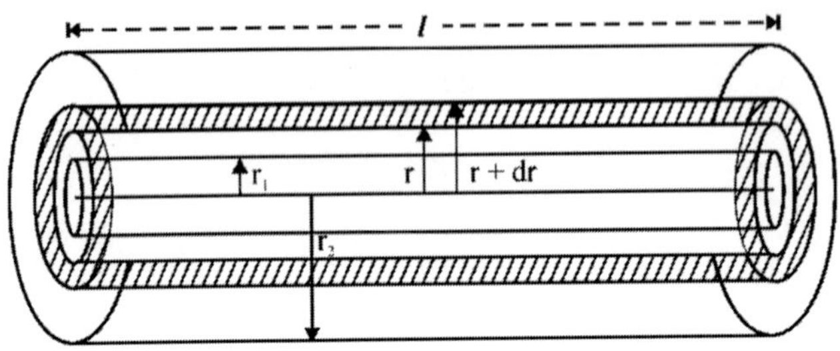

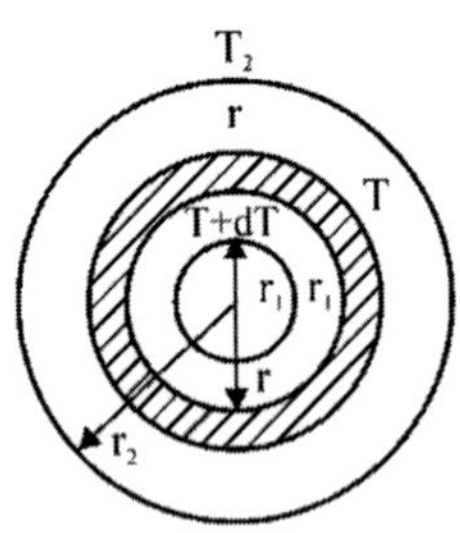

Fig. 4.5: Heat Flow Through Cylindrical Tube

Where The negative sign shows the T diminishes with an increase in r.

Rearranging equation (1) we have

$$Q\frac{dr}{r} = -2\pi \text{ k l dt} \quad (2)$$

In the steady state, Q must be same for all values of r. Hence we can integrate this equation between the limits r_1 and r_2 for and T_1 for T_2 Thus we have –

$$Q\int_{r_1}^{r_2}\frac{dr}{r} = -2\pi \text{ k l}\int_{T_1}^{T_2} dT$$

$$Q\log_e\frac{r_2}{r_1} = 2\pi \text{ k l}(T_1 - T_2)$$

$$k = \frac{Q\log_e\left(\frac{r_2}{r_1}\right)}{2\pi \text{ l}(T_1 - T_2)} \quad (3)$$

$$Q = \frac{2\pi\, l(T_1 - T_2)}{\log_e\left(\frac{r_2}{r_1}\right)} \qquad (4)$$

4.6 Heat Transfer by Convection

The equation of heat transfer through convection is given as

$Q = h\, A\, \Delta t$

where

h = Coefficient of heat transfer unit is $W/m^2\, K$

A = Area of heat transfer in m^2

Δt = Temperature difference causing heat flow.

The value of `h' depends upon physical properties of fluid like density, thermal conductivity, specific heat, viscosity and its dynamic characteristic like density. For simplifying the calculation, following equations are used in convective heat transfer

N_u = Nusselt Number = $\frac{hD}{k}$

R_e = Reynolds Number = $\frac{\rho D}{\mu}$

P_r = Prandtl Number = $\frac{\mu C_p}{k}$

G_r = Grashof number = $\frac{l^3 \rho^3 \beta g \Delta t}{\mu^2} = \frac{g\beta \Delta t L^3}{\gamma^2}$

Where h is heat transfer coefficient

D is plate spacing (L)

k is thermal conductivity

β is volumetric coefficient of expansion of air.

Δt is temperature difference between plates

ν is kinematic viscosity

μ is dynamic viscosity

C_p is specific heat

ρ is fluid density

γ is kinematics viscosity

For Natural Convection

For $G_r P_r$ ranging from 10^5 to 10^{18}

For vertical planes $N_u = 0.56\ (G_r.P_r)^{0.25}$

For $G_r.P_r$ exceeding 10^8

$N_u = 0.12\ (G_r \times P_r)^{0.33}$

For Forced Convection

Flow over the plate laminar flow ($Re < 10^5$)

$N_{u\ average} = 0.664\ Re^{0.5}\ Pr^{0.33}$

Flow over the plate, turbulent flow ($Re > 10^5$)

$Nu = 0.023\ P_r^{0.66}\ R_e^{0.8}$

For flow through pipe

$N_u = 0.023\ P_r 0.66 \times R_e 0.8$

4.7 Combined Conduction and Convection Heat Transfer

The heat transfer from flat plate placed in air, then the heat will be transferred from air to plate and then air by convection (Fig. 4.6).

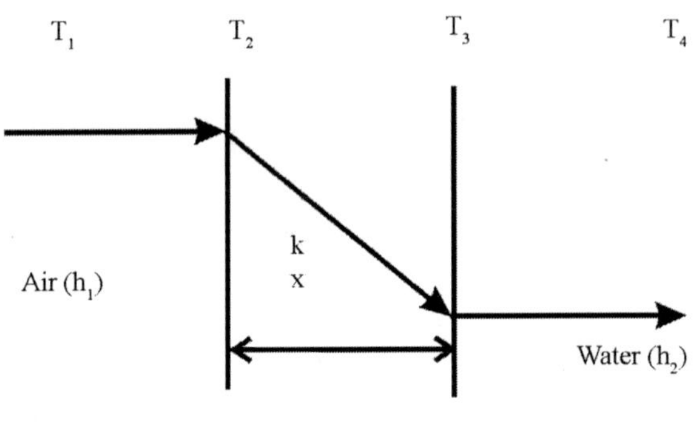

Fig. 4.6: Combined Effect of Conduction and Convection Heat transfer

Consider a flat plate of thickness x and having thermal conducting K, placed with hot air on one side and cold water to the other. Let temperature of air is T_1 and that of waters T_4. Temperature of plate faces be T_2 and T_3. Then heat transfer equations are

$$Q = h_1 A (T_1 - T_2)$$

$$= \frac{kA(T_2 - T_3)}{x} = h_2 A (T_3 - T_4)$$

Combining these equations

$$Q = \frac{T_1 - T_4}{\frac{1}{h_1} + \frac{x}{k} + \frac{1}{h_2}} A$$

Let U be overall heat transfer coefficient, then

$$Q = U\,A\,(T_1 - T_4)$$

Where

$$\frac{1}{U} = \frac{1}{h_1} + \frac{x}{k} + \frac{1}{h_2}$$

4.8 Radiation Incident on a Surface

The radiation incident on a body is partly absorbed, party transmitted and remainder is reflected. The Fig 4.7 illustrates such a situation.

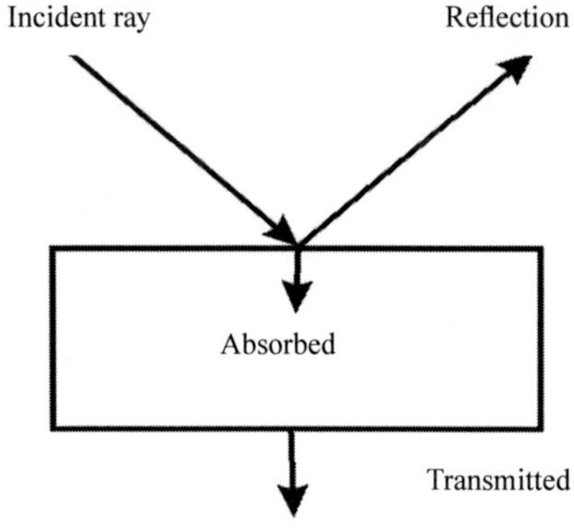

Fig. 4.7: Radiation Incident on a Surface

From the above definition

Reflectivity + Absorptivity + Transmissivity = 1

or $\rho + \alpha + \lambda = 1$

For a perfect reflection $\rho = 1$

For a non-reflective surface $\rho = 0$

A body with $\alpha = 1$ is called a black body and for a perfect absorber $\alpha = 1$

Also $\alpha = 0$ is non absorbing surface also $\lambda = 0$ represents air opaque surface

Further $\lambda = 1$ represent a perfectly transparent surface

also $\lambda = 0$ represents an opaque surface.

ReflectivityIt is the ratio of reflected energy to the incident energy. It is denoted by ρ (Rho).

Absorptivity It is the ratio of absorbed energy to the incident energy. It is denoted by α.

Transmissivity It is the ratio of transmitted energy to the incident energy. It is denoted by λ (gamma).

4.9 Newton's Law of Cooling

Consider the case of a body at temperature T_1 placed in surrounding at temperature T_2. It has been observed that the rate of loss of heat i.e., rate of cooling dQ/dt is directly proportional to the temperature difference (T_1T_2) or

$$\frac{dQ}{dt} \mu \left(T_1 - T_2\right)$$

Where dQ is the quantity of heat lost in time dt.

$$\frac{dQ}{dt} = -k\left(T_1 - T_2\right) \qquad (1)$$

This is known as Newton's law.

Thus, according to Newton's law, the rate of cooling is directly proportional to the excess of temperature.

We know that heat lost by a body depends upon its heat capacity. If m be the mass of the body and s its specific heat, then

$$\frac{dQ}{dt} = m\,s\frac{dT}{dt} \qquad (2)$$

This equation with Newton's law [eq. (1)] can be used to find the specific heat capacity of the liquid.

From eqns. (1) and (2),

$$\frac{dT}{dt} = -\frac{k}{m\,s}\left(T_1 - T_2\right)$$

or

$$\frac{dT}{dt} = -k\left(T_1 - T_2\right)$$

4.10 Stefan's Law / Heat Transfer by Radiation

According to Stefan's law, the total radiant energy E emitted per second from unit surface area of a black body (A perfectly black body is one which completely absorbs all the radiation falling on it irrespective of the wavelength) is proportional to the fourth power of its absolute temperature T. Thus

$E \alpha T^4$

or $E = \sigma T^4$

Where σ is known as Stefan's constant ($5.67 \times 10^8 W/m^2 K^4$)

When a black body at absolute temperature T_1 is surrounded by another black body at temperature T_2, then the net loss of energy per second per unit area of the former is given by

$E = \sigma (T^4_1 T^4_2)$

In case the body is not a perfectly black body, then amount of heat energy radiated per unit time per unit area will be

$\Delta E = e \sigma (T^4_1 T^4_2)$

Where e = Emmisivity of the surface, whose value depends upon the nature of surface

If A is the surface area of the body, then

$\Delta E = e A \sigma (T^4_1 T^4_2)$

4.11 Emissive Power, Absorptive Power and Kirchoff's Law

Emissive Power. The emissive power of a body, at a given temperature and for a particular wavelength is defined as the radiant energy emmitted per second by unit surface area of the body per unit wavelength range.

Absorptive Power. The absorptive power of a body, at a given temperature and for a particular wavelength is defined as the radiant energy absorbed per second by unit surface area of the body to the total energy falling per unit time on the same area.

Kirchoff's Law. This law states that the ratio of the emissive power to the absorptive power for radiation of a given wavelength is the same for all bodies at the same temperature and is equal to the emissive power of a perfectly black body at that temperature.

Wein's Displacement Law. The wavelength corresponding to maximum energy is inversely proportional to the Kelvin temperature and is given by

$\lambda m \, \alpha \frac{1}{T}$

or $\lambda m \, T$ = constant

The value of this constant is found to be 2.93 x 10^3 mK.

The knowledge of heat transfer through conduction, convection with special reference to radiation is important for design of solar energy devices. The Stefan's law is basically used to design black body used in solar devices.

Example. 4.1: Evaluate the heat transfer by conduction through a copper rod (R=300W/mK) of 20 cm length and cross sectional area of 1 cm^2. One end of the rod is maintained at 160°C, while the other at 60 °C. The curved surface of the rod is insulated.

Solution

$$\frac{Q}{t} = \frac{k\,A\left(T_{hot} - T_{cold}\right)}{d}$$

where K = 360 W/m K

A = Area of cross section = 1 cm^2

= 1 x 10^{-4} m^2

d = 20 cm = 0.2 meter

So. Q = [(300x1x10 – 4(160 – 60)]/0.2

= 15 W

Example 4.2: Calculate the heat transfer in water from a 25 mm diameter pipe in an solar water heater, if it is maintained at a temperature of 77 °C. Assume ambient water temperature is at 27 °C

Solution

T_1 = 77 + 273 = 350 K

T_2 = 27 + 273 = 300 K

Now, heat exchange by radiation from a black body to its surrounding is given by

$$H = \sigma A(T_1^4 - T_2^4)$$

$$H = 5.67 \times \frac{\pi}{4}\left(\frac{25}{1000}\right)^2 \left[\left(\frac{T_1}{100}\right)^4 - \left(\frac{T_2}{100}\right)^4\right]$$

$$= 5.67 \times 6.25 \times 10^{-4} \times \frac{\pi}{4}[150 - 81]$$

= 1919 x 10^{-4} Watt = 0.19 W

5

Solar Thermal Energy

5.1 Introduction

Energy in the form of heat is one of the main energy requirements in domestic, agricultural, industrial and commercial sector of our economy. From an enduse point of view, solar thermal energy finds its application in these areas. Solar thermal energy devices convert radiant energy of the sun into thermal energy for different productive works. In fact, solar energy consists of infrared radiation, it is characteristics of infrared radiation that whenever it falls on any object, it converts into thermal heat energy.

The conversion of sun light into thermal energy is easily or conveniently achieved by means of a metallic or plastic cover, painted black with ordinary black board paint or having selected coating over it and covered with one- or double-glazing cover for transmitting solar radiation inside it. This metallic cover is known as absorber, below it either channels or pipes or passages for allowing entry of water or air are provided, through which heat energy is transferred to working medium either air or water. The absorber plate is covered with suitable thickness of insulating material on its back or sides for preventing loss of heat from the absorber plate. This complete system is enclosed in an airtight box. When exposed to solar radiations, the blackened metallic surface absorbs solar energy and converts into heat. This heat could be used for cooking of food, heating of water and air, evaporation of moisture form the grains etc. In fact, absorber works on the principle of black body, which absorbs maximum and a good absorber, is a good emitter. Hence generated heat could be utilized for different applications.

5.2 Solar Energy Collection

The collection of the sun's heat is by means of collectors. The various types of collectors are used in solar energy utilization (Fig 5.1).

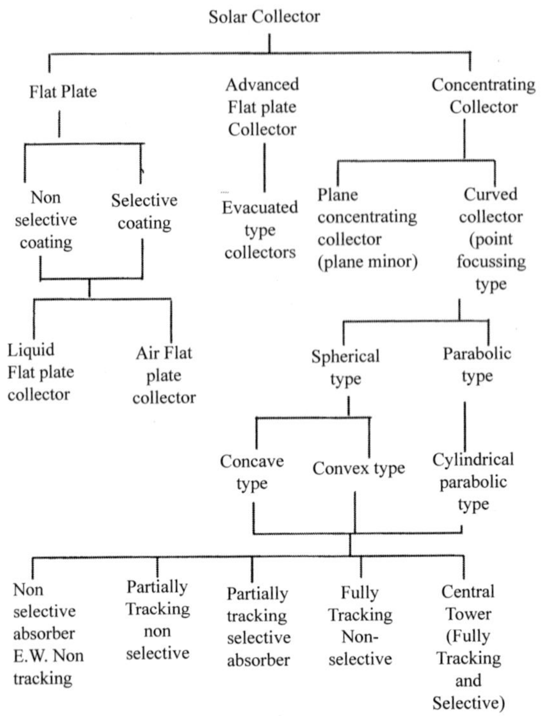

Fig. 5.1: Various Types of Collector

It has been observed that for entire temperature range required for domestic, agriculture and industrial sectors could be covered with the available technologies of conversion of solar energy into thermal energy. Also, thermal energy can be converted into electricity. The available technologies for solar thermal conversion can be categorized in Table 5.1.

Table 5.1: Application and System for use of Solar Energy

S.No.	Type of Application & Uses	Type of Conversion Device
1.	Low temperature applications (40-100 °C), For water heating drying, cooking & distillation	Use of flat plate collectors
2.	Medium temperature applications (100-350 °C) for heating, drying, steaming, industrial process heat and frying/ baking/ roasting	Use of concentrating parabolic collectors, evacuated tube collectors
3.	High temperature applications (350° to more than 1000°C) for electricity generation and industrial process heat.	Use of highly concentrating reflector system and point focusing collector and step type or multi facet type parabolic reflector.

There are large numbers of focusing concentration type collectors which may be used in solar energy-based devices. Some of these are as follows.

1. Parabolic Reflector
2. Step Reflectors or Fresnel Reflector
3. Spherical Reflector
4. Multi facet spherical reflector
5. Cylindricalparabolic reflector

Depending on types of application and desired temperature range these concentrators are used. The names of these concentrator itself explain their working.

5.3 Flat Plate Collector

A basic flat plate collector is the simplest and most widely used means to convert the sun's radiation into useful heat for meeting essential energy requirement by heating air, water or liquid. A basic flat plate collector essentially consist of following elements (Fig 5.2):

(a) An absorbing plate essentially flat made of metal upon which solar radiation falls and is absorbed and charged into heat or thermal energy. In fact, it is a black body, which absorbs and simultaneously emits the heat radiation for productive work .

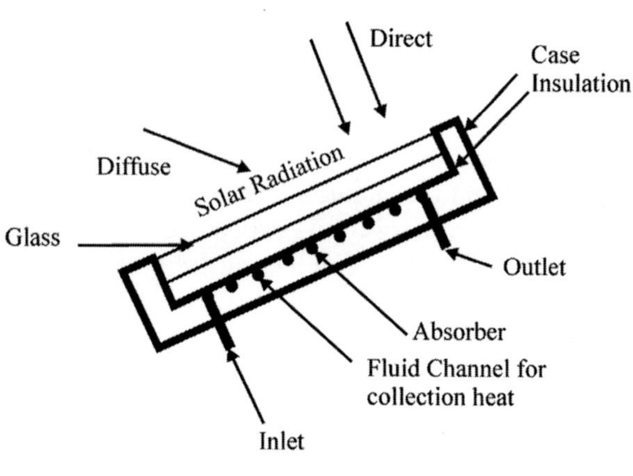

Fig. 5.2: Schematic of Flat Plate Collector

(b) Tubes, channels or passages connected to the flat absorbing plate used for circulation of fluid into, so that thermal or heat energy of absorbing plate may be transferred to fluid for further application or utilization.

(c) An insulating material sufficiently packed at the back and side of the absorbing plate so that conduction heat losses from the absorbing plate may be minimized and hence to maximize efficiency of the system. A perfect insulation of absorber plate can reduce cost of collector, which otherwise is required to meet desired capacity.

(d) A transparent cover (one or two sheets) of glass on transparent plastic for reducing upward convection and radiation heat loss from the absorber plate. Single glazing or double-glazing depending on requirement may be recommended. Double-glazing may additionally permit air as insulation in between it, hence provide barrier to heat radiation out flow from the black body.

(e) A weather tight container/ enclosure, which encloses all above components of the collector and provide good stability against weather conditions.

These solar collectors are classified into two main types based on the type of fluid used for heat transfer. These are known as liquid collectors and air collectors. The main difference between them is in the design of the passage / tube / channels for transfer of heat through fluid.

5.3.1 Liquid Flat Plate Collector

These types of collectors are generally at temperature less than 100°C and may be used in systems for supplying hot water in domestic, agricultural and industrial applications. There are three basic types of collectors' plates which may be used for liquid heating. This classification is based on the extent of wetted surface area relative to the absorbing surface area. These are as follows (Fig 5.3).

(a). **Pipe and Fin type:** Where liquid flows only in the pipe and hence has comparatively less wetted area and low liquid capacity. It has property of good corrosion resistance; however, it is quite expensive as compared to other. Here heat transfer takes place by conduction from absorber to pipe. Hence high thermal conductivity is required.

(b). **Sandwich type:** In this case rectangular or cylindrical sandwich type system in used, where the wetted area and water capacities are high.

(c). **Semisandwich type:** This is also known as Role bond type system. It is comparatively cheaper and lighter than other type. However, very susceptible to internal salt deposition.

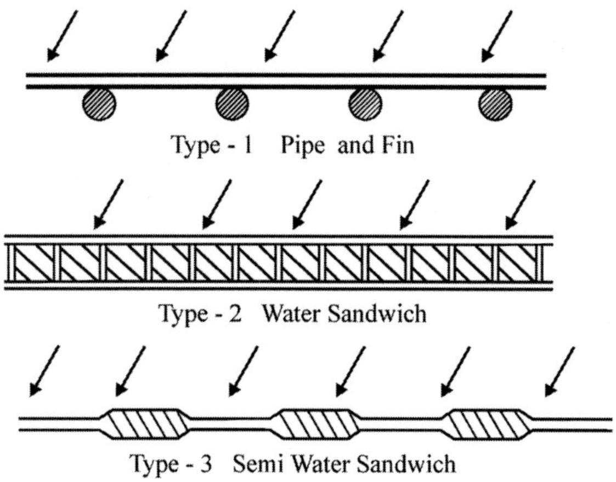

Fig. 5.3: Basic Water Collector Plate Type

5.3.2 Air Flat Plate Collector

The air is also used as a medium of heat transfer for many applications related to work of domestic, agricultural & industrial sectors. The flat plate collectors are also used for heating air, but in contract to liquid solar collectors. The difference between liquid flat plate collector and air collectors, is the mode of heat transfer between the absorber plate and the heated fluid. Generally, fin tube configuration is used for liquid flat plate collector. Where heat absorbed is transferred to the water tubes by conduction. Therefore, high thermal conductivity is required for improving efficiency of liquid flat plate collectors. While in air collectors, the air stream can be in contact with the complete absorbing surface, the plate conductivity becomes of small importance. Further, there is no corrosion of the absorber plate and hence light gauge metal sheets can be used in air collectors. Further air collectors do not require pressurized system and make plumbing work easy and also leakage is not big problem. Viewing to this relative simplicity, air collectors are easy to maintain and repair.

The working principle of air collector is virtually the same as that of the liquid flat plate collector, here also, air is circulated in contact with a black radiation absorbing surface which is having one or two transparent covers for heat loss reduction. Beneath the glazing, metallic absorber surfaces are used. Where air is circulated for getting heat of solar energy. The air heat plate collectors are classified into six categories, based on the type of absorbing surface (Fig 5.4).

(a) Simple flat plate collector

It is simple form of air collector, which is composed of one or two glazing or transparent cover over a absorber flat plate packed by proper insulation on the back and side of the collector. Here the airflow path may be either above or below or both above and below the absorber plate.

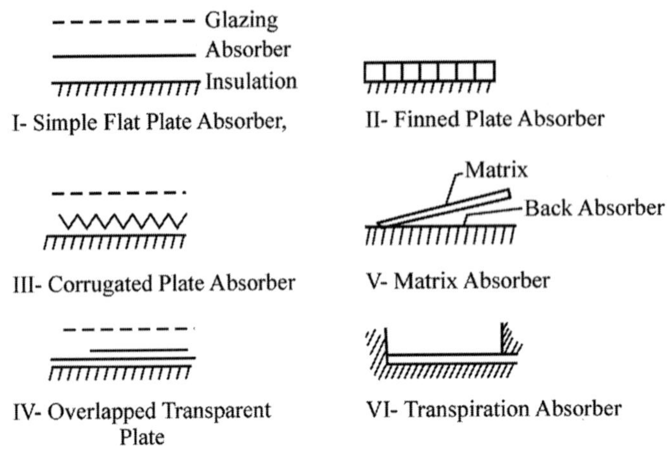

Fig. 5.4: Classification of Air Heaters based on Absorbing Surface

(b) Finned plate collector

In this type of system, the heat transfer area is enhanced by providing fins on the flat plate absorber, which are located in the air flow passage, hence more efficiency of solar energy collection and distribution.

(c) Corrugated plate collector

As the name indicated, in this case the absorber in corrugated rounded or Vtype triangles. This increases the heat transfer area and may make the surface directionally selective. However, more costly on account of more metallic sheet requirement for corrugation.

(d) Overlapped transparent plate type collector

This type of collectors is consisted of a staggered array of transparent plates, which are partially blackened. Here the airflow path is between the overlapped plates.

(e) Matrix type collector

In this case, an absorbing matrix is placed in the airflow path between the glazing and absorber plate. Through this, high heat transfer to volume ratio is obtained. Also, it offers low friction losses depending on the design. The material of matrix may be cotton gauze or loose packed porous material.

(f) Transpiration collector

It is modification of matrix type of collector. Hence matrix material is closely packed and the back-absorber plate is eliminated. Further, in this type of collector, air enters just under the inner most covers and flows down ward through the porous bed and into the distribution duct.

5.4 Solar Cooking

Cooking food with solar energy is a simple application of solar thermal technology. Solar cooker works on sun's energy and need no fuel, emit no smoke, no soots spoil the cooking utensils as well as kitchen walls and it keeps the environment cleans. Solar cooker is a device which demonstrate all fundamental of solar energy including transmittance, absorption, reflectance and black body concept.

The principle methods of cooking food are boiling, frying, roasting and baking. In conventional food habit the boiling technique is employed for preparation of pulse, vegetable and rice etc. During this process the temperature of food being cooked is about 100°C. For other methods of cooking higher temperatures are required. For frying and boiling, heat is supplied from all the sides of the material being cooked and heat to the food is being transferred through convection and radiation. In most of food, water is already present in raw material. Further, some water is also added to get homogenous mixture of food for cooking. In such cases, once the boiling temperature (Close to 100°C) is reached, not much heat is required except the heating rate should be equal to the rate of thermal loss from the vessel. The thermal losses from the cooking vessels are: evaporation loss from the food, radiation and convection loss from the surface of the cooking vessel etc. Hence in the design of solar cooker attention be given to following points:

1. It should provide sufficient heat for quick boiling of water so that food may be cooked as soon as possible.
2. Heat losses are minimized by incorporating insulation on the sides of vessel and keeping vessel covered with a lid.
3. The temperature requirement for cooking food by boiling is about 100°C, but to have high heat transfer rates, the temperature of the heat source should be high. Therefore, a solar collection of sufficient size be adopted for supplying heat higher than 100°C or so.

There are three broad categories of Solar Cookers available for cooking of food. The classifications of solar cooker are given in Fig. 5.5.

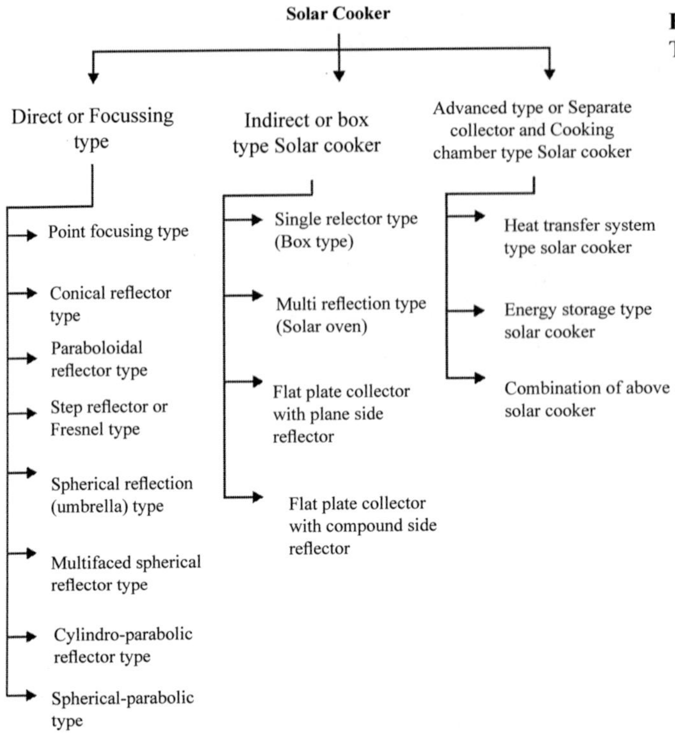

Fig. 5.5: Different Type of Solar Cooker

5.4.1 Brief History of Solar Cookers

Before the age of civilization, the cooking of food was unknown. People ate food in the condition in which they found it. Solar energy was the first time of energy used to heat wafers, to create a food source that is extremely healthy for the human body. The first known person to build a box to cook food using solar energy was Horace de Saussure, a Swiss naturalist who published his work in 1767. He cooked fruits in a primitive solar box cooker that reached temperatures of 88 °C. He became known as the grandfather of solar cooking. In the same era, in India, a British soldier patented a fairly sophisticated solar cooker that looked a lot like the Solar Chef. In 1894, China opened a restaurant in which solar cooked food was served. The present design of the solar cooker started evolving in the 1950s. A number of top engineers, scientists, and researchers were hired to study different aspects of solar cooking designs. These studies concluded that properly constructed solar cookers not only cooked food thoroughly and nutritiously, but were quite easy to make and use. In 1945, Sri M. K. Ghosh designed the first box-type solar cooker as a commercial product. Indian scientists designed and manufactured a number of commercial solar ovens and solar reflectors in 1950, but they were not readily accepted, partly because lower-cost alternatives were still available.

5.4.2 Direct or Focusing Type Solar Cooker

As the name indicates in this type of solar cooker same kind of solar energy concentrator (reflector) is used, which ultimately reflect sun's rays to a common point (focus or focal point) on which a cooking pot is placed. In this type of solar cooker, the cooker utilizes the direct radiation only.

Presently, there are two commercially available solar cookers in this category. Which are as follows:

(a) Dish Solar Cooker : It is concentrating type parabolic dish solar cooker with aperture dimension of 1.4 meter and focal length of 0.28 meter. It is suitable for individual and also for small establishments like dhabas, tea shops etc. The reflecting material used for fabrication of this cooker is aniodized aluminium sheet, which has a reflectivity of over 75%. The reflecting bowl is paraboloid dish made of reflecting sheets supported on suitable rings for holding them in fixed position. The sheets are joined in such a way that they automatically form the parabolic shape. The structure and frame of the bowl is strong enough so that the reflectors do not get deformed while turning it is various directions.

Second important part of this cooker is reflector stand, which is made of mild steel having powder coating for better durability. The stand is such that it can rotate reflector 360° around the horizontal axis passing through the focus and the center of gravity. It is also rotated around the vertical axis so it can adjust the cooker in the direction of sun. The tracking of the cooker is manual and has to be adjusted in 15 to 20 minutes during cooking time. The cooker has a delivering power of about 0.6 kw which can boil 2 to 3 liters of water in half an hour. The temperature achieved at the bottom of the vessel is around 350 to 400°C, which is sufficient for roasting, frying and boiling. This cooker is having a thermal efficiency of around 40% and can meet the needs of around 15 people and used from one hour after sunrise to one hour before sunset on clear days (Fig. 5.6)

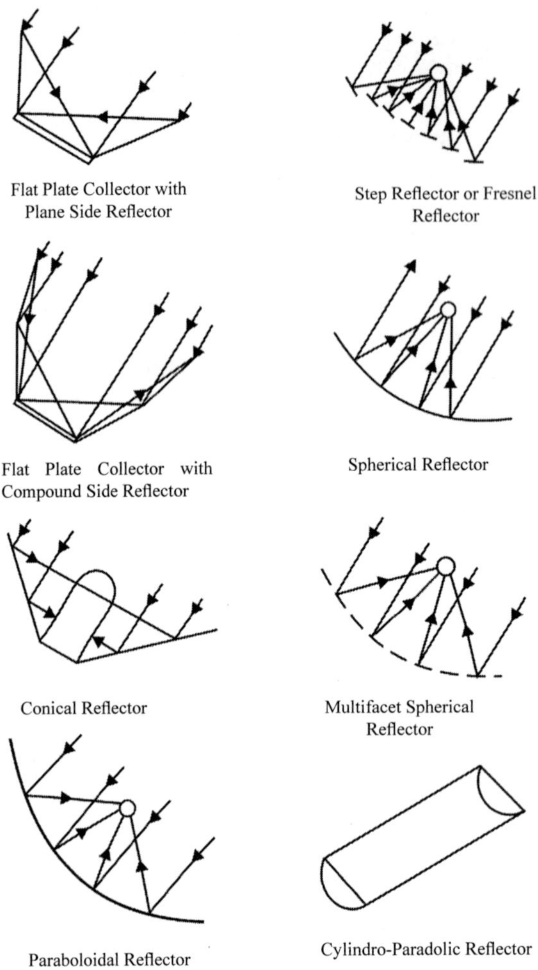

Fig. 5.6: Various type of Focusing Reflectors used in Direct type Solar Cooker

(b) Paraboloid Type Community Solar cooker for Indoor Cooking : It is possible to cook food within the kitchen with the help of this cooker. In this cooker a 7 sq.m large reflector standing outside the kitchen reflects the solar rays into the kitchen through an opening in its North wall while a secondary reflector further concentrates the rays on to the bottom of the pot painted black.

It is possible to cook the food of about 40 to 50 percent with the help of this cooker. The temperature attained is so high (400 °C) that the food could be cooked in a shorter time unlike box type solar cooker.

The major components of this cooker are as follows: Reflector frame, stand, rotating support, tracking device and secondary reflector.

The reflector frame, which is a primary reflector, is placed on a parabolic frame made out of metallic square tubes. It is a rigid structure of mild steel material. This reflector is made out of aluminium facet (plates)/rods on which reflecting mirrors of glass/ acrylic fixed which reflects the light towards secondary reflector. The secondary reflector is essentially a reflecting sheet so curved in a shape that it reflects the incoming rays towards bottom of the vessel.

This type of solar cookers is also having a rotating support, which is meant to rotate the frame along with the sun so that the solar rays are always directed on the focal point within the kitchen. The movement of the rotating support on which the frame is fixed is done with the help of a tracking device which is pendulum clockwork made out of air blower and cycle parts. The cooker is to set once a day is the morning and thereafter for rest of the time the clockwork keeps on rotating the reflector automatically.

5.4.3 Indirect or Box type solar cooker

Box type solar cooker is a typical example of solar energy devices, which demonstrate application of good reflector, absorber and transmitter.

In this type of solar cooker direct interception of solar radiation through glazing and indirect interception through reflection from plane mirror is made on the black body (insulated hot box) where raw material is placed for cooking. The insulated hot box may be square, rectangular or cylindrical in shape, which is painted black from inside with double-glazing for direct interception of solar radiation. The indirect entry of solar radiation through reflection is provided by single or multiple reflectors. Here the adjustment of cooker toward the sun is not so frequently required as in the case of direct type of solar cooker. It is slow cooker and takes long time for cooking and many of the dishes require roasting and baking can not be prepared with this cooker.

The commercially available box type solar cooker is illustrated in the Fig. 5.7. It has following important parts.

(a) The outer box : The outer box of a solar cooker may be made of wood, iron sheet or fiber reinforced plastic having suitable dimensions, which accommodate black body (inner box) and insulating pads.

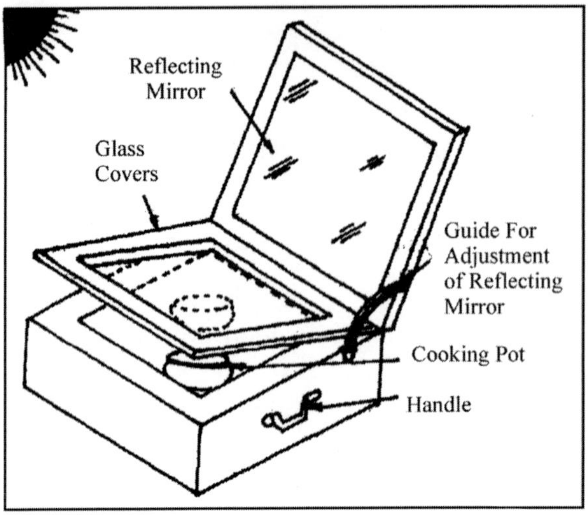

Fig. 5.7: Simple Box type Solar Cooker

(b) **The inner box :** The inner box may be made from G.I. sheet or aluminum sheet. All the four sides and the bottom of the inner box, which are exposed to the sun, are coated with black board paint for absorbing maximum amount of solar radiations. In fact, it is acting as black body, which absorbs energy directly intercepting through glazing and secondary reflecting form the plane mirror.

(c) **Insulation :** The hot box must be thermally insulated so that heat gain through solar radiation be effectively used for cooking purpose. Therefore, the space between outer box and inner box must be filled with a quality insulating material such as glass wool, thermocole etc. A suitable thickness of good quality insulating materials maintained the temperature of inner box.

(d) **Double Glazing :** A double glass cover is provided on the top of the inner box. These covers have length and breadth slightly greater than the inner box and can be fixed in a wooden frame maintaining a small spacing between the two glasses. This air cavity between glazing act as a insulator, which prevent heat losses from the inside box to top of the surface.

(e) Plane Mirror : A plane mirror is attached to the cooker, so that it enhances the entry of solar radiation by about 50 per cent in the inner box. In fact plane mirror acts as reflector and it increases the radiation input on the absorbing surface.

(f) Cooking containers : These containers with covers are made of aluminum or stainless steel and having dull black paint on their outer surface so that maximum amount of radiation can directly absorbed.

5.4.4 Multi Reflector Type

Another box type solar cooker with a threestep reflector, consisting of plane mirrors, is also available in India. The folding three-step asymmetric reflector box cooker retains the essential features of a box cooker and fitted with a threestep asymmetric reflector. As the reflector is adjusted relative to the box window, the distance of the mirror from the box window changes. Therefore, this design will not give optimum performance in different seasons of the year. Optimum performance can be obtained if the mirrors also are hinged together so that their relative angles can also be changed. This arrangement is done in a twostep reflector box cooker. In this design the two reflecting mirrors are hinged with each other and the width of each mirror is about 1.25 times the width of transparent window. This folding twostep asymmetric system is designed to give a concentration of 1.89 in summer and 2.0 in winter.

A solar oven is similar to a hot box cooker using an insulated box and multiple reflectors. The area of the box is kept as small as possible while the reflector area is large. The major difference is in the mechanism of directing solar radiation to the cooking area. In the hot box cooker solar radiation penetrates directly through the glass window while in a solar oven additional radiation after reflection also penetrates through the glass window. Due to enhanced solar radiation penetration and decreased cooker area for heat losses, the temperatures inside solar ovens can attain quite high temperature sufficient for cooking all varieties of foods. The mirrors are generally fixed at certain angles relative to the glass cover and the whole oven is oriented and tilted in the direction of the sun.

5.4.5 Advanced Type of Solar Cooker

It is a separate collector and cooking chamber type solar cooker, where solar energy is collected at a separate place with the help of either a flatplate or focusing collector and then this stored heat is transferred to the cooking vessel placed either at a separate place or indoor. Further, here the cooking in some cases can either be done with stored heat or the solar heat is directly transferred to the cooking vessel in the kitchen.

Now a day solar steam cooking system, which employs automatic tracking solar, dish concentrations are also available. These cookers can convert water into high-pressure steam. The steam thus generated is being used for cooking purposes.

5.4.6 Masonry Type Solar Cookers

The masonry type solar cooker was developed by Department of Renewable Energy Engineering, College of Technology and Engineering, Udaipur (27^0 42' N, 75^0 33 E). The body of the cooker was made of bricks and cement, while the hot box was made of an aluminium sheet. The dimensions of the hot box were 560 mm × 560 mm × 170 mm, and the overall dimensions were 920 mm × 920mm × 180 mm above the ground. One booster mirror, provided on the top, was attached using an adjustable arm. The overall dimension of the booster mirror was 560 mm x 560 mm. The absorber was painted with blackboard paint. Two glasses covers (each of 5 mm thick) – one on a removable iron angle and other on a wooden frame were provided over it, as shown in Fig. 5.8. The basic objective of use of masonry type solar cooker was to prepare animal feed, which require solar containers heating to maintain nutritional quality of feed.

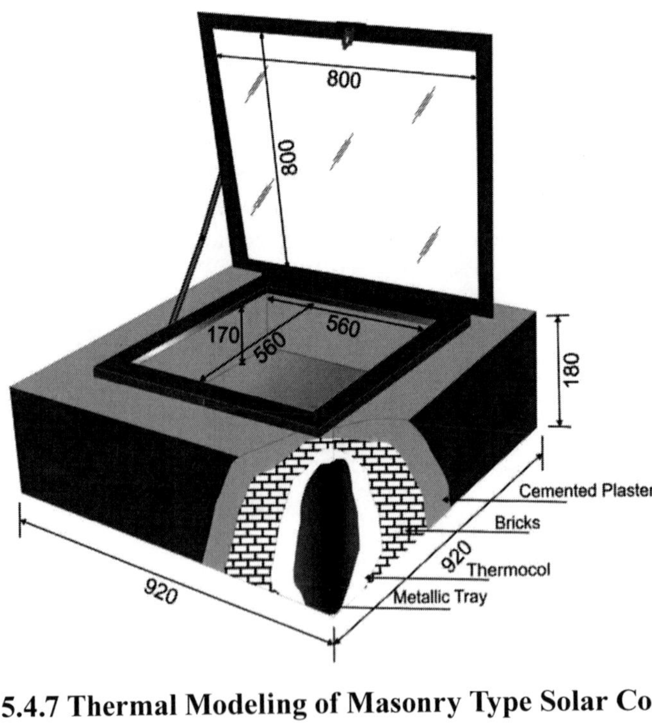

Fig. 5.8: Schematic of Animal Feed Solar Cooker (All dimensions are in mm)

5.4.7 Thermal Modeling of Masonry Type Solar Cooker

The masonry type cooker is similar to conventional hot box-type solar cookers. The various heat transfer routes in such a cooker are illustrated in Fig. 5.9. To develop a mathematical model of such an masonry type cooker, possible heat transfer routes are identified using an energy balance equation for heat transfer routes. The following assumptions are made to simplifying the model:

1. The reflective heat transfer between cooker sidewalls and the cooking vessel is negligible.
2. Appropriate thermal contact occurs between the cooking vessel and the absorbing surface of the cooker.
3. The heat exchange due to air inside the lid-covered vessel is negligible.

Energy balance at upper glass cover

$$(MC_p)_{gu}\frac{dt}{dt} = a_{gu}\,A_{gu}\,S_o + (Q_c + Q_r)_{gl\rightarrow gu} - Q_{c,gu\rightarrow am} - Q_{r,gu\rightarrow sky}$$

Where

$$Q_{c,gl\rightarrow gu} = h_{c,gl-gu} A_{gu}\left(T_{gl} - T_{gu}\right)$$

$$Q_{r,gl\rightarrow gu} = h_{r,gl-gu}\,A_{gu}\,(T_{gl} - T_{gu})$$

$$Q_{c,gu\rightarrow am} = h_{c,gu-am}\,A_{gu}\,(T_{gu} - T_o)$$

$$Q_{r,gu\rightarrow sky} = h_{r,gu-sky}\,A_{gu}\,(T_{gu} - T_{sky})$$

$$(MC_p)_{gu}\frac{dT_{gu}}{dt} = \alpha_{gu}\,A_{gu}\,S_o + (h_{r,gl-gu} + h_{c,gl-gu})\,A_{gu}\,(T_{gl} - T_{gu}) - h_{c,gu-am}\,A_{gu}\,(T_{gu} - T_o) - h_{r,gu-sky}\,A_{gu}\,(T_{gu} - T_{sky}) \quad (5.1)$$

Energy balance at lower glass cover

$$(MC_p)_{gl}\frac{dT_{gl}}{dt} = \tau_{gu}\,\alpha_{gl}\,A_{gl}\,S_o + Q_{c,a\rightarrow gl} + Q_{r,v\rightarrow gl} + Q_{r,ab\rightarrow gl} - (Q_c + Q_r)_{gl\rightarrow gu}$$

Where;

$$Q_{c,a\rightarrow gl} = h_{c,a-gl}\,A_{gl}\,(T_{air} - T_{gl})$$

$$Q_{r,v\rightarrow gl} = h_{r,v-gl}\,nA_{vb}\,(T_v - T_{gl})$$

$$Q_{r,ab\rightarrow gl} = h_{r,ab-gl}\,(A_{gl} - nA_{vb})\,(T_{ab} - T_{gl})$$

$$Q_{c,gl\rightarrow gu} = h_{c,gl-gu}\,A_{gl}\,(T_{gl} - T_{gu})$$

$$Q_{r,gl\rightarrow gu} = h_{r,gl-gu}\,A_{gl}\,(T_{gl} - T_{gu})$$

$$(MC_p)_{gl}\frac{dT_{gl}}{dt} = \tau_{gu}\,\alpha_{gl}\,A_{gl}S_o + + h_{c,a-gl}\,A_{gl}\,(T_{air} - T_{gl}) + h_{r,vc-gl}\,nA_{vb}\,(T_v - T_{gl}) + h_{r,ab-gl}\,(A_{gl} - nA_{vb})\,(T_{ab} - T_{gl}) - (h_{r,gl-gu} + h_{c,gl-gu})\,A_{gl}\,(T_{gl} - T_{gu}) \quad (5.2)$$

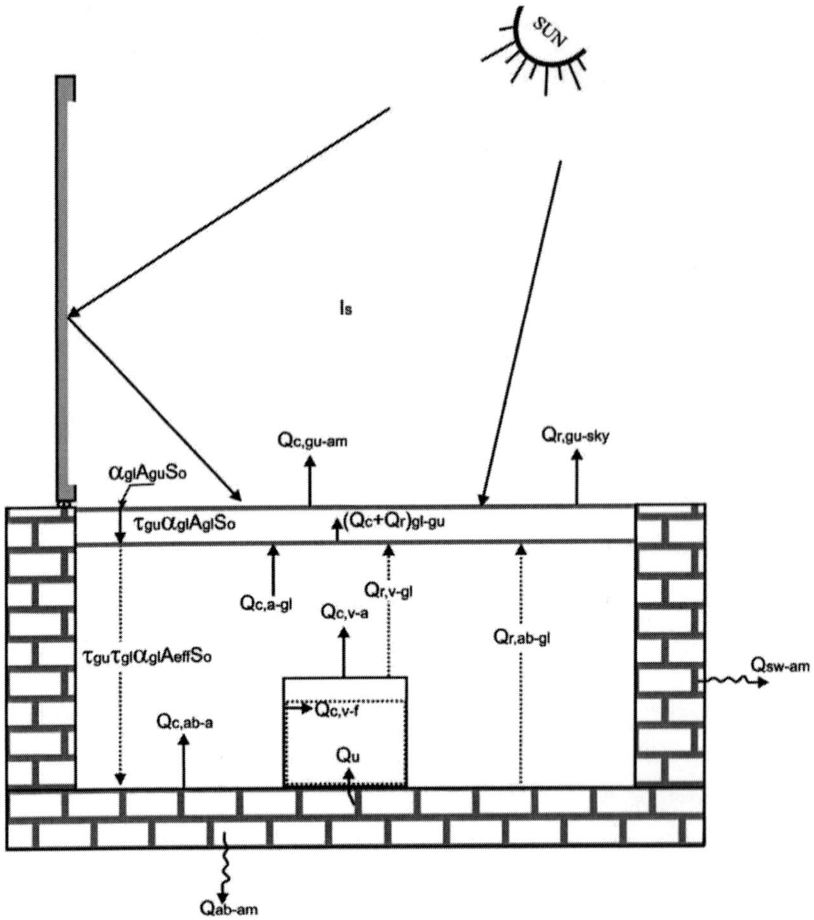

Fig. 5.9: Various mode of heat transfer in box type solar cooker

Energy balance of air inside the cooker

Where;

$$(MC_p)_a \frac{dT_a}{dt} = Q_{c,ab\rightarrow a} + Q_{c,v\rightarrow a} - Q_{c,a\rightarrow gl}$$

$$Q_{c,ab\rightarrow a} = h_{c,ab\text{-}a}\,(A_{ab} - nA_{vb})\,(T_{ab} - T_{air})$$

$$Q_{c,v\rightarrow a} = h_{c,v\text{-}a}\,nA_v\,(T_v - T_{air})$$

$$Q_{c,a\rightarrow gl} = h_{c,a\text{-}gl}\,A_{gl}\,(T_{air} - T_{gl})$$

$$(MC_p)\_a \frac{dT_a}{dt} = h_{c,ab\text{-}a}\,(A_{ab} - nA_{vb})\,(T_{ab} - T_{air}) + h_{c,vc\text{-}a}\,nA_v\,(T_v - T_a) - h_{c,a\text{-}gl}\,A_{gl}\,(T_{air} - T_{gl}) \quad (5.3)$$

Energy balance at absorber base

$$(MC_p)_{ab}\frac{dT_{ab}}{dt} = \tau_{gu}\,\tau_{gl}\,\alpha_{ab}\,A_{eff}\,S_o - Q_{r,ab\to gl} - Q_{c,ab\to a} - Q_{ab\to am} - Q_{sw\to am} - Q_u$$

$$Q_{r,ab\to gl} = h_{r,ab\text{-}gl}\,(A_{ab} - nA_{vb})\,(T_{ab} - T_{gl})$$

$$Q_{c,ab\to a} = h_{c,ab\text{-}a}\,(A_{ab} - nA_{vb})\,(T_{ab} - T_{air})$$

$$Q_{ab\to am} = U_{ab}\,A_{ab}\,(T_{ab} - T_o)$$

$$Q_{sw\to am} = U_{sw}\,A_{ab}\,(T_{ab} - T_o)$$

$$Q_u = U_{vb}\,nA_{vb}\,(T_{ab} - T_v)$$

$$(MC_p)_{ab}\frac{dT_{ab}}{dt} = \tau_{gu}\,\tau_{gl}\,\alpha_{ab}\,A_{eff}\,S_o - h_{r,ab\text{-}gl}\,(A_{ab} - nA_{vb})\,(T_{ab} - T_{gl}) - h_{c,ab\text{-}a}\,(A_{ab} - nA_{vb})$$
$$(T_{ab} - T_{air}) - U_{ab}\,A_{ab}\,(T_{ab} - T_o) - U_{sw}\,A_{ab}\,(T_{ab} - T_o) - U_{vb}\,nA_{vb}\,(T_{ab} - T_v) \qquad (5.4)$$

Energy balance inside vessel fluid

$$(MC_p)_f\frac{dT_f}{dt} = Q_{c,v\to f} = h_{c,v\text{-}f}\,nA_{vf}\,(T_v - T_{cf}) \qquad (5.5)$$

Energy balance at vessel

$$(MC_p)_v\frac{dT_v}{dt} = \tau_{gu}\,\tau_{gl}\,\alpha_v\,S_o\,nA_{vb} + Q_u - Q_{c,v\to f} - Q_{r,v\to gl} - Q_{c,v\to a}$$

$$Q_u = U_{vb}\,nA_{vb}\,(T_{ab} - T_v)$$

$$Q_{c,v\to f} = h_{c,v\text{-}f}\,nA_{vb}\,(T_v - T_{cf})$$

$$Q_{r,v\to gl} = h_{r,v\text{-}gl}\,nA_{vb}\,(T_v - T_{gl})$$

$$Q_{c,v\to a} = h_{c,v\text{-}a}\,nA_v\,(T_v - T_{air})$$

$$(MC_p)_v\frac{dT_v}{dt} = \tau_{gu}\,\tau_{gl}\,\alpha_v\,S_o\,nA_{vb} + U_{vb}\,nA_{vb}\,(T_{ab} - T_v) - h_{c,v\text{-}f}\,nA_v\,(T_v - T_{cf}) - h_{r,v\text{-}gl}$$
$$nA_{vb}\,(T_v - T_{gl}) - h_{c,v\text{-}a}\,nA_v\,(T_v - T_{air}) \qquad (5.6)$$

Net solar energy available on upper glass cover of the cooker consisting of radiation incident of glass cover pulse radiation reflected from booster mirror, hence the solar available at upper glass cover of the cooker can be written as:

$$S_o = I_s + I_b\,F_{mr}\,A_{mr}\cos\theta \qquad (5.7)$$

Where is view factor, which is taken unity and is angle of incident from mirror to upper glass cover which is taken as 45°.

(a) Heat Transfer Coefficients

The radiative heat transfer coefficient for heat transfer from surface i to j can be found from following relation

$$h_{i\to j}=\frac{\varepsilon\left(T_i+T_j\right)\left(T_i^2+T_j^2\right)}{\frac{1-\varepsilon_i}{\varepsilon_i}+\frac{1}{F_{ij}}+\frac{\left(1-\varepsilon_j\right)}{\varepsilon_j}\frac{A_i}{A_j}} \quad (5.8)$$

The convective coefficients $h_{c,v\text{-}f}$ and $h_{c,v\text{-}a}$ from the vessel may be calculated by treating the vessel walls for a vertical cylinder and its base or its cover treated as circular disks separately. These coefficients may be calculated using the following correlation [Holman 1981]:

$$h_{c,v-f}=\frac{0.686k_f}{L}\left(Gr_{v-f}\rho_r\right)^{0.25}\left[\frac{\rho_r}{\left(1+1.05\rho_r\right)}\right]^{0.25}+\frac{0.54k_f}{0.9d}\left(Gr_{v-f}\rho_r\right)^{0.25} \quad (5.9)$$

$$Gr_{v-f}=\frac{B_f^3\rho_f^2\left(T_v-T_f\right)\beta_{v-f}g}{\mu_f^2}$$

$$h_{c,v-a}=\frac{0.686k_a}{L}\left(Gr_{v-a}\rho_r\right)^{0.25}\left[\frac{\rho_r}{\left(1+1.05\rho_r\right)}\right]^{0.25}+\frac{0.54k_f}{0.9d}\left(Gr_{v-a}\rho_r\right)^{0.25}$$

$$Gr_{v-a}=\frac{B_{vL}^3\rho_a^2\left(T_v-T_a\right)\beta_{v-a}g}{\mu_a^2}$$

The convective heat transfer coefficient from the upper glass cover to the ambient ($h_{c,gu\text{-}a}$) is calculated by using correlation suggested by Duffie and Beckman (1991)

$$h_{c,gu\text{-}amb}=5.7+3.8\,V_w \quad (5.10)$$

The convective heat transfer coefficient between lower and upper glass cover can be estimated by using the Holland (1976) correlation:

$$Nu_{gl-gu}=1.0+\left[1.44\left(1.0-\frac{1708}{Ra_{gl-gu}}\right)\right]^{+}+\left[\left(\frac{Ra_{gl-gu}}{5830}\right)^{\frac{1}{3}}-1\right]^{+} \quad (5.11)$$

$$Ra_{gl-gu}=\frac{g\alpha_a\Delta T_{gl-gu}d^3}{\nu_a\kappa_a}$$

$$Nu_{gl-gu}=\frac{h_{c,gl-gu}d_{gl-gu}}{\kappa_a}$$

$$h_{c,gl-gu}=\frac{Nu_{gl-gu}\kappa_a}{d_{gl-gu}}$$

In the enclosure formed between the absorber base and lower glass cover, most of the space is occupied by the inside vessel. It is assumed that such an enclosure is heated from bottom sides; the heat lost from the upper side stems from the lower glass cover. The convective heat transfer coefficient in this model can be estimated by using the Holland (1976) correlation. The air resistance between the lower glass cover and absorber base is considered to be equal. Therefore, the values of $h_{c,ab-a}$ and $h_{c,a-gl}$ are twice that of the transfer coefficient across the enclosure.

$$h_{gl-gu} = h_{c,ab-a} = h_{c,a-gl} \tag{5.12}$$

(b) Energy Analysis

The energy input into solar cooking devices depends on the collector size and solar intensity, and be written as:

$$E_i = I_s A_{sc} \tag{5.13}$$

The energy output from a solar cooker depends on the quantity of water or food stuffs and temperature differences. The energy output can be expressed as:

$$E_o = \frac{m_w . c_{pw} \left(T_{fw} - T_{iw}\right)}{\Delta t} \tag{5.14}$$

The first law of thermodynamics states that the energy efficiency is the ratio of energy output to the energy input. It can be written as follows:

$$ç = \frac{E_o}{E_i} = \frac{m_w . c_{pw} \left(T_{fw} - T_{iw}\right) / \Delta t}{I_s A_{sc}} \tag{5.15}$$

(c) Exergy Analysis

The availability (exergy) of a solar flux with both beam and diffuse components can be represented by superposition as:

$$\Xi_i = I_b \left[1 - \frac{4T_o}{3T_s}\right] + I_d \left[1 - \frac{4T_o}{3\dot{T}_s}\right] \tag{5.16}$$

The idea of estimating the exergy of solar radiation as the exergy input into the solar cooker was expressed by Petela (2003) in order to determine the available energy flux into the solar cooker as follows:

$$\Xi_i = I_s \left[1 - \frac{4}{3}\left(\frac{T_o}{T_s}\right) + \frac{1}{3}\left(\frac{T_o}{T_s}\right)^4\right] A_{sc} \tag{5.17}$$

The sun's blackbody temperature of 5,762 *K* results in a solar spectrum concentrated primarily in the 0.3–3.0 mm wavelength band. Although the surface temperature of the sun (T_s) can be vary on the earth' surface due to the spectral distribution, the value of 6,000 *K* has been considered for the T_s.

The thermal exergy of heat energy at temperature *T* for non-isothermal processes is

$$\in = \int_{T_o}^{T} mc_p \left(1 - \frac{T_o}{T}\right) dQ$$

The thermal exergy content of water at temperature T_i can be calculated by

$$\in_w (T_{iw}) = m_w c_{pw} \left[(T_{iw} - T_o) - T_o \ln \frac{T_{iw}}{T_o} \right]$$

Water is heated by acquiring solar energy, and its temperature increases from T_{iw} to T_{fw}. Hence, there is an exergy gain that can be expressed as

$$\Delta \in_w = \in (T_{fw}) - \in (T_{iw})$$

The exergy output from the solar cooker can be estimated as

$$\Xi_o = m_w c_{pw} \left[(T_{fw} - T_{iw}) - T_o \ln \frac{T_{fw}}{T_{iw}} \right] \Big/ \Delta t \qquad (5.18)$$

Exergy efficiency is defined as a ratio of the exergy transfer rate associated with the output to the exergy transfer rate associated with the necessary input (Kotas, 1984). Kaushik and Gupta (2008) defined exergy efficiency for the solar cooker as the ratio of the cooker's output exergy (increase of exergy of water due to temperature rise) to the exergy input (exergy of solar radiation). Thus, the exergy efficiency is given by

$$\Psi = \frac{\text{Exergy output}}{\text{Exergy input}} = \frac{\Xi_o}{\Xi_i} = \frac{m_w c_{pw} \left[(T_{fw} - T_{iw}) - T_o \ln \frac{T_{fw}}{T_{iw}} \right] \Big/ \Delta t}{I_s \left[1 - \frac{4}{3}\left(\frac{T_o}{T_s}\right) + \frac{1}{3}\left(\frac{T_o}{T_s}\right)^4 \right] A_{sc}} \qquad (5.19)$$

5.5 Solar Desalination

There is an important need for clean, pure and potable or drinking water in many developing countries. Often water sources are brackish (i.e. contain dissolved salts) and contain harmful bacteria and therefore cannot be used for drinking purpose. In addition, there are many coastal locations where seawater is abundant but potable water is not available. Pure water is also useful for batteries and in hospitals or schools.

Distillation is one of many processes that can be used for water purification. This requires an energy input, as heat. Solar radiation can be the source of energy for this purpose. In this process, water is evaporated and which is further condensed as pure water thus separating water vapour from dissolved matter,

Solar water distillation is a technology in which water is first evaporated because of partial pressure difference between air &water and subsequently condense to form pure potable water. There are number of other approaches to water purification and desalination, such as photovoltaic powered reverse-osmosis, for which small-scale commercially available equipment are available. In addition, if treatment of polluted water is required rather than desalination, slow sand filtration is a good option.

5.5.1 Energy requirements for water desalination

The energy required to evaporate water is equal to the latent heat of vaporization of water. This has a value of 2260 kilojoules per kilogram (kJ/kg). This means that to produce 1 liter (i.e. 1kg since the density of water is 1kg/liter) of pure water by distilling brackish water requires a heat input of 2260 kJ.

It should be noted that, although 2260 kJ/kg is required to evaporate water where as to pump a kg of water through 20m head requires only 0.2kJ/kg. Therefore, energy requirement for desalination is quite more, therefore normally desalination is considered only where there is no local source of fresh water that can be easily pumped or lifted.

5.5.2 Working of simple solar still operates

Fig. 5.10 shows a single-basin still for distillation purpose. The main features of operation are the same for all solar stills. The incident solar radiation is transmitted through the glass cover and is absorbed as heat by a black surface in contact with the water to be desalinated. The water is thus heated and gives off water vapour. The vapour condenses on the glass cover, which is at a lower temperature because it is in contact with the ambient air, and runs down into a gutter from where it is fed to a storage tank.

Fig. 5.10: Solar Still

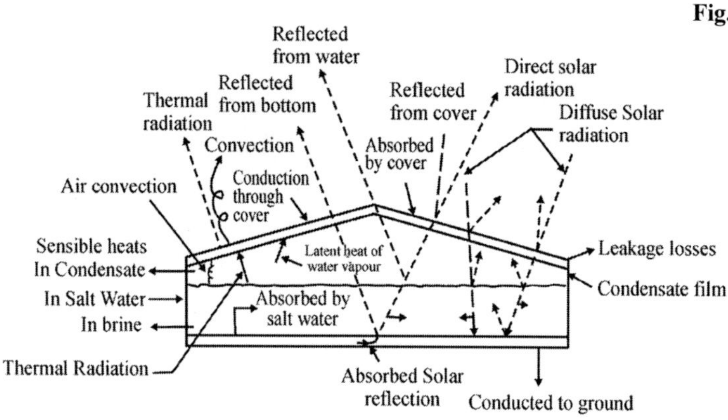

5.5.3 Design objectives for an efficient solar still

For high efficiency the solar still should maintain:

- a high temperature in feed (salinated) water
- a large temperature difference between feed water and condensing surface
- low vapour leakage.

A high temperature in feed water can be achieved if:

- a high proportion of incoming radiation is absorbed by the feed water as heat. Hence low absorption glazing and a good radiation absorbing surface are required
- heat losses from the floor and walls are kept low
- the water is shallow so there is not so much to heat.

A large temperature difference can be achieved if:

- the condensing surface absorbs little or none of the incoming radiation
- condensing water dissipates heat which must be removed rapidly from the condensing surface by, for example, a second flow of water or air, or by condensing at night.

5.5.4 Types of solar still and their performance

Single-basin stills are generally used as their behaviour is well understood. Efficiencies of 25% are typical for single basin type of solar still. Daily output as a function of solar irradiation is greatest in the early evening when the feed water is still hot but when outside temperatures are falling.

Material selection is very important. The cover can be either glass or plastic. Glass is considered to be best for most long-term applications, whereas a plastic (such as polyethylene) can be used for short-term use. However, more care is required in glass operation.

Sand concrete or waterproofed concrete are considered best for the basin of a long-life still if it is to be manufactured on-site, but for factory-manufactured stills, prefabricated ferro-concrete is a suitable material.

Multiple-effect basin stills have two or more compartments for evaporation & condensation as per position of sun. The condensing surface of the lower compartment is the floor of the upper compartment. The heat given off by the condensing vapour provides energy to vaporize the feed water above. Efficiency is therefore greater than for a single-basin still typically being 35% or more but the cost and complexity are correspondingly higher.

Wick stills - In a wick still, the feed water flows slowly through a porous, radiation-absorbing pad (the wick). Two advantages are claimed over basin stills. First, the wick can be tilted so that the feed water presents a better angle to the sun (reducing reflection and presenting a large effective area). Second, less feed water is in the still at any time and so the water is heated more quickly and to a higher temperature. The rise in height of wick is due to basic fundamental of surface tension.

Simple wick stills are more efficient than basin stills and some designs are claimed to cost effective than a basin still of the same output.

5.5.5 Hybrid designs

There are a number of ways in which solar stills can usefully be combined with another function of technology. Three examples are given:

- Rainwater collection. By adding an external gutter, the still cover can be used for rainwater collection to supplement the solar still output.
- Greenhouse-solar still. The roof of a greenhouse can be used as the cover of a still.
- Supplementary heating. Waste heat from an engine or the condenser of a refrigerator can be used as an additional energy input.

5.5.6 Output of a solar still

An approximate method of estimating the output of a solar still is given by:

$$\frac{\times G \times A}{2.3}$$

where:

Q = daily output of distilled water (liters/day)

η_o = overall efficiency

G = daily global solar irradiation (MJ/m^2)

A = aperture area of the still ie, the plan areas for a simple basin still (^{2})

In a typical country the average, daily, global solar irradiation is typically 18.0 MJ/m^2 (5 kWh/m^2). A simple basin still operates at an overall efficiency of about 30%. Hence the output per square meter of area is:

$$\text{Daily output} = \frac{0.30 \times 18.0 \times 1.0}{2.3}$$

$$= 2.3 \text{ liters (per square meter)}$$

The yearly output of a solar still is therefore approximately one cubic meter per square meter.

The cost of pure water produced depends on:

- the cost of making the still
- the cost of the land
- the life of the still
- operating costs
- cost of the feed water
- the discount rate adopted
- the amount of water produced.

5.6 Solar Water Heater

Water heating is one of the simplest applications of solar energy. Hot water is requirement of domestic activities as well as industrial activities. Hot water in household in not only required for taking both, but also used for cleaning utensils, washing cloths and floors etc. It has been observed that 100-liter solar water heater can saves about 2000 units of electricity annually if used for different application in a house hold or in industrial unit. Solar water heater offers a number of advantages such as:

1. Simple in construction and installation.
2. Almost no maintenance and operating cost.
3. It saves time and also energy for heating water.
4. Easily retrofittable to existing houses as per requirement.
5. Economically viable as compared to electric water heater.
6. Moderate temperature required, domestic affairs could easily achieved.

There are two basic types of solar water heaters: -

(a) Collector coupled to storage tank.

(b) Collector cum storage system.

First type of solar water can extract heat from solar collectors in two ways, namely through natural convection (also known as thermosyphon effect) and through a forced flow of water using an electrically operated pump. Collector coupled to storage tank type of water heater is used, as domestic solar water heater where the maximum temperature required is not more than 70°C. A typical system consists of solar collector (front glazing), metallic absorber, back insulation and collector box, insulated storage tank (with or without heat exchanger), piping, controls and pump.

Here heating is accomplished by collection of solar radiation with flat plate collector mounted on south facing roof or walls. The collector is usually placed below the storage tank, cold water from the storage tank flows down to the inlet of the collector, gets heated in the collector and rises to the tank by thermosyphonic effect. A density difference created by the temperature gradient causes the fluid to flow up in the collector by the thermosyphonic effect (Fig. 5.11). it has been observed that top portion of collectors should be alteast one feet or more than one feet lower bottom portion of storage tank in order to create thermosyphonic effect.

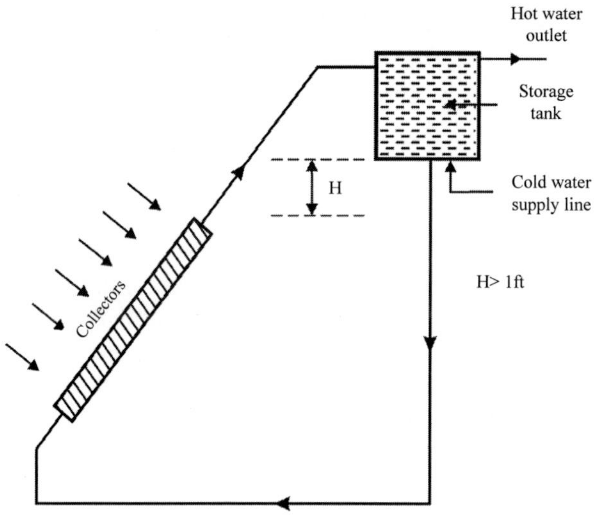

Fig. 5.11: Thermosyphon type Solar Water Heater (Direct Natural Circulation type)

The collectors are usually oriented in south with an inclination angle equal to latitude of the place. For round the year use, the inclination of the collectors from the horizontal should be equivalent to the latitude. However, for winter use, it is preferable to keep angle of inclination at latitude + 15°C. The required collector area for 100-liter hot water demand at 50°C is about 2 m^2.

In the cold climates, it is essential to mix anti-freeze materials in the heat extracting fluid circulating through the collector channels. In this case a heat exchanger may also be used and which prevent antifreeze fluid to flow directly through the collectors. It is known as indirect natural circulation type solar water heater (Fig.5.12).

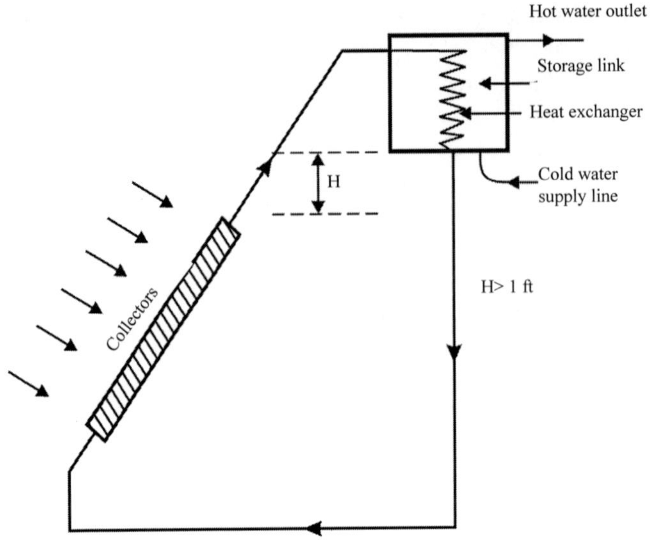

Fig. 5.12: Indirect Natural Circulation Solar Water Heater

In the case of collector cum storage type solar water heater, all the three functions i.e. solar energy collection, storage and control are combined into a single unit. Essentially, a typical collector cum storage water heating system consists of three main components front glazing (solar flat plate collector), absorber sheet and insulated storage tank. Sun's ray passes through the front glazing and gets absorbed by the absorber which further heats the water. The storage tank acts as a collector cum storage for hot water. A schematic diagram of system is shown in Fig. 5.13.

The utility profile of hot water by a user is an important factor to decide about the viability of solar water heater. This could be evaluated by using kilowatt meter, flow meter or energy meter for fuel consumption. If the load is continuous and when peak solar energy is available then solar water heater could be a good proposition. But to find out optimum size of the solar water heater, the load requirements on daily and on seasonal basis need to be known. This would help in deciding the collector area and also the storage capacity. This again would be limited by the open sunny space available for mounting collector field.

The effectiveness of solar water heater depends on proper matching of solar collector with hot water demand.

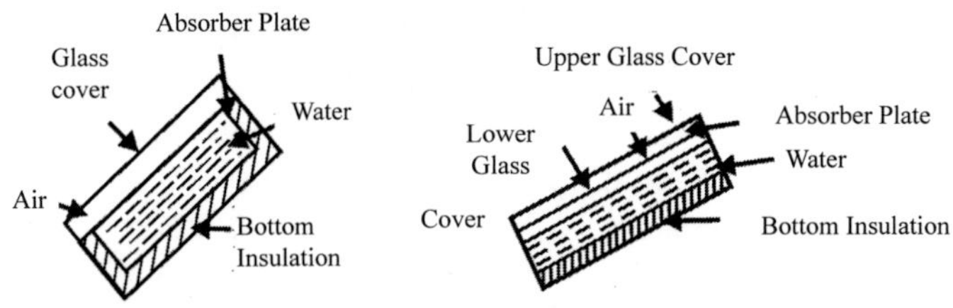

(A) System with Single Glazing (B) System with Double Glazing

Fig. 5.13: Collector-Cum-Storage Solar Water Heater

Normally the solar water heaters are either installed on the ground or on the rooftop. For system to be installed on the roof top roof construction and load bearing capacity, orientation of building accessibility, avoiding shadow due to nearby structures, (plants/ accessories/tall trees) and availability of cold water must be ensured. Similar care is taken for the systems to be installed on the ground. However, on ground chances of shade over solar collector are more hence efficiency reduces. Therefore, it is always recommended to have on the rooftop.

It is also possible to integrate the solar water heaters with the building design in the new construction. The solar collectors could either be made a part of the south-facing wall of the building or could be installed on the window/ door overhangs etc. Thus, the hot water requirement in the building could be supplemented even without affecting the aesthetics of the building. Several such systems are in operation in the country.

Normally four types of collectors are being used in solar water heaters for different temperature requirements. These are:

1. Unglazed polymer collectors for heating of swimming pool water.
2. Metallic collector with black painted absorber (e.g. copper tube/ copper sheet and copper tube aluminum sheet etc.) for temperature applications upto 60°C.
3. Collector with selectively coated metallic absorber for temperature upto 80 – 85°C, and
4. Evacuated tubular collector for temperature above 80°C.

Further, the collector has to be chosen on the basis of the requirement and temperature of the hot water.

In the thermosyphon type of solar water heater, the flow of water from bottom of the tank through the collector of the top of the tank takes place through temperature gradient created in the collector under the sunlight.

This configuration is being used in almost all the low capacity solar water heaters including those for domestic applications. Solar water heaters of capacity upto 5000 lpd [liter per day] have been installed in the country based on this principle and these have been successfully operating. Since there is no moving part in such a system, the chances of failure are very rare.

In the forced flow type of configuration pumps with controls are added at appropriate place in the system to ensure the flow of water through all the collectors. This type of system can either be a single pass system where the water enters the collector array at one end and leaves at the other or a recirculation type system. Where water is taken from the storage tank circulated through the collector array and returns back to the tank repeatedly. Since in a single pass system the collector operates at comparatively lower temperature than in recirculation type. The system efficiency is better in the previous case.

Collector interconnection must be able to keep up expansion and contraction between the collectors, prevent leakage at collector couplings, and prevent galvanic corrosion, compatible with heat transfer fluid (water) and resistant to weather and outside exposure. Brass flanges with neoprine gaskets are normally being used.

Normally B class GI pipe is being used for piping. Whenever this type is connected to copper header, a corrosion protector (gasket) is to be used to prevent galvanic corrosion. The insulation on the pipe must be protected from water absorption and UV deterioration. Sealants are to be used over the insulation layer. The insulated pipe is to be supported properly.

The capacity of solar water heating system can be boosted by increasing collector area. For large capacity solar water heaters, the collectors are to be connected in either parallel series arrangement or in series parallel arrangement. The advantages in parallel series arrangement are that the pipe length in series is reduced and also the thermal capacity of the system is reduced. This arrangement, however, has to have venting provision during initial filling and also for release of dissolved air. Whereas in the series parallel arrangement, the piping length is the least and so is the thermal capacity but it must be ensured that all the collectors running full with water, venting is effective and the flow rates are adequate to avoid over heating and blockage. (Fig 5.14). However, for domestic hot water requirement solar collectors must to be connected in series and proper matching with requirement of hot water is made.

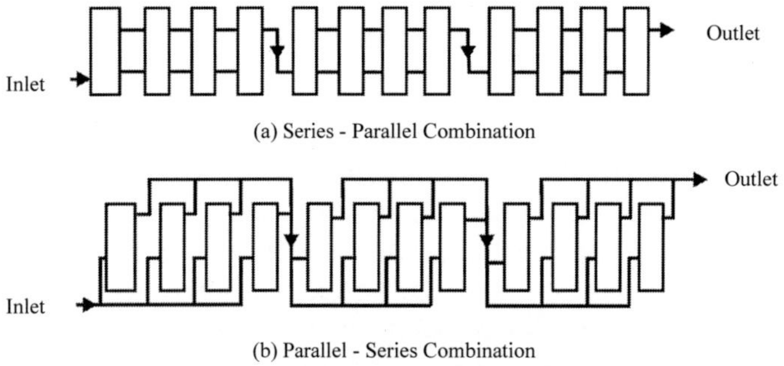

Fig. 5.14: Combination Techniques of Solar Collector

Heat exchanger must be used in case where the quality of water is such that the collectors are subjected to scale formation and also where freezing may occur during night. Counter flow heat exchangers and coil in storage tank type heat exchangers are normally being used in solar water heater.

For areas where night temperature drop below the freezing point, double loop systems with antifreeze mixture in the primary loop are to be used. The composition of the anti-freeze mixture depends on the minimum ambient temperature of the site. One can also use single loop design with drain off facility. In such a design, the collector field is drained off completely when the minimum temperature is likely to fall below the freezing point.

Solar water heaters normally require minimum maintenance. This is particularly true for systems based on the thermosyphon principle. However, cleaning of the collector glazing once in a week is desirable for better performance of the system. In areas where scale formation is a problem, de-scaling need to be done once in a year. In case of black painted collector, repainting may be required after a period of two to three years, whereas this may not be required in case of selectively coated collector. In case of forced circulation system, the conventional components of the system such as the thermal sensors, relays and pumps might need regular checkup and maintenance. The average cost of domestic solar water heater of 100-lpd capacity in the country is about Rs. 15,000. Assuming 5 kWh as daily average solar insolation and 300 clear sunny days in a year, a domestic solar water heater can supply upto 1800 kWh of thermal energy per annum. Similarly, the cost of commercial and industrial solar water heater varies from Rs. 70 to Rs. 120 per liter based on the site conditions, capacity and temperature requirements. The saving again would be 400 kWh of thermal energy per meter square of the installed collector area per annum.

5.7 Solar Ponds

With its vast potential of virtually untapped energy, living off the sun promises to be an attractive for the future. However, because of the intermittent and diffused nature of solar radiation, the problem of the high cost of collecting and storing solar energy neglects the economic viability of large-scale solar thermal applications. For such applications, it is more effective and economical to devise a solar system that combines both the collection and storage functions in place of the conventional flatplate solar collector and hot water storage tanks. Large expanses of water bodies like ponds, lakes or oceans are natural potential solar collectors with builtin storage characteristics. Experience shows that the water in such a pond heats upto only a few degrees, because of the natural convection currents that are set into medium as soon as heat is absorbed at the bottom of the pond.

In a mass of water of low depth, the solar radiation falling on the water will penetrate and be absorbed at the bottom, raising the water temperature. But buoyancy will immediately cause this heated water to rise to the surface and the heat will be rapidly dissipated to the surroundings. If the water in the lower regions of the pond of lake could be made heavier than at the top, it could stay at the bottom and retain the absorbed heat, thereby yielding larger temperature differences between the bottom and the surface layers of water.

In fact, this effect has been observed in some natural lakes whose water is found to contain a concentration of salt that increases with the depth. The resulting `salinity gradient' (saltslope) gives the lake a vertical density difference so that heated `heavy' water remains at the bottom. A significant temperature rise of 40°C to 50°C has been observed in such waters.

Solar Ponds is an artificially constructed water pond in which significant temperature rises are caused to occur in the lower regions by preventing convection currents. The more specific terms "salt gradient solar pond" or "nonconvecting solar ponds" are also often used. At the moment solar ponds are mostly being studied on an experimental basis in a number of places. However, they appear to be economical for large area applications and it is likely that they will be used extensively in the future as problems connected with their operation and maintenance are resolved.

The solar pond, which is actually a large area solar collector, is a simple technology that uses water pond between one to four meters deep as a working material for three functions: a) collection of solar radiant energy and its conversion to heat energy (upto 95°C), b) storage of heat, and c) transport of thermal energy out of the system. It possesses a thermal storage capability spanning the seasons. The surface area of the pond affects the amount of solar radiation it can collect. The bottom of the ponds is generally lined with a durable plastic liner made from material such as black polyethylene and hypalon reinforced with nylon mesh. This dark surface at the bottom of the pond increases the absorption of solar radiation. Salts like magnesium chloride, sodium chloride or sodium nitrate are dissolved in the water, the concentration being highest/densest at the bottom (20 percent to 30 percent) and gradually decreasing to almost zero at the top (Fig 5.15).

Typically, a saltgradient Solar Pond consists of three zones:

(a) an upper convective zone of clear `fresh' water that acts as the solar collector/receiver and which is relatively the shallowest in depth and in generally close to ambient temperatures.

(b) a gradient which serves as an insulating layer by reducing heat losses in the upward direction. This nonconvective zone is much thicker and occupies more than half the depth.

(c) a lower convective zone with the densest salt concentration, which serves as the heat storage zone. Almost as thick as the middle nonconvective zone, salt concentration and temperatures in this zone are nearly constant. The Solar Pond is filled in stages. To obtain the difference in salt density through this stepped concentration gradient, the pond is filled with atleast 3 successive layers of salt solution (10 to 20) cm thick), one on top of another, each less dense than the layer below, so that the top layer is fresh water or nearly so, while the bottom layer contains the most salt. Naturally this kind of stepped concentration is not likely to remain stable for long, and would eventually disappear because of diffusion and evaporation. In order to maintain the stability, concentrated bring is introduced at the bottom periodically while the surface is “washed” with fresh water. If the pond contained no salt, the bottom layer (which is denser in the solar pond) would be less dense than the top layer because hot water expands. The less dense hot water would then rise to the surface and the layer would mix. Therefore, the purpose of the salt density difference is to keep the different `layer's of the solar pond separate.

When solar radiation strikes the pond, most of it is absorbed by the surface at the bottom of the pond. The temperature of the dense salt-water layer is therefore higher than the fresh water layer near the surface of the pond. The denser, salt water at the bottom prevents the heat being transferred to the top layer of fresh water by natural convection currents.

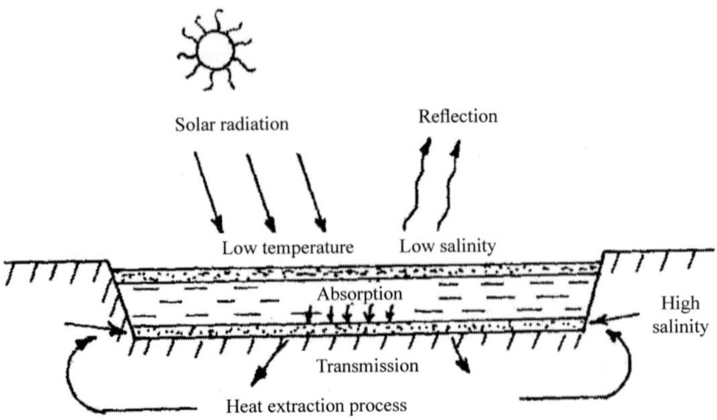

Fig. 5.15: Cross Section of Solar Pond

In order to extract the energy stored in the bottom layer, hot water is removed continuously from the bottom, passed through a heat exchanger and returned to the bottom to generate electricity, heat stored in the salt water is piped to an evaporator. Liquid freon in the evaporator is heated and changed into a gas. The pressure generated by the gas spins a turbine and electricity is produced from the generator. Freon gas is then cooled and recycled to be used again.

5.8 Solar Furnaces

Solar furnaces are optical system, which are used for obtaining high temperature through concentration of solar radiation into a small area. Viewing to their output, they are also known as solar energy concentrations, where area may vary from as small as 1m^2 paraboloid to several thousand of m^2 of reflectors known as heliostats. Basically, solar furnaces offer following advantages, which are as follows:

1. It is comparatively simpler device for obtaining high temperature for various heating applications.
2. It is ideal tool to study the chemical, optical, electrical and thermodynamic properties of materials at high temperature.
3. It is a device for getting high temperature controlled atmosphere which is capable for yielding continuous observations or observations in the absence of any electromagnetic fields.

However, solar furnaces are costly, especially their initial capital cost, and they are suitable only for intermittent operations due to intermittent nature of sunshine. Further, very high temperature in solar furnace can be obtained only on a very small area and the temperature gradient in the focal spot is also quite high.

Essentially solar furnace consists of three main components, such as

1. Concentrator (Either paraboloidal or spherical reflector)
2. Heliostat for directing solar radiation parallel to the optical axis of the concentrator.
3. Sun tracking mechanism (Manual tracking or by astronomical method or by servo systems).

The solar furnaces are classified on the basis of their temperature requirement (Fig. 5.16). In the direct type solar furnace, the heat flux density obtained is highest but in this case both the lens and reflectors and the target moves with the sun, making the system impractical and most inconvenient. Various configuration involved in the design of solar furnace is illustrated in Fig. 5.17

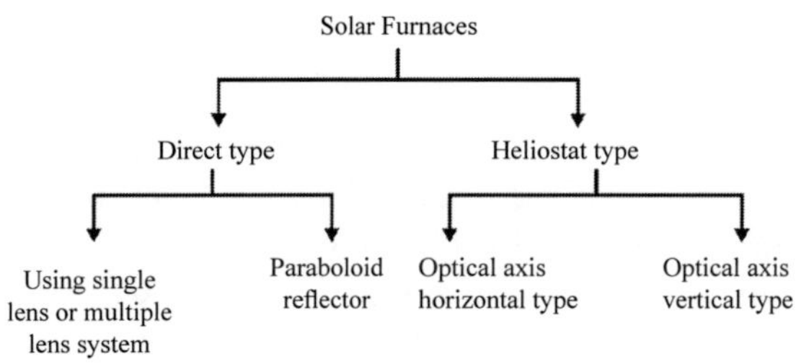

Fig. 5.16: Classification of Solar Furnaces

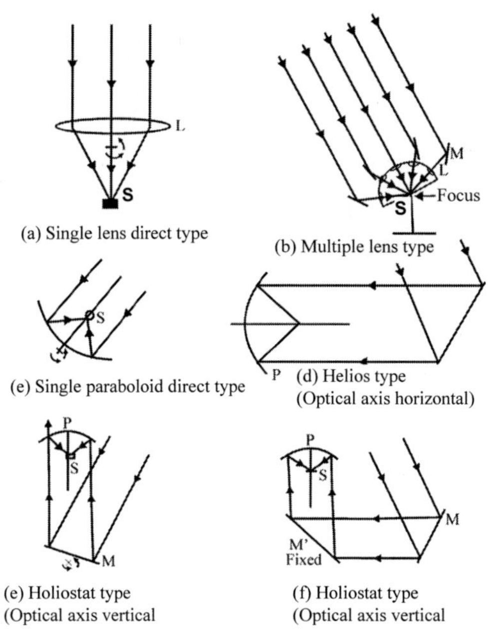

Fig. 5.17: Types of Solar Furnaces

5.9 Solar Thermal Power Generation

In the solar thermal power generation route, solar energy is collected on a collector's surface and then transferred to an absorber. A fluid flowing through the absorber transfers heat to a heat power generating machine i.e. turbine. Where heat energy is converted into mechanical energy and in turns a generator (attached with turbine) where mechanical energy is converted into electrical energy. The thermodynamic heat cycle here works exactly in the same way as in the case of conventional steam based power station. Here only difference is that the heat source is absorber of the solar collector instead of a boiler or a combustion chamber. Two types of solar thermal power generation systems are available:

(i) Centralized system (Tower concept)

(ii) Distributed system (Farm concept)

In the tower concept, a field of heliostats, which follows the direction of sun, concentrates direct solar radiation into a central receiver mounted on the top of a tower. Fig. 5.18 gives a schematic diagram of a solar thermal tower power plant. A heat transfer fluid removes heat from the receiver and in turn carries it to another heat storage unit. This heat source is used to operate an electrical power plant. The advantages of the farm concept are its modularity and relatively less maintenance.

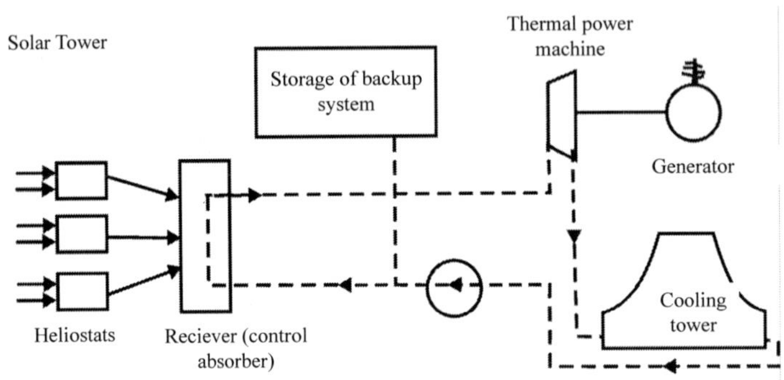

Fig. 5.18: Solar Tower Concept of Electricity Generation

Here the heliostats consist of glass mirror mounted in segments in a steel frame. The mirrors are backed by formed core / steel sheet laminates. The heliostats are usually supported by stainless steel pipes on reinforced concrete foundation.

The performance of the heliostats depends mostly on the following parameters:

(i) **Selected mirror construction** It should ensure that deformation which may result from wind, temperature, etc. remain within the prescribed limits.

(ii) **Surface treatment of mirrors** to maintain the reflectivity constant during their lifetime.

(iii) **Stability of the supporting and tracking mechanism** It is used for frequent automotive adjustment of the mirrors according to sun.

The other important component in the receiver, which is usually designed on a cavity to minimize the radiation losses, is essentially a boiler or steam generator. It is located at the top of the tower, and it receives the sun's energy directed on it by the heliostats. An array of spaced tubes forms the boiler surface. The heat transport media (water, gas or medium depending upon the chosen heat cycle) flows through these tubes. As temperature of 1000°C or more can be reached, the use of suitable materials and its design are crucial. Here on tower concept, higher temperature and hence higher thermal power is possible. It is more suitable for central utilization. Maximum possible power generation from tower concept is about 100 MW.

In a farm concept (Fig. 5.19) flat plate or parabolic dish or trough collectors may be used for generation of electricity. Each collector is connected to the forward flow tubes and the return flow tubes of the central thermal power machine. A parabolic trough collector consists of an axial parabolic reflector at the focus of which a heat receiver in the form of a tube is located. Parabolic troughs are usually made of glass segments hot moulded to shape, conventionally silvered on the back face. After silvering, the edges are often sealed by a special substance to prevent moisture and other particles from entering between the glass and silvered layer and spoiling the mirror effect. The segments are then provided with a backing surface, which could be metal, fiberglass or concrete. Concentration ratios of between 10 and 100 can be achieved by such collectors resulting in heat transport medium temperature of 150 to 350°C. The reflectance surface is generally mirrored to achieve reflection of the order of 87 per cent. The focal line of the concentrate is usually made to coincide with the axis of rotation of the supports. A number of troughs are combined together to form module.

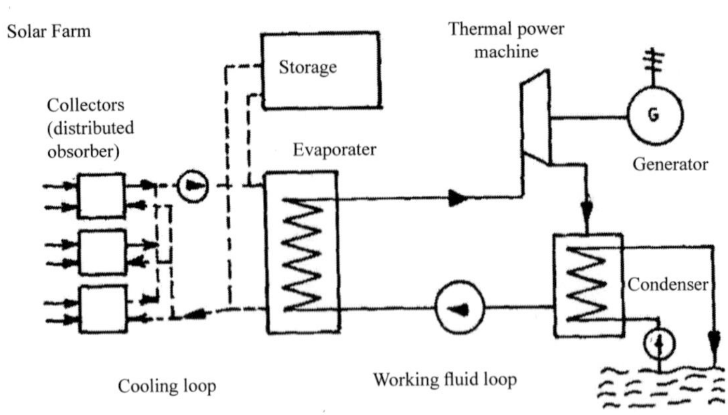

Fig. 5.19: Solar Farm Distribution System for Electricity Generation

A parabolic dish or point focusing solar collector is capable of achieving concentration from about 50 to several hundred, as a result of which temperature of 500°C or more can be reached in the central receiver. A parabolic dish is primarily a parabolic formed by rotation of a parabola along a vertical axis. In other words, cross section of parabolic dish in any direction will reveal a common parabola. Parallel sun's rays falling on the dish are focused on a point and as a result of which high temperatures can be reached at the focus. A receiver placed on the focus could be used in a prime mover for generation of electricity. In parabolic dishes the achievable temperature may range from 300 to 1600°C. Parabolic dishes may also be used for producing industrial process heat. A number of parabolic dishes can be coupled together to generate the required power.

Following types of heat engine are often used with parabolic dish system:-

(i) Organic Rankine cycle engine

(ii) Stirling cycle engine

(iii) Air Brayton cycle engine

In each of these cycles hot gas or vapor is expanded through an engine or turbine to produce work and is thereby cooled. The gas vapor is further cooled to reject heat and is finally returned to its initial state for getting energized by solar radiation and then complete the cycle.

In this distributed solar thermal power generating system, a hybrid concept has also been developed. In this approach, the working fluid in heated by solar energy to the maximum possible extent. Generally parabolic trough collector systems are being used in this hybrid concept.

In order to maximize the efficiency of the solar system and to ensure steady output during fluctuating solar insulation, a supplementary boiler that can be fired on multi-fuels such as biomass, etc. may also be added.

5.10 Design Considerations for Solar Thermal Power Plants

Electricity generation by use of solar energy through thermal route has been considered an important means of conversion of solar energy to usable form. Solar thermal power plant has different design consideration as compared to conventional power plant. All subsystem should be carefully integrated or method for getting maximum efficiency of generation and utilization. Following points are properly considered for design of solar thermal power plants.

(a) Site selection

Main considerations for site selection for electricity generation are good solar insolation level, preferred latitude between 20° and 35°, plain and wasteland having good drainage and having little economic value, preferably away from polluting industries, dry climate, well connected with transport routes, good availability of water for steam generation, mirror washing and cooling tower, etc. To reduce the transmission and distribution losses, it will be desirable that plant be located near the center and load curve of the area should preferably have a discernible daypeak.

(b) Requirement of land

While designing a solar thermal power plant for medium latitudes (20° to 35°), for maximum collection of solar energy, it is desirable that the distance between the two rows should be kept at approximately three times the aperture of the collectors. On this basis the estimated land requirement per MW works out to around 7 acres. Land area selected should be such as to minimize piping and ensure heat transfer fluid flow balancing in modules.

(c) Capacity selection

Large size solar thermal plants have obviously to be designed to operate in synchronous mode with the existing grid. It may, therefore, be argued that the capacity of the proposed solar thermal power project should be declared on the basis of grid capacity. In case the capacity of proposed project is decided at a level of higher than the local peak demand, part of generation can be fed to the existing grid for the nearby load centers.

(d) Orientation of collectors

It is also very important aspect for electricity generation through solar thermal route. The most favorable orientation of parabolic trough collectors for a solar thermal power plant is obviously NorthSouth axis. This orientation maximized the yearround electricity generation. The electricity generation almost follows the solar radiation curve and not necessarily the load curve. However, it may be possible to orient the collectors in a manner such that electricity generation is increased during offnoon hours though at the expense of perhaps greater reduction in its quantum around noon. This may be desirable if electricity during offnoon is valued far higher than the electricity around noon. Similarly, by orienting the collectors along EastWest axis, annual electricity generation for latitude around 26 °N is reduced by 15-20%.

(e) Selection of collector area and other components

Because of diurnal and seasonal variation of solar insolation, the collector field sizing at a particular level is of critical importance. Sizing of the collector field at lower insolation levels would mean that around noon hours, a part of the field would remain unused unless the turbine could take overloading. Sizing of the field at higher insolation levels would result in part load operation of turbogenerator set for most of the time.

To take care of yearly variation in solar radiation, the collector field should be sized on the basis of a Typical Meteorological Year (TMY). It is desirable to conduct technoeconomic feasibility studies of a solar thermal power plant on the basis of TMY data.

Other important considerations in determining size of the collector field are: frequency of insolation level at a particular hour; cost of collectors; long term marginal cost of power generation for each unit, increment in collector area, etc.

Pumping of the heat transfer fluid through absorber tubes and the heat exchangers is a major energy consuming auxiliary requirement in a solar thermal power plant. Therefore, proper selection of pumps to ensure continuous flow with highest possible efficiency is important.

The primary requirements are to overcome the system pressure drop at required flow rate and high efficiency.

(e) Heat exchangers

A Solar Thermal Power Plant has a number of heat exchangers such as heat exchangers to transfer heat from heat transfer fluid to water/steam (steam generator system comprising economizer, boiler, super heaters and preheaters), surface condenser, feed heaters, etc. The complete heat exchanger system should preferably be thermally and mechanically designed for about 110% of capacity, based on the turbinegenerator nameplate rating.

(f) Turbine

Selection of turbine for solar thermal power plant is another critical consideration. It has been estimated that one-point drop in turbine efficiency increases collector area by three points, which means highly efficient turbines are desirable.

(g) Back up capacity requirement

Backup capacity of the system should be optimized on the basis of load curve, duration of fluctuations in solar insolation, etc. If load demand during evenings is lower than nominal plant capacity and fluctuation in solar insolation is high. Installation of a partial back up is recommended. It is advisable to select backup capacity after careful analysis of cost effectiveness. To minimize the precipitation of carbon and sulphur oxides which may cause mirror and metallic corrosion, use of certain fuels having non-corrosive property and sulphur heavy stock oil (LSHS), etc. is recommended. In case load demand is high during evening and morning along with frequent fluctuations due to grid failure and cloudy weather and good quality fuel for back up is not available or is costly, a thermal storage is recommended.

5.11 Storage of Solar Energy

One of the limitations in the use of solar energy is that it is time dependent and intermittent in nature. Therefore, there is urgent need to store the solar energy during the period, when it is available in abundant quantity and use it, when it is not available. Use of solar energy in the form of air conditioning and cooling acquire a significant position, however the position become severe during winter, because of its low availability during this period. Thus, the use of solar energy be integrated with its optimum capacity of storage system. There are different methods of storage. The choice of media for energy storage also determined method of storage. In fact, the choice of a storage system depends on the various factors, such as.

1. The nature of device or system to be connected on the storage system.
2. Time and duration of availability of solar energy.
3. Duration during which it is expected to use.
4. The degree of reliability or percentage accuracy required.
5. The convenience and efficiency of the storage system.

6. Flexibility in the system i. e. provision of incorporation of auxiliary energy system.
7. Total cost involved in setting up the storage system, which include initial capacity investment and operating cost, if any.

5.12 Energy Storage

Solar energy storage system may be classified as shown in Fig. 5.20

Fig 5.20: Types of Solar Energy Storage

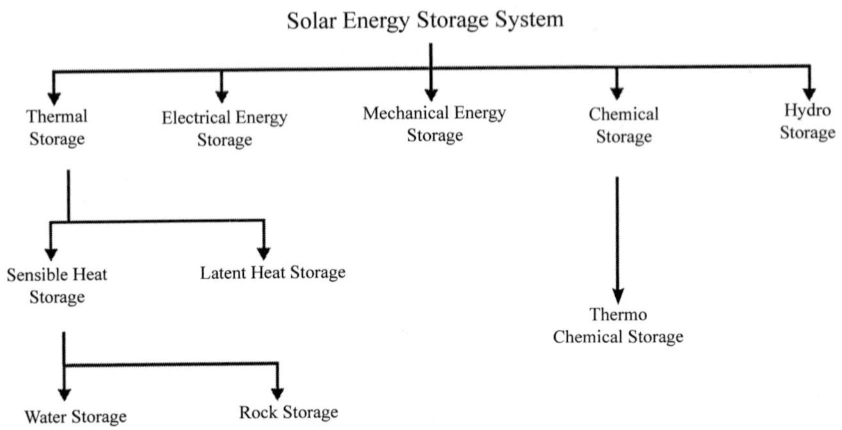

5.12.1 Thermal Storage

Energy can be stored by heating, melting or vaporization of material and then simultaneously energy can be recovered as heat when the process is reversed. Whenever any material is heated, it would develop following two types of changes;

(1) Sensible heat change

The rise in temperature of the material is there in sensible heating and as such it would not give any indication of phase change i.e. solid to liquid or liquid to vapour etc. The total energy responsible for this change can be represented as

$$Q_s = MC_p (T_2 - T_1)$$

Where

M = mass of material to be heated

C_p = specific heat of material

T_2 = final temperature of material (before boiling or melting)

T_1 = initial temperature of material at which heating is started.

(2) Latent heat change

This change indicate phase change i.e. conversion of solid into liquid or liquid into vapour through the action of heat. Here the energy responsible for this change can be represented as

$$Q_L = Q_s + M_w L$$

Where

M_w = mass of matter evaporated

L = Latent heat of vaporization.

The basic fundamental of above two heat changes can be incorporated for development a solar storage system based on thermal storage aspects. Thermal energy storage is essential for both domestic water and space heating applications and for the high temperature storage system needed for thermal power applications. Water and rock are two examples, in which solar energy can be stored on the basis of thermal storage aspects. The other materials, which can be used for these storages are iron shot, iron (Red iron oxide or iron ore), concrete and refractory material like magnesium oxide, aluminium oxide and silicon oxide.

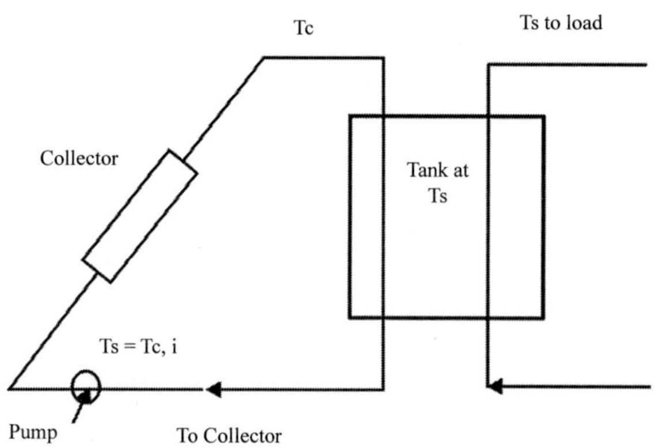

Fig 5.21: Sensible Heat Storage System

Water is most common material for storage of solar energy. The example of this approach is insulated water storage tank coupled with solar water heater, where energy is added by circulating water through collector and is removed by circulating water through load (Fig. 5.21)

The optimum tank size for the float plate collector system is usually about 70 lit/m2. Water has the following characteristics for storage medium

1. It is readily and easily available useful material.
2. It has higher thermal storage capacity
3. It permits energy addition and removal from the medium itself, thus eliminating any temperature drop between transport fluid and storage medium.
4. It includes pumping for transportation purpose. The pumping cost applicable to water is small.
5. The heat transfer can be made by natural convection as in domestic solar water heater or by forced convection with the help of pump, blower or fan etc.
6. It is superior storage medium as compared to rock, gravel or crushed stone, because of its lower material cost and lower volume required per unit of energy storage.

The rock, gravel or crushed stone can also be used as energy transport mechanism with air in sensible heat storage. The rock permits large heat transfer area but its thermal capacity, however, is only about half that of water. Therefore, rock storage volume will be about 3 times the volume of a water tank that is heated over the same temperature interval. The rock has following characteristics-

1. It can more easily contained than water.
2. It acts as its own heat exchanger, thus reduction in total system cost.
3. It permit storage at high temperature (above 100 °C or so), which cannot be used in liquid form without an expensive, pressurized storage tank.
4. The heat conductivity of rock storage depend on air flow. The conductivity of the bed is low when air flow is not present.
5. The cost of storage material is low.
6. The heat transfer coefficient between the air and solid is high.

A packed bed storage unit is shown in Fig. 5.22 it essentially includes a container, porous structure to support the bed and air distributions. In operation, the flow of heat is maintained in one direction through the bed and in opposite direction during removal of heat.

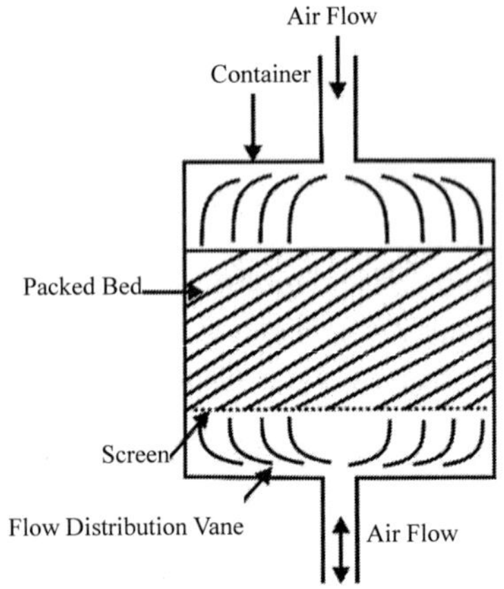

Fig. 5.22: Packed Bed Storage Unit

Similarly, in the thermal storage rout, latent heat storage units are also available. Materials that undergo change of phase in a suitable temperature are adopted for this approach. Few examples are Glauber's slat ($Na_2SO_4.10H_2O$), $Fe(NO_3).6H_2O$, organic compounds, Salt Eutectics and water. The material used in latent heat storage units have following characteristics-

1 The phase change take place at a high latent heat effect. It store large quantities of heat.

2 The phase change process is reversible over a very large number of cycles without serious degradation.

3 It accompanied with limited supper cooling

4 It has a mean for transfer heat into it and out of it.

5 The phase change occurs close to its actual melting temperature.

6 The material used for this is essentially harmless i.e. non toxic, non-inflammable, non-combustible and non-corrosive.

7 The preparation of phase changing material for use is relatively simple.

5.12.2 Electrical energy storage

Capacities and inductors are used for storing electrical energy. The capacitors can store large amount of electrical energy at high voltage and low current for long period, however, inductors store electrical energy at low voltage and high current.

Theoretically total energy stored in capacitors is given as

$$H_{cap} = \frac{1}{2} V\varepsilon E^2$$

Where

V = is the volume of the dielectric

ε = is a constant and

E = is the electric field strength

The electric energy store in capacitor depends upon the break down strength of dielectric. Best example of dielectric material is mica. Presently the electrical energy stored in capacitor is not economical, because it can store energy upto 12 hours duration. Further, most of capacitor is costly. Since the conductivity of dielectric is nil, therefore there will always be losses due to leakage. As a result of it the storing period is limited in capacitor.

The energy stored in an indicator is given as

$$H_{ind} = \frac{1}{2} V \mu H_m^2$$

Where μ is permeability of the material and H_m is the magnetic flux density. Therefore, in order to have more electric energy storage (H_{ind}), both μ and H_m should be large. Further for H_m to be more, high magnetic field are necessary, which required strong supporting structure. In addition to this, at the time of releasing of electrical current from the inductor, the circuits carrying large current should be open. The opening of circuit carrying large currents is a big problem in present setup.

Electrical energy can also be stored in the primary cell, which comprises two electrodes (anode and cathode) and an electrolyte (an ionic conduction). The primary cell is very simple device for storing electrical energy. Since it has no moving parts, therefore, it works at high efficiency and always the output is in the form of electrical energy. However, these cells are expensive, because of their very low storing capability for a given volume of cell. Examples are mercury, silver-zinc cell, mercury-carbon, lead acid, nickel-iron and nickel-cadmium batteries.

5.12.3 Chemical energy storage

The chemical energy storage of solar energy consists of a battery, which is charged photo chemically and discharged electrically whenever needed. The photochemical reaction takes place in the presence of solar radiation. Some of the reactions in type of storage are:

$$2NOCL + \text{Photons} \rightarrow 2NO + Cl_2$$

$$AgCl(s) + \text{Photons} \rightarrow Ag(s) + \frac{1}{2} Cl_2$$

$$NO_2 + \text{Photons} \rightarrow NO + \frac{1}{2} Cl_2$$

$$H_2O + \frac{1}{2} O_2 \rightarrow H_2O_2$$

It has been observed that all in above reactions, the rate of reverse reaction is slowed down, when one of its products is precipitate, also it have side reaction and back reaction at faster rate. These are the reason why chemical storage of solar energy is not getting popular.

Hydrogen energy can be generated through electrolyzation of water in presence of solar radiation. It is also possible to electrolyze water with solar generated electrical energy, store oxygen and hydrogen and recombine in a fuel cell to regain electrical energy.

$$2H_2O \leftrightarrow 2H_2 + O_2$$

Further, solar energy can also be used for anaerobic fermentation of solar algae into combustible fuel gas i. e. methane (CH_4). The methane is an excellent fuel for meeting out much energy related activities; it is lighter than air and can be easily stored at room temperature. The stored energy of CH_4 can be released into thermal energy whenever it burns in presence of oxygen.

Stored energy of $CH_4 + O_2 \rightarrow$ Thermal energy (20 MJ m^{-3})

It has been observed that solar energy can be converted into methane gas with 2% efficiency. Thus, one square kilometer of algae filed could produce an amount of methane storing 4 MW of converted solar energy.

Photosynthesis is also a method of solar energy storage in chemical form.

$$6CO_2 + 6H_2 O \rightarrow C_6 H_{12} O_6 + 6O_2$$

During this reaction, in presence of chlorophyll and solar radiation, atmospheric CO_2 and H_2O are combined and produce carbohydrates. These carbohydrates can further release stored energy in thermal energy, at high temperature, whenever needed

5.12.4 Hydro storage

Hydro storage is a reasonable efficient method of storing solar energy by storage water into an elevated reservoir during period of solar energy radiation and recovering the stored energy by running water through a turbine when energy is needed (Fig. 5.23).

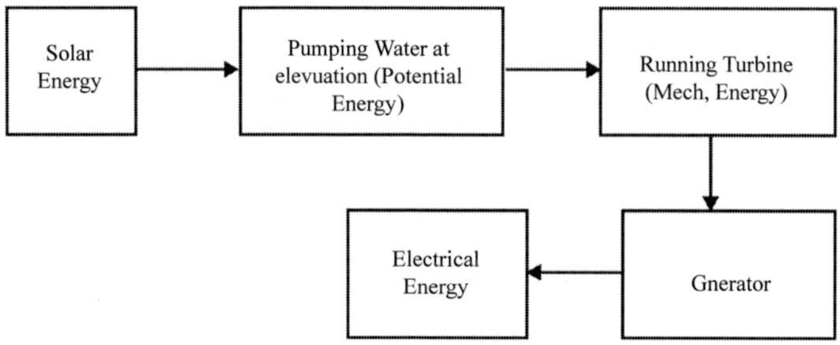

Fig. 5.23: Hydro Storage System

Pumped water at elevated will have potential energy

Where m is mass of water 7 h is height of tank.

If this potential energy is converted into kinetic energy to run a turbine, then

Kinetic Energy (K.E.) of water during running $= \frac{1}{2} mV^2$

Therefor $mgh = \frac{1}{2} mV^2$

$V = \sqrt{2gh}$

This velocity will be used to design turbine coupled with generator.

6

Solar Drying

6.1. Introduction

Solar crop drying has been demonstrated to be cost effective and an effective alternative to traditional and mechanical drying systems, especially in location with god sunshine during the harvest season. Considerable energy savings can be made with open sun drying since the source of energy is free and sustainable. However, this method of drying is extremely weather dependent and has the problems of contamination, infestation, microbial attacks, etc., thus affecting the product quality. Additionally, the drying time required for a given commodity can be quite long and results in post-harvest losses. Solar drying of agricultural products in enclosed structures by natural convection is an attractive way of reducing post harvest losses and low quality of dried products associated with traditional sun-drying methods. The crop is vulnerable to damage due to hostile weather conditions. The crop is also susceptible to re-absorption of moisture, if it is kept on the ground during periods of no sun, which reduces its quality. This type of problem can be solved by solar dryer. Many dryers have been developed and used to dry agricultural product in order to improve shelf life .Most of these either use expensive sources of energy such as electricity or combination of solar energy and some other form of energy which is not cost effective and dependent on regular supply of fuel.

Principally, drying requires temperature for removing moisture from product, air flow rate to remove moisture from drying chamber and optimum relative humidity within the drying chamber to allow fast drying. These all three requirements for drying can easily be achieved from solar energy. Solar drying is one of the direct applications of solar heat, which involves removal of moisture from the products through the application of solar energy. The removal of moisture requires only low temperature heating, which can be met easily by absorbing solar energy by the surface. This increases the temperature of the air inside the dryer. The moisture produced from the drying product is usually carried out along with the exhaust air. The characteristic of solar energy is good for the drying at low temperature i.e. high flow rates with low temperature rise. The intermittent nature of solar radiation can not affect the drying performance at low temperature. Further, solar energy is available at the site of use and saves transportation cost.

Significant developments of the past in the area of solar drying are reviewed. Increasing fuel prices stimulated research on solar drying in highly mechanized agricultural systems. In solar drying, solar-energy is used as either the sole source of the required heat or as a supplemental source. To dry various agricultural products at farm level, natural circulation solar dryers are preferred over forced circulation solar dryers because of non-availability or erratic power supply in rural areas. For large scale drying applications, a large number of direct, indirect, natural circulation solar dryers have been developed. Presently, drying of agricultural commodities in the food processing industry is being made with the help of various types of mechanical dryers. There may be about 2500–3000 of such dryers in operation.

6.2. Working Principle

Solar drying systems are classified primarily according to their heating modes and the manner in which the solar heat is utilized. Further solar dryers can broadly be categorized into direct, indirect and specialized solar dryers. The three modes of drying are: (a) open sun drying, (b) direct (natural solar dryer) and (c) indirect in the presence of solar energy. The working principle of these modes mainly depends upon the method of solar-energy collection and its conversion to useful thermal energy.

6.2.1 Open sun drying

Open sun drying is a traditional method practiced widely in tropical climates for drying agricultural products. Despite the rudimentary nature of the processes involved, such techniques still remain in common use. Because the power requirements (i.e. from the solar radiation and the air's enthalpy) are readily available in the ambient environment, and as little or no capital cost is required and running costs low (often labour only), these are frequently the only commercially viable methods for drying agricultural produce in developing countries. Considerable saving can be made with this type of drying since the source of energy is free and sustainable. In open sun drying, the crop is spread in a thin layer on the ground and exposed directly to solar radiation, wind and other ambient conditions. The working principle of open sun drying is shown in Fig. 6.1. The solar radiation falling on the crop surface is partly reflected and partly absorbed. The absorbed radiation and surrounding heated air heat up the crop surface. A part of this heat is utilized to evaporate the moisture from the crop surface to the surrounding air. The part of this heat is lost through radiation (long wavelength) to the atmosphere and through conduction to the ground surface. The open sun drying method is associated with various problems, which include:

(i) Rapid reconstitution of the crop due to sudden and unexpected rainfall.

(ii) High ambient relative humidity, especially in tropical climates.

(iii) Severe damage and loss of crop due to insect infestation.

(iv) Animal and human interference.

(v) Dust and debris carried by the wind.

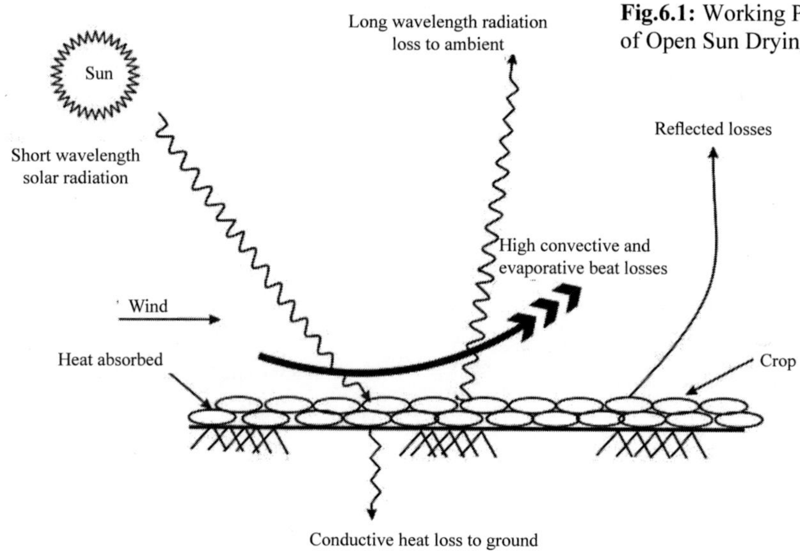

Fig.6.1: Working Principle of Open Sun Drying

6.2.2 Natural convection dryer

Natural convection type solar dryer essentially consists of an enclosure for keeping the products to be dry with a transparent cover placed over the enclosure. The internal surfaces of the enclosure are painted black. The evaporation of moisture from the product takes place due to direct absorption of solar radiation by the product as well as transfer of heat by the internal surface, which get heated by the radiation incident on them. Here, ultimately removal of moisture takes place through naturally created draft.

Application of natural convection type dryer solely depends on availability of solar energy. In the beginning solar-heated air is circulated through the produce by bouyancy forces or as a result of wind pressure, acting either singly or in combination. These dryers are often called ``passive" in order to distinguish them from systems that employ fans to convey the air through the crop. The latter are termed ``active" solar dryers. Natural-circulation solar-energy dryers appear the most attractive option for use in remote rural locations. They are superior operationally and competitive economically to natural open-to-sun drying. The advantages of natural-circulation solar-energy tropical dryers that enable them to compete economically with traditional drying techniques are:

(i) They require a smaller area of land in order to dry similar quantities of crop that would have been dried traditionally over large land areas in the open.

(ii) They yield a relatively high quantity and quality of dry crops because fungi, insects and rodents are unlikely to infest the crop during drying.

(iii) The drying period is shortened compared with open air drying, thus attaining higher rates of product throughput.

(iv) Protection is afforded the crop from sudden down pours of rain.

(v) Commercial viability, i.e. their relatively low capital and maintenance costs because of the use of readily available indigenous labour and materials for construction.

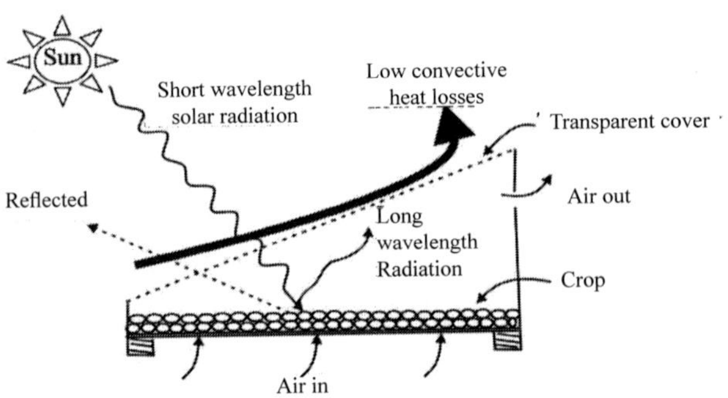

Fig 6.2: Working Principle of Direct Solar Cabinet Drying

The principle of direct solar crop drying is shown in Fig.6.2. This is also called cabinet dryer. Solar Cabinet dryer mainly consist of a drying cabinet. One side of the cabinet is glazed to admit solar radiation, which is converted in to low grade thermal heat thus raising the temperature of the air, the drying chamber, and the produce. As solar radiation incidence on glass cover part of it reflect back to atmosphere and remaining is transmitted inside cabin dryer. Part of radiation transmitted in to dryer again reflected back from the surface of the produce.

The remaining part transmitted radiation is absorbed by the surface of the produce. Due to the absorption of solar radiation, produce temperature increase and it starts emitting long wavelength radiation which is not allowed to escape to atmosphere due to presence of glass cover unlike open sun drying. Thus, the temperature above the produce inside chamber becomes higher. The glass cover server one more purpose of reducing direct convective losses to the ambient which further become beneficial for rise in produce and chamber temperature respectively. However, convective and evaporative losses occur inside the chamber from the heated produce. The moisture is taken away by the air entering into the chamber from below and escaping through another opening provide at the top.

6.2.3 Forced convection dryer

The crop is not directly exposed to solar radiation to minimize discolouration and cracking on the surface of the crop. The forced convection dryer air is heated in collector and is forced on to the drying material through external means like fans or pumps. These dryers are often called ``active mode" solar dryers.

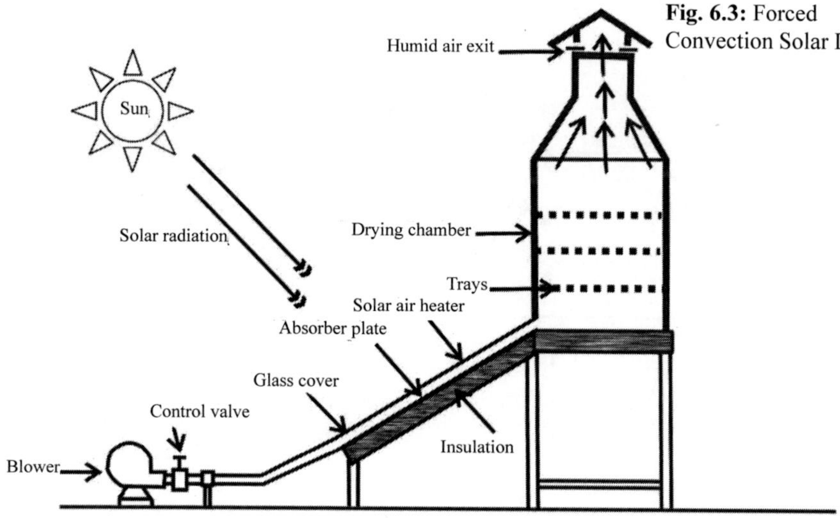

Fig. 6.3: Forced Convection Solar Dryer

Fig.6.3 shows a schematic of the major components of forced convection type solar dryer. The solar air heater or collectors is placed at an appropriate angle to optimize solar-energy collection. Tilting the collectors is more effective than placing them horizontally because more solar energy can be collected when the collector surface is nearly perpendicular to the sun's rays and by tilting the collectors, the warmer, less dense air rises naturally into the drying chamber. In forced convection dryer, the solar-heated air flows through the solar drying chamber in such a manner as to contact as much surface area of the produce as possible. Leafy agricultural produce and thinly sliced food materials are placed on perforated drying trays. As heated air forced to drying chamber through several layers of produce on trays, it becomes moisture laden. This moist air is vented out through the outlet port. Fresh air is then taken in to replace the exhaust air. Forced convection type solar dryers are known to be suitable for drying higher moisture content foodstuffs such as papaya, kiwi fruits, brinjal, cabbage and cauliflower slices and such dryers are comparatively efficient, faster and can be used for drying large agricultural products.

Dryers are also classified as direct as well as indirect type of solar dryer. In the direct type solar dryer, the material to be dry is put on trays in the dryer and direct entry of sun through a suitable transparent cover is made and finally moisture after getting evaporated goes out. Where as in the indirect solar dryer some air heaters are used through which air gets heated and passes over to the drying material. Similarly, direct cum indirect type of solar dryer, directly sun radiations are allowed inside the tray and also heated air from outside air collectors is used for boosting output of the dryer. There are varieties of each of these dryers. Few solar dryers are shown schematically in Fig. 6.4.

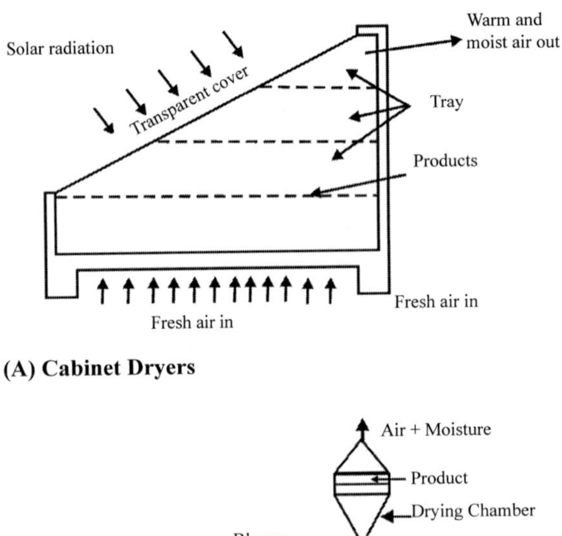

(A) **Cabinet Dryers**

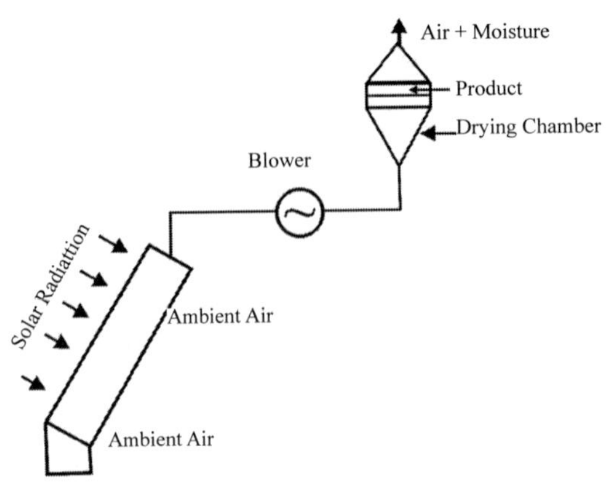

(B) Indirect Forced Convection Solar Dryer

Fig. 6.4: Various type of Solar Dryer

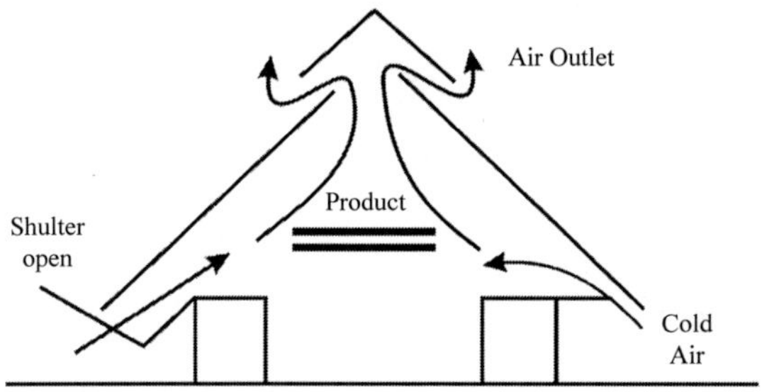

Greenhouse Solar Dryer
(Natural Convection Type)

(C) Greenhouse Solar Dryer

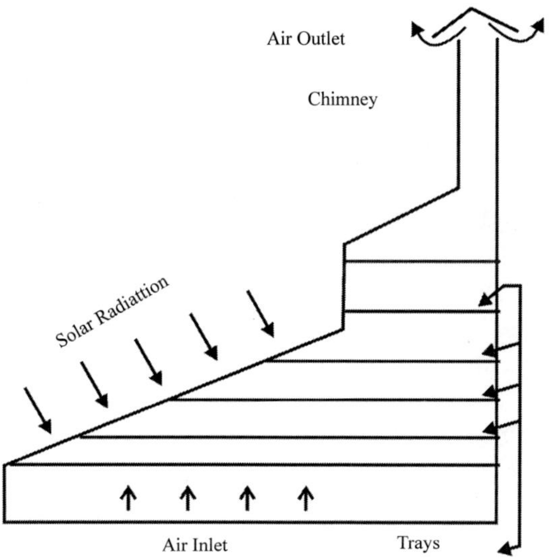

(D) Natural Convection Direct cum Indirect Solar Dryers

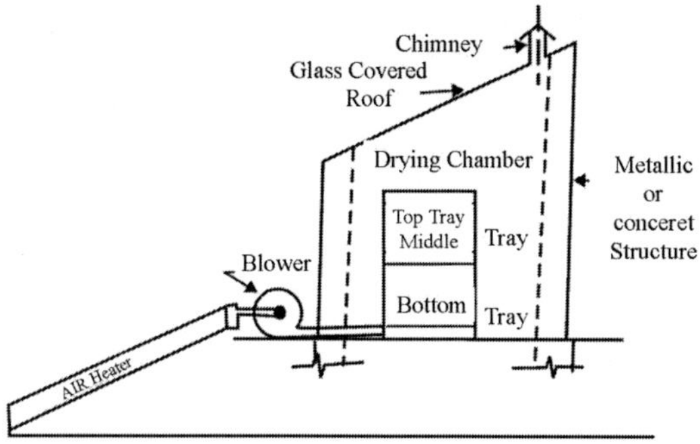

(E) Forced Convection Direct cum Indirect Solar Dryer

6.3. Classification of Drying Systems

There are numbers of way to classify the drying system but it broadly classified according to their operating temperature ranges into two main groups of high temperature dryers and low temperature dryers. Further it also classified on the basis of their heating sources into fossil fuel dryers and solar-energy dryers. At industrial level where high temperature is required for drying high temperature dryer are used and its heat source is fossil fuel. Agricultural produced mostly drying in the range of 60-75 °C used low temperature dryer usually powered by solar energy. Solar-energy drying systems are classified primarily according to their heating modes and the manner in which the solar heat is utilized.

6.3.1 High temperature dryers

In industries, dryers are used to make product suitable for marketing. A majority of industrial dryers used steam/ electricity for heating operation required for removal of moisture. Some dryers also used petroleum-based fuels like kerosene, LPG and light diesel oil (LDO) for heating the air which is to remove the moisture. Dryers operated in these fuels are in position to generate high temperature that is why it is called high temperature dryers. These types of dryers are necessary when very fast drying is desired. These types of dryers usually used when products require a short exposure to the drying air. High temperature dryers are usually classified into batch dryers and continuous flow dryers. In batch dryers, the products are dried in a bin and subsequently moved to storage. Thus, they are usually known as batch-in-bin dryers. Continuous flow dryers are heated columns through which the product flows under gravity and is exposed to heated air while descending.

6.3.2 Low temperature dryers

The drying of agricultural products is primarily a low temperature operation as higher temperatures are likely to result in destruction of nutrients and flavour of the food. Recommended values of drying temperatures for various agricultural products are given in Table 6.1. In low temperature drying systems, the moisture content of the product is usually brought in equilibrium with the drying air by constant ventilation. Thus, they do tolerate intermittent or variable heat input. Low temperature drying enables crops to be dried in bulk and is most suited also for long term storage systems. Thus, they are usually known as bulk or storage dryers. Their ability to accommodate intermittent heat input makes low temperature drying most appropriate for solar-energy applications.

Table 6.1: Moisture Contents of Solar Drying of Various Agricultural Produces

Product	Moisture content		Maximum allowable temperature (°C)
	Initial (%)	Final (%)	
Apples	80	24	65-70
Apricots	85	18	65
Bananas	80	15	70
Cabbage	80	4	65
Cabbage	80	4	55
Carrots	70	5	65-75
Cassava	67	17	70
Cauliflower	80	6	50
Chilies	80	5	65
Corn	24	14	50
Cotton	50	9	75
Cotton seed	50	8	75
Fabric	50	8	75
Figs	70	20	70
French bean	70	5	75
Garlic	80	4	75
Garlic flakes	80	4	55
Grapes	80	15-20	70
Green beans	70	5	75
Green peas	80	5	60
Guavas	80	7	65
Maize	35	15	60
Mango	85	12	20
Mangoes	84	28	70
Millet	21	4	45
Mulberries	80	10	65
Nutmeg	80	20	65
Okra	80	20	65
Onion flakes	80	10	55
Onion rings	80	10	55
Onions	85	6	55
Paddy, parboiled	30-35	13	50
Paddy, raw	22-24	11	50
Peaches	85	18	65
Pineapple	80	10	65
Potato chips	75	13	70
Potatoes	75	13	75
Prunes	85	15	55
Red lauan	90	20	75
Rice	24	11	50

Sorrel	80	20	65
Spinach	80	10	75
Sweet potato	75	7	75
Tomatoes	95	7	60
Wheat	20	16	45
Yams	80	10	65

6.4. Classification of Solar-Energy Drying Systems

Solar dryer is a device for drying agricultural product under controlled conditions. The controlled drying means controlling the drying parameters like drying air temperature, humidity, drying rate, moisture content and air flow rate. The dryer are classified in several categories depending upon the mode of heating or the operational mode of heat derived from the solar radiation, and its subsequent use of removal of moisture from the wet product as shown in Fig. 6.5. Broadly solar dryer can be classified into two categories

a) Active solar dryer (forced circulation)

b) Passive solar dryer (natural circulation).

6.4.1 Active solar dryer

In active solar drying system, hot air is generated, outside the drying chamber. These dryers incorporate external means, like fans or pumps, for moving the solar energy in the form of heated air from the collector area to the drying beds. In an active dryer, the solar-heated air flows through the solar drying chamber in such a manner as to contact as much surface area of the food as possible. Thinly sliced foods are placed on drying racks, or trays, made of a screen or other material that allows drying air to flow to all sides of the food. Once inside the drying chamber, the warmed air will flow up through the stacked food trays. The drying trays must fit snugly into the chamber so that the drying air is forced through the mesh and food. Active solar dryers are known to be suitable for drying higher moisture content foodstuffs such as papaya, kiwi fruits, brinjal, cabbage and cauliflower slices.

6.4.2 Passive solar dryer

A passive solar dryer is one in which the drying product is directly exposed to the sun's rays. Direct passive dryers are best for drying small batches of fruits and vegetables such as banana, pineapple, mango, potato, carrots and French beans. This type of dryer comprises of a drying chamber that is covered by a transparent cover made of glass or plastic. The drying chamber is usually a shallow, insulated box with air-holes in it to allow air to enter and exit the box. The food samples are placed on a perforated tray that allows the air to flow through it and the food.

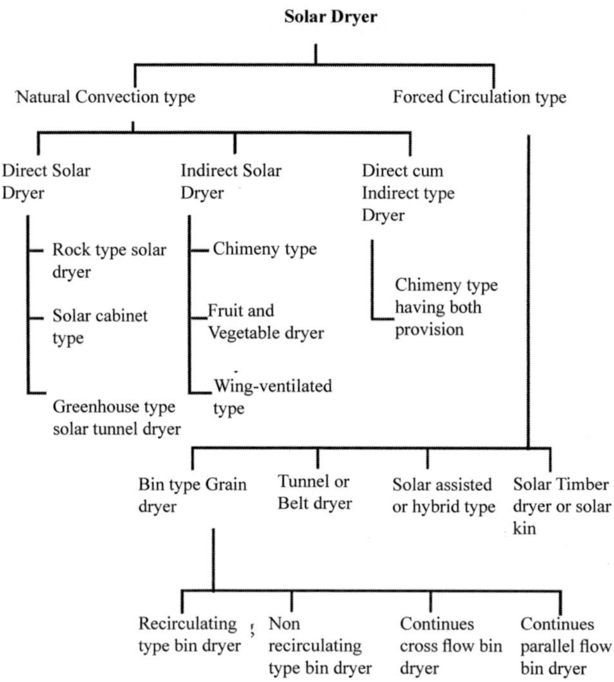

Fig. 6.5: Classification of Dryers and Drying Modes

6.5 Design of Tunnel Dryer

To carry out design calculation and size of the tunnel dryer, the following assumption are made as summarized in Table 6.2

Table:6.2: Assumption for Design of Tunnel Dryer

Items	Condition or assumptions
Location	Udaipur (27° 42 N. 75° 33' E)
Product	Surgical Cotton
Loading rate (M) in each batch	600 kg
Initial moisture content (m_i)	40%
Final moisture content (m_f)	5%
Ambient air temperature	32 °C (Average for April, May–June)
Maximum allowable temperature	65 °C
Sunshine hour	8h
Incident solar radiation (I)	5.5 kWh m^{-2}
Dryer efficiency (η)	30%
Thickness of plastic sheet	200 μ (UV stabilized)
Density of air at ambient (kg m^{-3})	1.252
Height of chimney (H)	2.40 m

The mass of water to be removed during drying, M_w kg

$$M_w = \frac{m_i - m_f}{100 - m_f} \times W$$

Mass of water removed per hour m_w, kg h-1

$$m_w = \frac{M_w}{t_d}$$

Total energy required Q kJ

$$Q = WC_p (T_d - T_a) + M_w \lambda$$

Energy required per hour Q_t KJ h^{-1}

$$Q_t = \frac{Q}{t_d}$$

Drying area

It has been observed that about 68% of the area of hemispherical shaped solar tunnel dryer toward south is able to receive sunlight whereas remaining 32% of the area toward north is shaded from the sun. Global solar radiation (I) for Udaipur region is 5.5 kWh m^{-2}. The floor area of dryer is calculated as follows:

$$\text{Area} = \frac{Q_t}{(I \times \eta \times 0.68)}$$

Size of solar tunnel required

Energy gainer area is about 68%

Area of hemispherical-shaped solar tunnel dryer = π r L

Diameter 3.75 m is kept constant for easy entry and other convenience

Design of North wall

Area shaded from sun is about 32%

Total area of north wall to be protected Ap = drying area x 32%

Perimeter of north wall (P) = π r

Since perimeter (P) cover diametrical length (Lp) = 3.75 m

Arc width of cover through which energy losses (w)=Ap/L

Arch width (w) will cover diametrical length (Lp_1)

$$L_{p1} = \frac{L_p \times w}{P}$$

Required length of protector (h_p)

$$h_p = \sqrt{(w)^2 - (L_{p1})^2}$$

Design of chimney

Quantity of air need to absorb m_w kg of water

$$Q_a = \frac{m_w \times \lambda}{C_a \times \rho_a (T_e - T_a)}$$

Now, Qa amount of moist air is needed to be removed in 8 h ($m^3 s^{-1}$)

$$= \frac{Q_a}{(8 \times 60 \times 60)}$$

Draft produce if we assume height of chimney by 2.40 m (four number of 0.60 m)

Di = H g (ρa – ρe)

But actual draft D_a

$D_a = 0.40 \times D_i$

Velocity of exit air (v)

$$v = \sqrt{\frac{2D_a}{\rho_e}}$$

Thus, if assumed this exit air is being carried out by four chimneys

$$Q_c = \frac{Q_a}{4}$$

Area of chimney

$$A = \frac{Q_c}{v \times K}$$

6.6 Thermal Modeling of Solar Tunnel Dryer

The present thermal modeling work was inspired by the work of Janjai et al. (2011). A mathematical model to assess the thermal performance of a tunnel dryer was developed. The following assumptions were made while developing the mathematical model:

a) There is no stratification of the air inside the tunnel dryer.

b) Drying computation is based on a thin layer drying model.

c) Specific heats of air, cover and product are constant.

d) Absorptivity of air is negligible.

e) Radiative heat transfers from the floor to the cover and from the floor to the product are negligible.

The schematic diagram of heat transfers of the solar tunnel dryer is shown in Fig. 6.6, and the energy flows through different components of the dryer as follows:

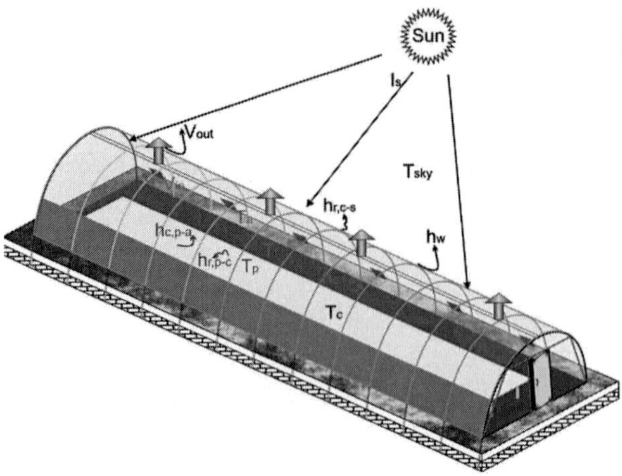

Fig. 6.6: Various Heat Transfer Mode in Solar Tunnel Dryer

Energy Balance of the Cover

The tunnel is covered with UV stabilized low density plastic sheet. Solar radiations falling on cover are allowed to be transmitting inside the tunnel. Energy balances on the tunnel cover are considered as follows:

The thermal energy accumulation in the tunnel cover = Convective heat transfer between air inside the tunnel and cover + Radiative heat transfer between the sky and the cover due to radiation + Convective heat transfer between cover and ambient air + Radiative heat transfer between the product and the cover + Solar radiation absorbed by the cover.

Energy balance of the cover is as follows:

$$Q_{th\text{-}c} = Q_{c,c\text{-}a} + Q_{r,c\text{-}s} + Q_{c,c\text{-}w} + Q_{r,p\text{-}c} + Q_{ab}$$

$$m_c C_{pc} \left(\frac{dT_c}{dt}\right) = A_c h_{c,c\text{-}a} (T_a - T_c) + A_c h_{r,c\text{-}s} (T_{sky} - T_c) + A_c h_w (T_o - T_c) + A_p h_{r,p\text{-}c} (T_p - T_c) + A_c \alpha_c I_s \quad (6.1)$$

Where;

Rate of thermal energy accumulated in the cover

$$Q_{th\_cover} = m_c C_{pc} \left(\frac{dT_c}{dt}\right)$$

Convective heat transfer between air inside the tunnel and cover

$$Q_{c,c\text{-}a} = A_c h_{c,c\text{-}a} (T_a - T_c)$$

Radiative heat transfer between the sky and the cover due to radiation

$$Q_{r,c\text{-}s} = A_c h_{r,c\text{-}s} (T_{sky} - T_c)$$

Convective heat transfer between cover and ambient air

$Q_{c,c\text{-}w} = A_c\, h_w\, (T_o - T_c)$

Radiative heat transfer between the product and the cover

$Q_{r,p\text{-}c} = A_p\, h_{r,p\text{-}c}\, (T_p - T_c)$

Solar radiation absorbed by the cover

$Q_{s,c} = A_c\, \alpha_c\, I_s$

Energy Balance of Air inside the Tunnel

Air inside the tunnel dryer gets heated because of the convective heat transfer among the floor, product and air. The energy balances for the same can be written as follows:

Thermal energy accumulation in the air inside the tunnel dryer = Convective heat transfer between product and air + Convective heat transfer between floor and air + Thermal energy gain of air from product due to sensible heat transfer + Thermal energy gain by air inside the dyer due to inflow and outflow of the air in the chamber + Overall heat loss from the air inside the dryer to ambient + Solar energy absorbed by the air inside the dryer

Energy balance of air inside the tunnel dryer is as follows:

$$Q_{th\text{—}air} = Q_{c,p\text{-}a} + Q_{c,f\text{-}a} + Q_{s,a\text{-}p} + Q_{t,gain} + Q_{l,am} + Q_{s,ab}$$

$$m_a\, C_{pa}\, \left(\frac{dT_a}{dt}\right) = A_p\, h_{c,p\text{-}a}\, (T_p - T_a) + A_f\, h_{c,f\text{-}a}\, (T_f - T_a) + D_p\, A_p\, C_{pw}\, \rho_p\, (T_p - T_a)\, \frac{dM_p}{dt}$$
$$+ \rho_a\, V_{out}\, C_{pa}\, T_{out} - \rho_a\, V_{in}\, C_{pa}\, T_{am} + U_c\, A_c\, (T_o - T_a) + [(1 - F_p)\,(1 - \alpha_f) + (1 - \alpha_f)\, F_p]$$
$$I_s\, A_c\, \tau_c \qquad (6.2)$$

Where;

Thermal energy accumulated in the air inside the dryer

$Q_{th\_air} = m_a C_{pa} \left(\dfrac{dT_a}{dt}\right)$

Convective heat transfer between product and air

$Q_{c,p-a} = A_p\, h_{c,p\text{-}a}\, (T_p - T_a$

Convective heat transfer between floor and air

$Q_{c,f-a} = A_f\, h_{c,f-a}\, (T_f - T_a)$

Thermal energy gain of air from product due to sensible heat transfer

$Q_{s,a-p} = D_p\, A_p\, C_{pv}\, \rho_p\, (T_p - T_a) \dfrac{dM_p}{dt}$

Thermal energy gain by air inside the dyer due to inflow and outflow of the air in the chamber

$$Q_{t,gain} = \rho_a V_{out} C_{pa} T_{out} - \rho_a V_{in} C_{pa} T_o$$

Overall heat loss from the air inside the dryer to ambient

$$Q_{l,am} = U_c A_c (T_o - T_a)$$

Solar energy absorbed by the air inside the dryer

$$Q_{s,air} = [(1\text{-}F_p)(1 - \alpha_f) + (1 - \alpha_f) F_p] I_s A_c \tau_c$$

Energy Balance of the Product

Convective and Radiative are major heat transfer mode while balancing energy of product inside the tunnel. The energy balances energy for the same can be written as follow:

thermal energy accumulation in the product = Convective heat transfer between product and air + Radiative heat transfer between product and cover + Thermal energy lost from the product due to sensible and latent heat transfer + thermal energy gain by the product

The energy balance on the product gives:

$$Q_{th\_pro} = Q_{c,p-a} + Q_{r,p-c} + Q_{lost} + Q_{s,pro}$$

$$M_p(C_{pp} + C_{pl} M_p) \frac{dT_p}{dt} = A_p h_{c,p-a} (T_a - T_p) + A_p h_{r,p-c} (T_c - T_p) + D_p A_p \rho_p L_p \frac{dM_p}{dt} + F_p \alpha_p I_s A_c \qquad (6.3)$$

Where;

Thermal energy accumulation in the product

$$Q_{thpro} = m_p (C_{pg} + C_{pl} M_p) \frac{dT_p}{dt}$$

Convective heat transfer between product and air

$$Q_{c,p-a} = A_p h_{c,p-a} (T_a - T_p)$$

Radiative heat transfer between product and cover

$$Q_{r,p-c} = A_p h_{r,p-c} (T_c - T_p)$$

Thermal energy lost from the product due to sensible and latent heat transfer

$$Q_{lost} = D_p A_p \rho_p L_p \frac{dM_p}{dt}$$

Solar thermal energy gain by the product

$$Q_{s,pro} = F_p \alpha_p I_s A_c \tau_c$$

Energy Balances on the Concrete Floor

Thermocol sheet was sandwiched between dyer floor and the ground to reduce the conductive heat losses. The energy balances on the concrete floor of the tunnel dryer can be considered as follows:

Thermal energy accumulation in the floor = heat transfer between air inside the tunnel and the floor due to convection + Solar radiation absorption on the floor.

The energy balance on the concrete floor of the tunnel dryer can be written as:

$$Q_{th\_floor} = Q_{c,f\text{-}a} + Q_{s,floor}$$

$$m_f C_{pf} \frac{dT_f}{dt} = A_f h_{c,f-a}(T_a - T_f) + (1 - F_p)\alpha_f I_s A_f \tau_c \quad (6.4)$$

Where;

Thermal energy accumulation in the floor

$$Q_{th\_floor} = m_f C_{pf} \frac{dT_f}{dt}$$

Convective heat transfer between inside dryer air and the floor

$$Q_{c,f\text{-}a} = A_f h_{c,f\text{-}a} (T_a - T_f)$$

Solar radiation absorption on the floor

$$Q_{s,floor} = (1 - F_p)\, \alpha_f\, I_s\, A_f\, \tau_c$$

Mass Balance Equation

Air temperature inside the tunnel dryer is higher than corresponding ambient temperature. It increases the ability to pick up moisture from the product and product gets dried.

The rate of moisture accumulation in the air inside tunnel dryer = Rate of moisture inflow into the dryer due to entry of ambient air + Rate of moisture outflow with exit air from the tunnel dryer + Rate of moisture removed from the product inside the dryer

The mass balance inside the tunnel dryer can be written as:

$$M_{in\_air} = M_{inflow} + M_{outflow} + M_{rem}$$

$$\rho_a V \frac{dH}{dt} = A_{in} \rho_a H_{in} v_{in} - A_{out} \rho_a H_{out} v_{out} + D_p A_p \rho_p \frac{dM_p}{dt} \quad (6.5)$$

Where;

The rate of moisture accumulation in the air inside tunnel dryer

$$M_{in\_air} = \rho_a V \frac{dH}{dt}$$

Moisture inflow into the dryer due to entry of ambient air

$$M_{inflow} = A_{in}\, \rho_a\, H_{in}\, v_{in}$$

Moisture outflow with exit air from the tunnel dryer

$$M_{outflow} = A_{out}\, \rho_a\, H_{out}\, v_{out}$$

Moisture removed from the product inside the dryer

$$M_{rem} = D_p\, A_p\, \rho_p \frac{dM_p}{dt}$$

Heat Transfer and Heat Loss Coefficient

Radiative heat transfer coefficient from the cover to the sky ($h_{r,c-s}$) was computed according to Duffie and Beckman (1991):

$$h_{r,c-s} = \varepsilon_c \sigma \left(T_c^2 + T_{sky}^2\right)\left(T_c + T_{sky}\right) \tag{6.6}$$

Radiative heat transfer coefficient between the product and cover ($h_{r,p-c}$) is computed as suggested by Duffie and Beckman (1991).

$$h_{r,p-c} = \varepsilon_p \sigma \left(T_p^2 + T_c^2\right)\left(T_p + T_c\right) \tag{6.7}$$

The correlation with sky temperature (T_s) and ambient temperature (T_o) is adapted from Duffie and Beckman (1991).

$$T_{sky} = 0.0552\, (T_o)^{1.5} \tag{6.8}$$

Convective heat transfer coefficient from the cover to ambient due to wind (h_w) is computed as (Watmuff et al., 1977):

$$h_w = 2.8 + 3.0\, V_w \tag{6.9}$$

Convective heat transfer coefficient inside the solar greenhouse dryer for either the cover or product and floor (h_c) is computed from the following relationship

$$h_{c,f-a} = h_{c,c-a} = h_{c,p-a} = h_c = \frac{Nuk_u}{D_h} \tag{6.10}$$

Where D_h is given by

$$D_h = \frac{4WD}{2(W + D)} \tag{6.11}$$

Nusselt number (Nu) is computed from the following relationship (Kays and Crawford, 1980):

$$Nu = 0.0158 Re^{0.8} \tag{6.12}$$

Reynolds number which is given by:

$$Re = \frac{D_h V_a \rho_a}{v} \tag{6.13}$$

As the wind speed outside the dryer and the speed of drying air inside the dryer are very low, we assumed that the overall heat loss coefficient (U_c) of heat transfer from air inside the dryer to ambient air was approximately equal to the conductive heat transfer coefficient of the cover. This coefficient can be written as:

$$U_c = \frac{k_c}{\delta_c} \tag{6.14}$$

Humidity Ratio (H)

$$H = 0.62198 \frac{P_{ws}}{\left[(100 \times P_{vp}) - P_{ws}\right]} \tag{6.15}$$

Saturation vapour pressure (P_{ws}) is calculated by Weiss (1977) expression

$$P_{ws} = 061078 \exp (17.2694\ T_a)/(T_a + 237.3) \tag{6.16}$$

Vapour pressure (P_{vp}) is calculated by Antoine expression as appears in Singh and Kaushik (2013).

$$P_{vp} = \exp\left(25.317 - \frac{5144}{T_a}\right) \tag{6.17}$$

Relative Humidity (Rh)

$$Rh = P_{vp}/P_{ws} \tag{6.18}$$

6.7 Energy and Exergy Assessment

6.7.1 Energy Analysis

Energy Input

$$E_{in} = I_s \times A$$

Energy Output

$$E_{out} = \frac{m_{da} C_{p_a} (T_a - T_o)}{\Delta t}$$

Energy Efficiency

$$\eta = \frac{m_{da} C_{p_a} (T_a - T_o) / \Delta t}{I_s \times A} \tag{6.20}$$

6.7.2 Exergy Analysis

Following section 2.6, the exergy of solar radiation input to dryer is as follows:

$$Ex_{in} = Ex_{rad} = I_s \times A\left[1 - \frac{4}{3}\left(\frac{T_o}{T_s}\right) + \frac{1}{3}\left(\frac{T_o}{T_s}\right)^4\right]$$

Exergy output given by equivalent work of energy input as follows:

$$Ex_{out} = E_{out} \times \left(1 - \frac{T_o}{T_a}\right)$$

Exergy utilization efficiency of the dryer to raise the inside air temperature thus given by:

$$\Psi_{solar} = \frac{Ex_{out}}{Ex_{in}} \tag{6.21}$$

Inside the tunnel hot air has the exergy potential to pick up the moisture from the product and exit through the chimney. This process is considered as the steady flow and exergy equation can be derived by the simplification of the general exergy equation as suggested by Akpinar (2011).

Exergy potential of air inside the dryer is as follows:

$$Ex_{inside} = C_{P_a}\left[(T_a - T_o) - T_o \ln\left(\frac{T_a}{T_o}\right)\right]$$

Exergy of output air through chimney is as follows:

$$Ex_{outlet} = C_{P_a}\left[\left(T_{out@ch} - T_o\right) - T_o \ln\left(\frac{T_{out@ch}}{T_o}\right)\right]$$

Exergy efficiency of drying air is as follows:

$$\Psi_{drying\ air} = \frac{C_{P_a}\left[\left(T_{out@ch} - T_o\right) - T_o \ln\left(\frac{T_{out@ch}}{T_o}\right)\right]}{C_{P_a}\left[(T_a - T_o) - T_o \ln\left(\frac{T_a}{T_o}\right)\right]} \tag{6.22}$$

The net exergy efficiency of solar tunnel dryer can be written as:

Net Exergy efficiency $\Psi_{net} = \Psi_{solar} \times \Psi_{drying\ air}$ (6.23)

6.8 Case Studies on Walk-in-Type Solar Tunnel Dryer for Methi (Fenugreek) Leaves Drying.

6.8.1 Background

In the present study Fenugreek leaves were dried in a walk-in type of solar tunnel dryer. India is the major producer and exporter of Fenugreek in the world. It is about 90% of the Indian total production of 36000 MT which is available in Rajasthan and Gujarat of which around 33% – 34% is exported. Fenugreek leaves are a kind of green leafy vegetable with a slightly bitter taste grown only during India's winter season. Most people dry the leaves to use in the off season. The dried leaves are known *kasuri methi*. Much of the literature has reported that, the leaves are good source of iron, therefore, it they are often used to cure anemia. Fenugreek leaves are also considered to be a rich source of Vitamin K, protein and nicotinic acid. In light of the importance of Fenugreek leaves, an experiment was conducted to assess the performance of a walk-in type solar tunnel dryer as well as analyze the energetic and exergetic performance of such system.

6.8.2 System Description

The walk-in type of solar tunnel dryer consist of a metallic type frame structure covered with a UV stabilized semi-transparent polyethylene sheet of 200 μm thickness as illustrated in Fig. 6.7. The technical specifications for the dryer are presented in Table 6.3.

Fig.6.7: Schematic of Solar Tunnel Dryer

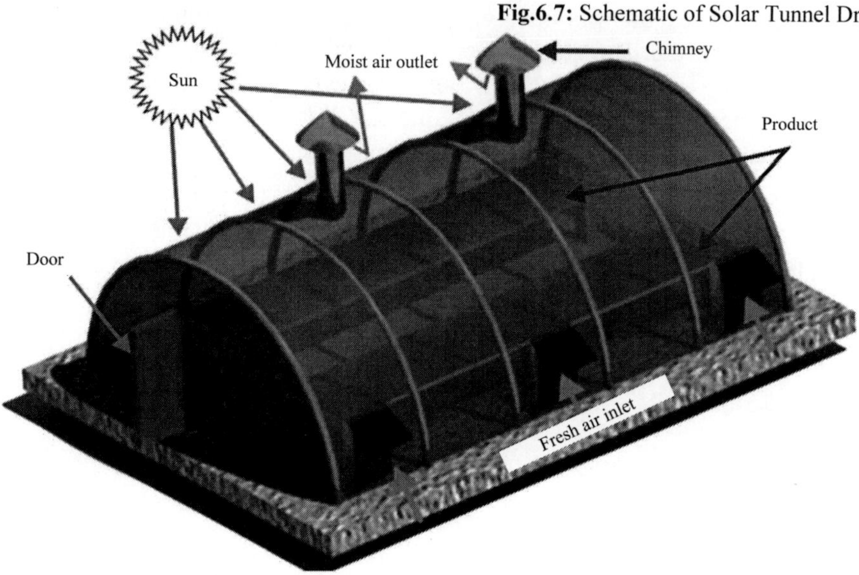

Table 6.3: Technical Specification of Solar Tunnel Dryer

Location	Udaipur (27' 42°N, 75' 33°E)
Floor area	18.75 m^2
Diameter of dryer	3.75 m
Length of dryer	5.0 m
Number of chimneys	2 nos
Height of chimney	0.44 m.
Diameter of chimney	0.20 m
Thickness of plastic cover	200 μ (UV stabilized)
Maximum allowable temperature	65 °C

6.8.3 Thin Layer Drying Kinetics Model

Moisture ratio plays a vital role in crop drying, so it was estimated by using the following formula:

$$MR = \frac{(M_t - M_e)}{(M_o - M_e)} \tag{6.24}$$

The moisture ratio (MR) was simplified to M_t/M_o by the scientists and researchers due to the frequent change in relative humidity of inlets for the drying air to the drying chamber. A thin layer of drying kinetics model equations is presented in Table 6.4, and they were tested to select the best model for describing the drying curve equation of *Fenugreek* leaves during the drying process in the walk-in type solar tunnel dryer.

The best fit was determined by using three parameters: Higher values for the coefficient of determination (R^2), the lowest of the reduced sum square errors (χ^2) and root mean square error (RMSE) equations 25-27, to achieve the best goodness of fit.

$$R^2 = 1 - \frac{\sum_{i=1}^{n}(MR_{exp,i} - MR_{pre,i})^2}{\sum_{i=1}^{n}(MR_{exp} - \overline{MR}_{pre,i})^2} \tag{6.25}$$

$$\chi^2 = \frac{\sum_{i=1}^{n}(MR_{exp,i} - MR_{pre,i})^2}{N - n} \tag{6.26}$$

$$RSME = \left[\frac{1}{N}\sum_{i=1}^{n}(MR_{exp,i} - MR_{pre,i})^2\right]^{\frac{1}{2}} \tag{6.27}$$

Table 6.4: Thin Layer Drying Kinetics Model

S.No.	Model name	Model
1.	Newton	MR = exp(-kt)
2.	Page	MR = exp(-kt^n)
3.	Modified Page	MR = exp(-kt)^n
4.	Verma et al.	MR = a exp(-kt) + (1 - a) exp(-gt)
5.	Wang and Singh	MR = 1 + at + bt^2
6.	Henderson and Pabis	MR = a.exp(-kt)
7.	Modified Henderson and Pabis	MR = a exp(-kt) + b exp(-gt) + c exp(-ht)
8.	Midilli et al.	MR = a exp(-kt^n) + bt
9.	Logarithmic	MR = a exp(-kt) + c
10.	Two term	MR = a exp(-k$_0$t) + b exp + (-k$_1$t)
11.	Two term exponential	MR = a exp(-kt) + (1 - a) exp(-kat)

6.8.4. Results and Discussion

(a) Thermal Performance

It is desirable to assess the temperature of the drying system at various positions. In this experimental study, as shown in Fig. 6.8, the air temperature inside the dryer was measured at four different locations. The average temperature inside the dryer, temperature at chimney outlet, and ambient temperature corresponding to solar radiation measurements are shown in Fig. 6.9, which reveals that average temperature inside the dryer increases as solar radiation increases. On the first day of drying, the air temperature inside the dryer was 54°C; on the second day, it was 55°C. The solar radiation varied from 593 to 911 Wm^{-2} and 621 to 911 Wm^{-2} during the first and second days of drying, respectively. The air temperature difference between inside the dryer and the ambient temperature varied from 13°C to 20°C and 15°C to 23°C during the first and second days of drying, respectively. The temperature at the chimney outlet was always lower than that of inside air temperature. This temperature varied from 1°C to 5°C and from 1°C to 3°C during the first and second days of drying, respectively, which is responsible for producing a draft and allowing the moist air to escape from the dryer through the chimney.

Fig. 6.8: Solar Tunnel Dryer with Temperature Data Logger

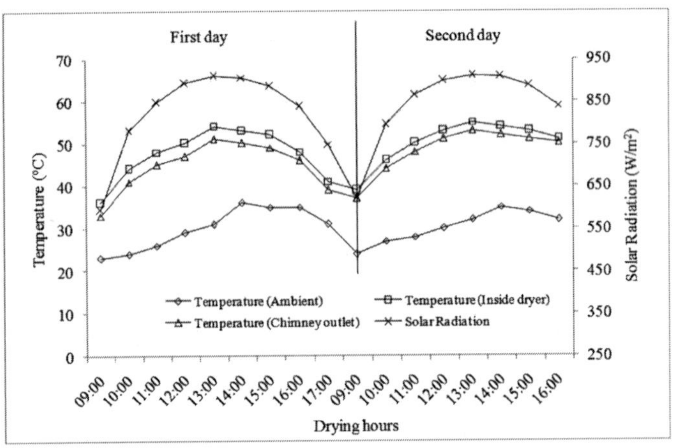

Fig. 6.9: Temperature Profile and Solar Radiation During Experimental Study

Fig. 6.9: Temperature Profile and Solar Radiation During Experimental Study

(b) Modeling of Drying Curves

Fresh *Fenugreek* (Methi) leaves were purchased from a local market and were washed thoroughly in fresh water and then the roots were removed (Fig.6.10). After washing and weighing the methi with stem just as farmer dries it in open field conditions was loaded in the tray as shown in Fig.6.11.

The variation in the moisture content ratio of *Fenugreek* leaves in the tunnel dryer with time is illustrated is Fig. 6.12. It reveals that, no constant rate drying has been observed; rather, it seems as though the drying rate fall. The total moisture ratio (d.b.) reduction during the first day of drying was ranged from 1.00 to 0.08%, and during second day it ranged from 0.07 to 0.01%.

Fig.6.10: Fresh Methi (*Fenugreek*) leaves

Fig.6.11: Inside View of Solar Dryer with Drying Product

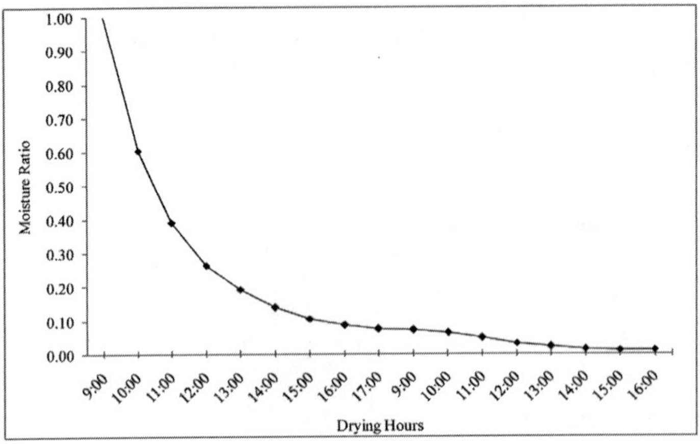

Fig. 6.12: Variation in Moisture Content with Time

The thin layer drying kinetics models listed in Table 6.5 were evaluated based on R^2, χ^2 and RSME parameters. Verma et al. (1985) model was found to be model best fit for thin layer drying of *Fenugreek* leaves in the walk-in type of the solar tunnel dryer. The value of statistical parameters for Verma et al. (1985) model were determined to be R^2=0.99932; χ^2 =0.0000534 and RSME= 0.00662.

The best fit model i.e., the Verma et al. (1985) was validated by comparing the predicated model-based moisture content with the experimentally measured moisture content of the product. It was found that the experimental moisture ratio data are closest to the straight line representing the predicated moisture ratio as illustrated in Fig. 6.13.

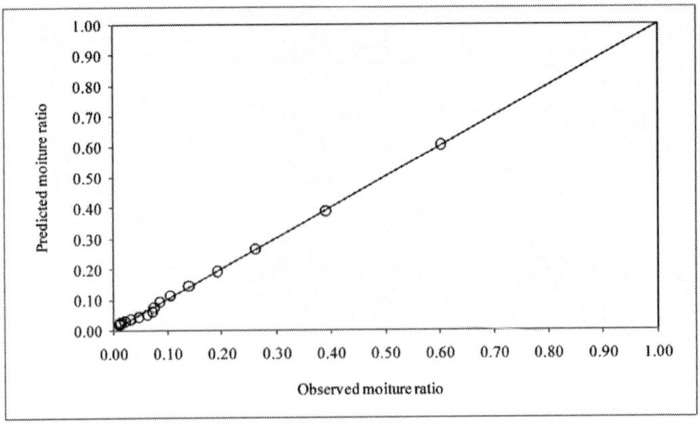

Fig.6.13: Relation Between Predicated and Observed Moisture Content

Table 6.5: Modeling of Moisture Ratio for Thin Layer Drying of Fenugreek Leaves

Model	Model constant	R^2	c^2	RSME
Newton	k=0.42387	0.98528	0.00101	0.03094
Page	k = 0.5444; n= 0.77088	0.99756	0.00017	0.01257
Modified Page	k=0.454391; n= 0.770883	0.99756	0.00017	0.012572
Verma et al.	k=0.18139; a= 0.29983; g=0.68158	0.99932	0.0000534	0.00662
Wang and Singh	a=-0.19827; b=0.00914567	0.74555	0.019051	0.129655
Henderson and Pabis	a=0.96325; k=0.40692121	0.98670	0.0009812	0.029425
Modified Henderson and Pabis	a=0.321084; k=0.406921; b=0.321084; g=0.406921; c=0.321084; h=0.406921	0.98670	0.001338	0.029425
Midilli et al.	a=0.984065; k=0.540733; n=0.828407; c=0.018314	0.99835	0.00014	0.01034
Logarithmic	a=0.947632; k=0.470443; c=0.035878	0.99534	0.000366	0.017373
Two term	a=0.481625; ko=0.406921; b=0.481625; k1=0.406921	0.98670	0.001132	0.029425
Two term exponential	a=1.00000; k=0.423877	0.98528	0.0010854	0.034102

(c) Energy and Exergy Analyses

Energy efficiency was calculated by using Equation 20. The efficiency variation with drying time is illustrated in Fig. 6.14. Total energy supplied by the sun to the dryer is the product of dryer surface area and solar radiation. The energy utilized to raise the drying air is considered as energy out, and it was small as compared to energy input that is why the energy efficiency of such a dryer is low as compared to other solar thermal devices extensively working at the elevated temperatures. On the first-day, energy efficiency and solar radiation varied across the range of 0.841 - 1.613 %, and 593 Wm^{-2} - 911 Wm^{-2} respectively. On the second-day of drying the value of solar radiation and energy efficiency was in the range of 621 Wm^{-2} - 911 Wm^{-2}, and 1.245 -1.609 % respectively. It was found that on the first-day product released, there is more moisture as compared to second day of drying. Owing to this, the temperature gradient during the first day of drying is comparably lower than that of second day of drying. That is why the energy efficiency on first day of drying is lower than the second day of drying and the same condition is applicable for the exergy efficiency. The net exergy efficiency of the solar tunnel dryer for drying methi (Fenugreek) leaves is the product of exergy efficiency of solar radiation (Eq.4.21) and the exergy efficiency of drying air (Eq. 4.22) as presented in Equation 4.23. The asset value of this varied from 0.018 -0.070 % during the first day of drying and from 0.059 – 0.102 % for the second day.

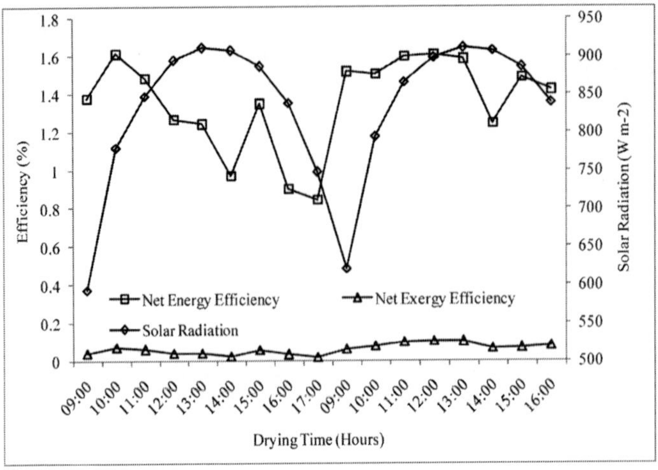

Fig. 6.14: Energy and Exergy efficiency of tunnel dryer during methi (Fenugreek) drying

6.9 Salient Features of The Solar Tunnel Dryer

- It is a modular walk-in-type hemi-cylindrical poly house type design.
- The dryer consists of three major components viz Poly house super structure having provisions of inlets for fresh air and outlets for the hot humid air, Cement concrete floor and insulated black wall on northern side of the super structure and Sub-system for spreading the product inside the tunnel for drying.
- The cross-section of the hemi-cylindrical super structure has been standardized. The inside diameter is 3.7 m and height 2 m. The length may vary from 5 m to 20 m depending upon batch size and type of product(s) to be dried.
- The dryer can be installed on ground surface or on roof top depending upon convenience and availability of the open space.
- Poly house is laid east-west direction so as to capture maximum amount of solar energy.
- The floor of the dryer is laid at a gradient of 2-5° along the length. It helps induce air flow through natural convection.
- Inside the poly house, an insulated metallic black wall is provided along the northern side of the tunnel to increase heat collection and reduce heat loss through the northern side.

- The length of the dryer, size of inlet airports and number of exhaust chimneys and exhaust fans are worked out based on the characteristics of the product to be dried and its initial and final moisture content, ambient conditions and the time period available for drying.
- On both sides of the tunnel, a door of convenient size is provided to facilitate loading and unloading of the product. An exhaust fan is provided at the upper end of the tunnel. Another exhaust fan may also be provided on the other side of the tunnel depending upon the requirements/design calculations. The humidity sensor(s) installed inside the tunnel dryer automatically control on/off operation of the exhaust fan(s).
- The tunnel dryers may be designed for removal of 175-200 kg of moisture from the product per day.
- The space inside the poly house is used for spreading the product so as to achieve maximum moisture removal. The hangers/trays and trolleys are designed depending upon type of the product.
- The super structure is covered with good quality UV-stabilized semi-transparent polyethylene sheets of 200 micron thickness, which may last up to 3 years.
- The electricity consumption is very low and maintenance requirements are also low.

The structural components of STD are mainly Hoops, Foundation, Floor and UV Stabilized Polythene Film. The fabrication detail of each component is as follows:

a) **Hoops: -**There are integral parts of structure of STD. The shape of the hoop is semi cylindrical & separated by equal distance from one another. Hoops are made by bending galvanized iron pipes for creating an opening door frame in the drying chamber. About 15 cm lengths of each hoop at the ends remain unbent so that it can easily be fitted in the foundation.

b) **Foundation pipes*:*** -They are used to provide a firm support to hoops in the structural G.I. pipes of 25mm diameter and 180cm length is used for this purpose.

c) **Drying Floor: -** Floor of STD is made of cement concrete of 1:2:4 ratio and painted black for better absorption of solar radiation.

d) **Cover Material: -** UV stabilized low density polythene film of 200 micron is used as a cover material. A single layer polythene film for the cover of STD is preferred due to material economy and easy handling.

6.10 Operation and Maintenances of Solar Tunnel Dryer (STD)

- STD is a simple technology and managed by employing people of low to medium skill.
- Product should be loaded into the STD by 9.00 AM to get full advantage of sun shine hours.
- Exhaust fan setting is made for desired maximum humidity level inside the dryer for maintaining high quality of the product and reducing drying time.
- Operating instructions vary depending upon the product being dried and should be adhered to the instructions.
- Major maintenance requirement is for change of the poly cover.
- Normally poly cover is required to be changed every 3-4 years depending upon the local environmental conditions.
- Good quality UV-stablized sheet of 200 microns is used for covering the structure.
- Exhaust chimneys, north wall, floor and other surfaces inside the tunnel have to be painted black as per the requirement.

6.11 Points to be kept in view while Using Solar Tunnel Dryer

- For faster drying rate, the air flow should be through the product with continuously rotating drying trays.
- Care must be taken to keep products in the drying trays always above the hole of fresh air inlets.
- In no any case product, be spread on the floor.
- While drying all trays must be filled with products, otherwise hot air will pass through least resistance path.
- Under load and over load will definitely disturb drying rate.
- Temperature of the STD can be attended about 60°C, which is actually required for the drying of products.
- Always keep tunnel door closed while drying is continued other day.
- In night door must be close if products are placed for second day drying.
- In order to maintain the colour and quality, the low temperature drying is recommended, which can easily be attended by STD.

- In summer if chances of temperature enhancement are more than 65°C, than it is advised to open the door during peak hours.
- Drying of agriculture products found uniform in STD because of intermediate heating of air in between the trays.
- If drying is continued for second day, it is recommended that products kept in bottom tray be moved to upper tray and material of second tray should be moved to next tray and top most tray material be moved to bottom drying in order to have uniform drying and prevention against over drying.
- Space between the two trays should be 30cm, which permit intermediate heating of air between the trays.
- No tray is kept empty while drying.
- In STD, maximum temperature is attended towards the south side followed by the center and north side.
- Thus, stand of trays should be kept at center to utilize the optimum solar input.
- Material should be loaded with maximum 30 mm bed thickness to provide the fast drying and good quality of product with uniform drying.
- Drying is found fast and uniform in the northern and southern trays due to rotation of the frame as compared to trays loaded on stationary frames.
- Care must be taken in maintaining the polythene; there must not be any tempering on polythene.
- If it develops cut at any place, it must be stitched immediately to avoid heat losses.
- Dust if any on the surface of polythene must be immediately clean, otherwise it will prevent entry of solar radiation inside it.
- Exhaust fan should only be used when more humidity is observed inside the dryer.
- In winter, poly cover must be covered with some insulating material in night, if drying is continued for next day.

7

Solar Greenhouse

7.1 Introduction

A greenhouse is quasi-permanent structure, covered with a transparent or translucent material, ranging from simple homemade designs to sophisticated pre-g\fabricated structures, wherein the environment could be modified suitable for propagation or growing of plants. Materials used to construct a greenhouse frame may be wood, bamboo and steel or even aluminum. Covering can be glass or various rigid or flexible plastic materials. Depending on the covering material, different terminology have been used in the context of greenhouse structures as mentioned below:

Glasshouse: A greenhouse with glass as the covering material is called glasshouse

Polyhouse: It is a greenhouse with polyethylene as the covering material.

7.2 Plant Environment And Greenhouse Climate

A plant grows best when exposed to an environment that is optimal for that particular plant species. The aerial environment for the plant growth can be specified by the following four factors:

1 Heat or temperature

2 Light

3 Relative humidity

4 Carbon dioxide

While plants have precise optimum environmental conditions for best growth, most are tolerant to variations in these conditions within some limits. However, permanent damage would occur when they are exposed to conditions outside these limits. At the same time plants are subjected to attack by pest and disease.

Greenhouse crop production protected against adverse environmental conditions and allows pest and disease to excluded or controlled. Besides providing protective enclosure, a greenhouse also acts as a 'Heat trap'. It admits solar radiation and converts this energy into heat by raising the temperature of the greenhouse air. While this is the basis of the greenhouse's ability to perform its tasks, it also affects others environmental factors.

Environmental conditions inside the greenhouse can be modified suiting to the potential growth of the plants. The extent of climate modification will, however, depend on the design of greenhouse and is generally related with it cost. Higher the capacity of greenhouse to modify its climate, higher is the cost of its construction.

The way in which a greenhouse gets heated exposed to sunlight is similar to heating of the earth's surface end its adjacent atmosphere. When solar radiation reaches the earth, small portion is reflected back into the space while the remainder is absorbed at the surface raising its temperature. In the same way, when solar radiation reaches to greenhouse cover- surface, a small amount (normally – 15-20%) is reflected back from the surface while the remainder transmitted to the interior. Plants, soil and other objects absorb most of this transmitted radiation and remainder is reflected as shown in Fig 7.1

The absorbed radiation raise the temperature of absorbing surfaces and objects with the heat energy being immediately transferred to the greenhouse air by convection and evaporation, thereby increasing the temperature and humidity.

Greenhouse lose heat in three ways

1 Warm air moves out the greenhouse and is replaced by cooler outside air.
2 Heat is transferred through the covering material itself.
3 Heat is lost through the soil/floor

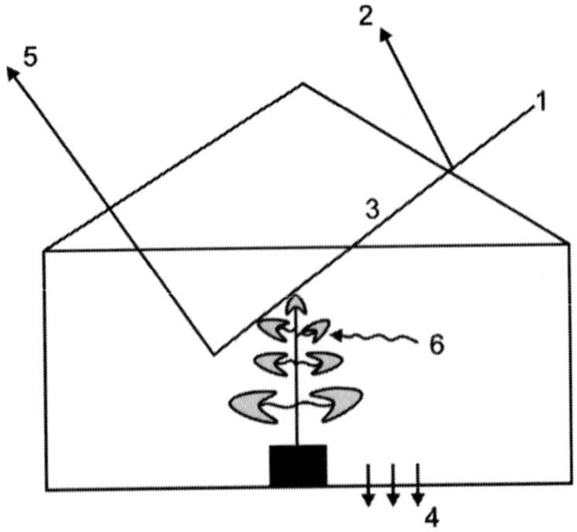

Fig.7.1 Solar Radiation Heats the Greenhouse

1. Solar Radiation Reaching the Greenhouse, **2.** Solar Radiation Reflected from Greenhouse Surface, **4.** Solar Radiation Transmitted to Interior of Greenhouse, **5.** Absorbed Radiant Heat Energy Conducted into Surface, **6.** Solar Radiation Reflected from Internal Surfaces, And **7.** Absorbed Radiant Heat Energy Transferred to the Air by Convection and Conduction and Evaporation)

Increasing the temperature of the growing environment in a greenhouse is unavoidable, but it is also the most important function of a greenhouse. Enhanced temperatures accelerate plant growth, and allow sustaining plant growth even when outside ambient temperatures are unfavorably low. However, during summers inside temperatures rise higher than the optimum level and, therefore, cooling/ ventilation provision are necessary.

Most plant grows better within 60-80% relative humidity (RH) of air. Low RH increases the evaporative demand on the plant, while high RH can depress this demand inhibiting the uptake of nutrients, particularly calcium. In general, the RH inside the greenhouse is higher than the outside, mainly due to transpiration load. Effective ventilation is required to control higher RH level.

In most part of the country, solar radiation is not a limiting factor for the plant growth. Light control inside the greenhouse can be affected conveniently either by shading or by supplementary lighting whenever required. Grower in northern India should, however be careful in monitoring light level in winters especially during prolonged foggy conditions. In peri urban areas, particulate pollutants get deposited on the plastic roof thereby reducing the light transmission significantly. This problem is compounded in winter. In such conditions, it is necessary to wash the roof frequently to maintained adequate light level inside the greenhouse.

Plants use carbon dioxide from the atmosphere for photosynthesis. Carbon dioxide concentration inside the greenhouse in the early morning is always higher the outside. With the onset of sun, this level quickly depletes and goes down the normal level during the day if adequate air exchanges are not maintained. Carbon dioxide enrichment is generally accomplished by burning suitable fuel like propone.

7.3 Type of Greenhouses

The greenhouse design and cost range from a simple plastic walk-in tunnel costing about Rs. 90/- per sq meter to a climatic controlled, saw tooth type of greenhouse with automatic heating, ventilation and cooling, costing more than Rs. 3000/- per sq meter. The selection of the greenhouse should be determined by the grower expectations, needs, experience, and above all its cost effectiveness in relation to the available market for the product. Obviously cost of greenhouse is very important and may outweigh all other considerations.

Greenhouses are classified in different shape, which also determine their cost, climate control and use in term of crop production. Commonly used structural designs are briefly described below and shown in Fig 7.2

A. Gable

This is the most basic structure similar to a hut like construction with glass as the covering material. The roof frame can be inclined at any angle to present an almost perpendicular face to sun angle to minimize losses due to external reflection. The structure also allows large openings in the side walls and at the ridge for high rates of natural ventilation. Modern gable shaped greenhouses are multi-span unit with bay widths of 6 – 12 meters.

B. Gambrel

These structures are similar to the gable but have high strength to withstand high wind loads during storms. This is more suitable where wood or bamboos are to be used for the greenhouse construction.

C. Skillion

In this kind of structure, the roof consists of single sloping surface. This is because the greenhouse is built as the southward extension of a building with a solid wall on the northern side. Such greenhouses have the advantage of low structure requirement.

D. Curved-roof

The semicircular tunnel greenhouse structure appeared with the introduction of polyethylene film as the covering material. These structures, besides being most simple and easy to construct, have the advantage of high strength with a relatively light frame due to inherent strength of the curved arch. But these structures have the disadvantage of poor ventilation efficiency since the curved roof is not amenable to the incorporation of ridge ventilation.

In an attempt to improve the ventilation efficiency of curved roof greenhouse, raised arch type of structure have been adopted. This design has vertical side walls, which permit high head room and improved ventilation due to wind velocity.

E. Saw-tooth

In these structures, the ventilation efficiency of curved roof greenhouse separately by a series of sloping surface, all of which are pitched at the same angle and facing in the same direction. The vertical surface consists entirely of ventilating area. These types of greenhouse are most efficient from ventilation point of view. Such greenhouses are also suitable for multi span structures. Orientation of saw tooth greenhouses can also be used as a means of maximizing natural ventilation. By facing the open vertical ventilation areas away from the wind, airflow over the greenhouse roof creates a negative pressure, which facilitated in sucking out warm greenhouse air. However, this air dynamic relies on the premise that there are large ventilation areas in the greenhouse walls on windward side.

F. Shade houses

Shade nets are perforated plastic material used to cut down the solar radiation and prevent scorching or wilting of leaves caused by marked temperature increases within the leaf tissue from strong sunlight. These nets are available in different shading intensities ranging from 25% to 75%. Leafy vegetable and ornamental greens are recommended to be grown under shade net whose growth rates are significantly enhanced to unshaded plants when sunlight is strong

G. Net houses

Shade house and net house are often synonymously used but more correctly a net house is enclosed with perforated screen primarily to act as a barrier for the entry of insect and pasts. Insect proof nylon nets are available in different intensities of perforations, ranging from 25 to 60 mesh. Net of 40 mesh are effective means to control entry of most flying insect and save crop from diseases.

7.4 Covering Materials of Greenhouses

As mentioned earlier, the purpose of a greenhouse covering is to allow sunlight to pass through it so that the energy is retained inside. Glass was the main covering material in the early greenhouses. With the introduction of plastic materials, there are now several alternatives available for greenhouse covering. A brief description of covering material is given below:

Fig. 7.2: Design of Greenhouses

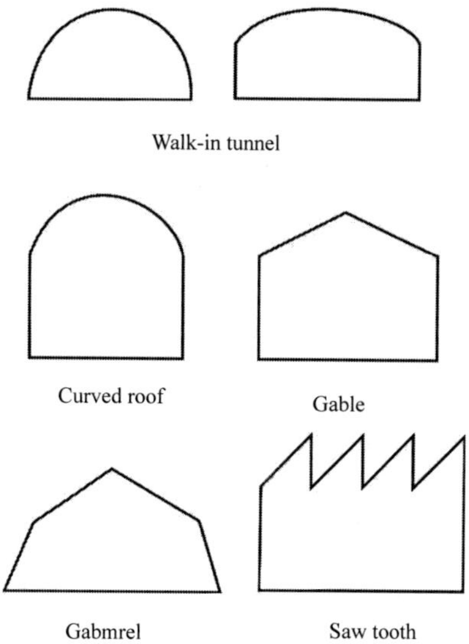

a. Glass

A clean, transparent provide the maximum light transmittance to the extend of 90%. However, being heavier in weight, it requires elaborate structure for adequate support. It is brittle and can break with minimum shock or vibration resulting in high maintenance costs.

b. Acrylic

This material has long service life, good light transmittance (80%), moderate impact resistance, but prone to scratches. It has high coefficient of expansion and contraction. Being inflammable and costly. It is not a preferred material.

c. Polycarbonate

It is available in single or double wall sheet of different thickness. A new polycarbonate sheet has good light transmittance of about 78%, but reduces with age. It has excellent impact resistance and low inflammable. High cost limits its use on large scale.

d. Fiberglass reinforced plastic panels (FRP)

These plastics consist of polyester resins, glass fiber stabilized etc. It has a limit transmittance of about 80% and has high impact resistance with a service life ranged from 6-12 years. Good quality FRP materials for greenhouse are not quite assured.

e. Polyethylene

A clear polyethylene sheet has about 88% light transmittance. Its higher strength and low cost have made it most popular replacement to glass. An ultra-violate (UV) stabilized plastic sheet can have a service life of 3 year. These sheets are generally available in 7- and 9-meter widths with 200-micron (0.2 mm) thickness.

7.5 Greenhouse Ventilation

Ventilation of an enclosed space is the process of replacing the air inside the space with the ambient or conditioned air. In case of open circuit ventilation, it is the ambient air replacing the enclosed air mass. In the case of closed-circuit ventilation, the air from the enclosure is drawn, conditioned and returned to the enclosure. A hybrid ventilation system, working as an open circuit system for a part of the time and as a closed circuit for the rest of the time, is the also used, especially plant growth chambers.

The need for greenhouse ventilation arises for the control of one or the following parameters:

1. Temperature.
2. Humidity

3. Carbon dioxide concentration
4. Air circulation around plants

It may be noted that ventilation is not the only method of controlling the above four parameters rather it is an important means to achieve the above mentioned parameters.

7.5.1 Ventilation requirements

Ventilation requirements would vary depending on whether it is for the control of temperature, moisture, CO_2, any other gas in the greenhouse air, or for maintaining air circulation around plants. Ventilation requirement may be determined either by rules of thumb or by analytical procedures. Some of the well accepted rule of thumb are as follows:

1. Provide an airflow rate of 0.03 – 0.04 cu.m/sq.m. or 6 – 8 cfm/sq ft of floor space.
2. Provide one air exchange per minute of ventilation for summer cooling.
3. Airflow rate in the greenhouse should be such that air velocity around the plants is maintained in the range of 0.5 to 1.0 m/s.

The above-mentioned norms are, in general, convenient and help in determining capacity of the ventilation system. However, for more precise estimation, analytical methods may be used.

7.5.2 Methods of ventilation

There are two methods of ventilation, namely, natural and forced. Natural ventilation refers to providing sufficient open area in the greenhouse structure so that ambient air by itself enters into the greenhouse after displacing an equal amount of greenhouse air. In the forced ventilation system, auxiliary power is used to move air through the greenhouse. A brief description of these methods is given below.

7.5.3 Natural ventilation

Many older greenhouses depend on natural ventilation for air movement. However, with increased concern about the high cost of energy required to operate greenhouses today, increased emphasis is again being placed on natural ventilation systems for greenhouses. In greenhouses employing natural ventilation systems, sidewall vents and ridge vents continuous for the full length of the building can be opened as far as desired to allow air to move through the house. To be ventilated satisfactorily, the house must have both sidewall and ridge vents. If a house has only side vents, then it can only be ventilated during periods of wind movement outside. Using ridge vents and side vents permits the greenhouse to be vented by both wind pressure and thermal gradients. Thermal gradients generally are created within the greenhouse by solar energy heating the materials inside, which in turn heat the air. As air is heated, it becomes lighter and rises through the ridge vents, with the makeup air coming from outside through the sidewall vents. If sidewall and ridge vents are properly sized, quite satisfactory ventilation rates can be achieved with some degree of temperature control. A natural ventilation system will not be as dependable or satisfactory as a mechanical ventilation system in terms of providing continuous, uniform greenhouse ventilation. Some newly-designed greenhouses with natural ventilation systems are capable of achieving a high degree of environmental modification for increased plant production. However, for those plant species that require air temperatures lower than outside air temperature, evaporative cooling must be used. Mechanical ventilation is an integral part of any evaporative cooling system.

7.6 Greenhouse Cooling System

The need to cool a greenhouse arises whenever the greenhouse air temperature crosses the upper limit of the crop tolerance. Failure to bring down the temperature effectively may result in either partial or total crop failure within only a very short period of time. Design of appropriate cooling system for greenhouse operation in most parts of India is essential because the cooling season may last upto to eight months annually.

7.6.1 Fan and Pad evaporative cooling systems

Fan and pad systems consist of exhaust fans at one end of the greenhouse and a pump circulating water through and over a porous pad installed at the opposite end of the greenhouse (Fig 7.3)

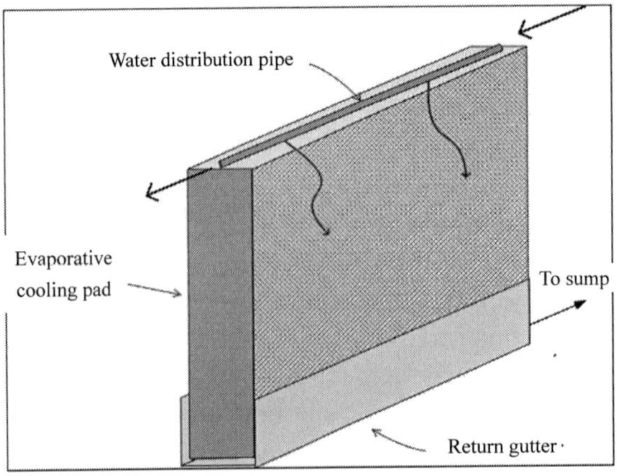

Fig. 7.3: Evaporative Cooling Pad.

Water flows along the distribution pipe and drains down into the pad material. The sump should be large enough to hold all run-off when the pump is turned off.

If all vents and doors are closed when the fans operate, air is pulled through the wetted pads and water evaporates. Removing energy from the air lowers the temperature of the air being introduced into the greenhouse.

The air will be at its lowest temperature immediately after passing through the pads. As the air moves across the house to the fans, the air picks up heat from solar radiation, plants, and soil, and the temperature of the air gradually increases. The resulting temperature increase as air moves down the greenhouse produces a temperature gradient across the length of the greenhouse, with the pad side being coolest and the fan side warmest.

7.7 Greenhouse Control Systems

Once a greenhouse structure is built various techniques, devices, etc. must be added in order to control the environment. Control systems include those for lighting, heating, cooling, relative humidity and carbon dioxide enrichment.

7.7.1 Light

(a) **Importance:** Maximum light transmission, of the appropriate quantity and quality (photosynthetically active radiation, 400-700 nm), through the greenhouse structure to the plants is crucial for optimum photosynthesis, growth and yield.

(b) **Structural considerations:** Large sections of glazing material (glass, polyethylene, polycarbonate, etc.), held in place by few supports, results in higher light levels and less shading. Minimize other opaque structures above the crop that would cause shading such as heaters, carbon dioxide generators, opaque vents, etc.

(c) **Too much light**

Shade paint/white wash: A mixture sprayed on the outside of the greenhouse.

This will either wear off by the end of the summer or it can be washed off.

External shade cloth: Fabric cloth, placed on the outside of the greenhouse, made of varying degrees of mesh size to exclude specific amounts of light(ex.: 30%, 40%, 50% shade).

Internal shade cloth: Fabric cloth, as above, hung inside the greenhouse.

(d) Too little light

Occurs above/below 300 north/south latitudes during the "winter".

White reflective ground covers: These are now in common use in commercial greenhouses in all locations and can significantly increase light levels tothe plant canopy.

7.7.2 Heating

Each plant species has an optimum temperature range. Heating devices will maintain the temperature within that range during periods of cold weather.

(a) Types of heat loss from a greenhouse

Conduction: Heat transfer either through an object or between objects in contact. Conduction depends on area, path length, temperature differential and physical properties of the object(s).

Example: Heat loss through the glazing material on the greenhouse.

Convection : Heat transfer by the movement of warm gas or liquid to a colder location. Convection depends on temperature differential.

Example: Movement of warm air near the plants upward toward the roof.

Radiation: Heat transfer between separated objects. Radiation occurs from all objects and depends on the areas, temperatures and surface characteristics of the objects involved.

Example: Heat transfer from all objects in the greenhouse.

(b) The basic heating system :

Consists of a **fuel burner**, **heat exchanger**, **distribution system** and **controls**.

Heat delivery to the crop is by **convection** and **radiation**.

The fuel - usually burn **natural gas**, but can also use oil, coal, wood, etc.

(c) Heating by hot water or steam

Hot water or steam can be produced using boilers fired by natural gas, etc. The hot water or steam is then transported throughout the greenhouse in pipes. The pipes can end in a heat exchanger where a fan distributes heated air. The pipes can run along the floor and also be used as cart rails between aisles. Heat will then rise upward through the crop by convection. Heat pipes can also be positioned within the crop to steer plant growth . Heated tubes can create "bottom heat" for propagation or growing.

(d) Heating by hot air

Fuel is burned to heat air that is then distributed by fans around the greenhouse. Horizontal air flow (HAF) fans circulate warm air above the crop. Fan jet systems, with unit heaters or heat exchangers and perforated polyethylene tubes, distribute warm air and improve air movement and ventilation throughout the greenhouse.

(e) Moveable nighttime insulation: Insulating material (cloth or film curtains) can be positioned above the crop or near the roof to retain heat near the crop. The insulating material used during the night can be the same material used for shading during the day.

7.7.3 Cooling

High temperatures can be detrimental to plant growth. High temperatures can cause such problems as

- Thin, weak stems or, as in tomatoes, stick trusses (thin, weak truss stems)
- Reduced flower size or, as in tomatoes, flower fusion and boat formation
- Delayed flowering and/or poor pollination/fertilization and fruit set
- Flower and bud/fruit abortion

(a) Cooling requirements and calculations

According to the National Greenhouse Manufacturer's Association 1993 standards about 8 cubic feet per minute/square feet of greenhouse floor area is required OR 1 full greenhouse volume exchanged per minute in warm climates.

CFM = height x width x length (i.e., volume)

Example: Using the greenhouse dimensions in the heat calculation example:

CFM = volume lower section + volume triangular top

= (8 x 24 x 48) + (6/2 x 24 x 48)

= 9216 + 3456

= 12,672 cubic feet per minute => size fans/pads accordingly

(b) Passive ventilation systems

Shading: Shade cloth or shade paint/white wash, besides regulating the light intensity, can also help to cool the greenhouse.

Ridge Vents: Vents in the roof of a greenhouse that allow hot, interior air to escape. The area of the vents should be 25% of the floor area.

Roll-up Side Walls: Can be used in flexible glazing (polyethelene film) single bay greenhouses where the side walls can be rolled up several feet allowing a natural horizontal flow of air over the plants. As with ridge vents, the area of the side wall vents should be 25% of the floor area.

Cooling Towers: Water cooled pads at the top part of tall towers cool the surrounding air which then drops displacing warmer air below.

Removable Roof: Recent greenhouse designs can include a roof that retracts completely for natural ventilation. This would allow for adaptation of greenhouse grown plants to outside conditions prior to movement outside.

(c) Active cooling systems

Fan and Pad: "Evaporative cooling" where air from the outside is pulled through porous, wet pads (usually cellulose paper). Heat from the incoming air evaporates water from the pads, thereby cooling the air. Evaporative cooling will also help to increase the relative humidity in the greenhouse.

Fogging Systems: Uses evaporative cooling like the fan and pad but incorporates a dispersion of water droplets that evaporate and extract heat from the air.

This system gives better uniformity since the fogging is distributed throughout the greenhouse and not just near one a pad end as with the fan and pad system. The smaller the droplet size, the faster each droplet evaporates and therefore the faster the cooling. Mist droplets = 1000 microns in diameter.

7.7.4 Relative Humidity (RH)

High or low relative humidity can be detrimental to plant growth.

(a) **Effects on transpiration** – When RH is too high, transpiration (the movement of water from inside the leaf to the outside) is reduced along with movement of mineral nutrients. When RH is too low, transpiration may be increased significantly resulting in plant wilt.

(b) **Effects on pollination** – When RH is too high, the pollen can clump on the stigma causing cat facing or the pollen may not be released from the anthers at all. When RH is too low, the normally sticky stigma can dry out and the pollen may not stick to it's surface, decreasing pollination.

(c) **Ways of controlling RH in the greenhouse-** Relative humidity can be increased by running the cooling pads or by fogging. Relative humidity can be decreased by running the heaters or simply venting.

7.7.5 Carbon dioxide enrichment

The rate of photosynthesis is dependent upon the availability of carbon dioxide. Carbon dioxide enrichment is most important during the winter months in the morning. The sun has risen and photosynthesis has begun. The plants can reduce the levels of carbon dioxide from the ambient level of about 330 ppm (higher in cities due to industry and vehicles) to around 220 ppm. Lowered carbon dioxide levels will reduce growth and can cause flower and fruit drop reducing overall yields.

Ways of controlling carbon dioxide levels in the greenhouse

Ventilating (bringing air in from the outside) may provide sufficient carbon dioxide during the Spring, Summer and Fall months. Ventilating during the Winter months, or anytime in cold climates, will, however, result in cold outside air being brought into the greenhouse. Heating will then be needed to maintain the proper temperature which may become uneconomical. Therefore, carbon dioxide generation is a typical and effective way to increase levels in the greenhouse during the winter or in cold climates. Carbon dioxide generators can burn various types of fuel including natural gas (most economical)or propane. Carbon dioxide levels above 800 ppm, even as high as 1200 ppm, have been shown to be beneficial to plant growth.

7.7.6 Air circulation

One reason for having a greenhouse is to create a "controlled environment" for all of the plants. Further, each plant within the greenhouse should receive the same conditions. However, especially during times when the heating and cooling systems are not in operation, pockets of high or low temperature, relative humidity or carbon dioxide may develop which can be less than optimal for plant growth or flower/fruit development. Therefore, proper air circulation is required with the enclosure.

(a) Ways of improving air circulation

Horizontal air flow (HAF) fans can be placed in the rafters of the greenhouse to circulate air above the crop. This helps to minimize pockets of warm or cold air and high or low humidity or carbon dioxide within the greenhouse. HAF fans can be used in conjunction with hot air heating systems to circulate warm air throughout the greenhouse. HAF fans can also be used at anytime to enhance air mixing in the greenhouse.

8

Solar Photovoltaic Technology

8.1 Introduction

Solar energy can be directly converted into electrical energy. Energy conversion devices which are used to convert sun light into electricity by use of the photovoltaic effect are called solar cells. The advantages of solar cells are:

1. They are reliable, convenient and durable.
2. It is green, sustainable and pollution free technology.
3. Directly solar radiation can be converted into electricity (conversion of global solar light by flat plate photovoltaic).
4. Maintenance cost is quite low therefore suitable even in isolated and remote area, since no moving parts are involved.
5. Solar cells are quite compatible with almost all environments, respond instantaneously with solar radiation.
6. Considerable expected lifetime i.e. 20 years or even more.
7. No distribution system is required, because they are decentralized source of energy and located at the place of use.
8. No moving parts in energy generation and conversion, portable and easily assembled at any place in a short time.
9. No distribution and transmission losses.
10. They don't create pollution as compared to other conventional sources of energy.
11. Modular nature in which desired capacity can be designed by simple integration method.
12. Ability to function unattended for long periods.
13. Compatible and integratable with any other power sources.

14. Remote monitoring and control feasible.
15. Recyclable material.
16. Large specific power per unit weight, particularly in case of thin film solar cells.
17. Ideally suited for small stand-alone power systems particularly in remote and gridless areas.
18. Taking life cycle into account, PV is economically viable as also a more favourable solution for many small / medium power applications.
19. As a sun-based energy production & distribution industry, it is expected to be a significant drive of world economics and generation of job.

However, solar cells, have same disadvantages, which are as follows:

1. Solar energy has intermittent nature, therefore, can be used only in the presence of sun.
2. Present costs of solar cells are very high.
3. Presently no adequate, reliable, efficient and convenient energy storage system is available.
4. The conversion efficiency is also relatively low, which is limited to 25 per cent.

8.2 SWOT Analysis of Solar Photovoltaic System

The SWOT analysis provides an analysis of the strength, weaknesses, opportunities and threats of SPV. This analysis defines the role of SPV in meeting the electricity needs for several applications.

Strengths: (1) Cost effective: in terms of life cycle (2) Reliable (3) Low maintenance (4) Environmentally benign (5) Sunshine is free and SPV is fuel independent (6) Locally generated power greater energy security and control of access to energy (7) Flexible sizemodular construction (8) Transportabilitymodular construction.

Weaknesses: (1) Sun dependent (sunlight is highly diffused, intermittent and requires backup storage) (2) High initial cost (3) System maintenance (4) Disposal (batteries) (5) Power type requires inverter (6) Vulnerability (the modular and portable characteristics leave them vulnerable to theft and venations.

Opportunities: (1) SPV is an appropriate and cost-effective source of electricity for basic services and amenities in an environmentally sensitive manner (2) Best option for decentralized and day light energy requirement (such as irrigation).

Threats: (1) Cannot stand if low cost environmentally friendly technology is developed (2) other renewable energy sources are locally available.

8.3 Semiconductors

It has been observed that certain substances like germanium, silicon etc. have resistivity (10^4 to 0.5 ohmmeter) between good conductors like copper having resistivity 1.7 x 10^8 ohmmeter and insulators like glass having resistivity 9 x 10^{11} ohmmeter. These substances are known as semiconductors. Thus, a substance, which has resistivity in between conductors and insulators in known as semiconductor. Here it is natural to think that why a semiconductor cannot be regarded as resistance material? The answer is that nichrome, one of the highest resistance materials has resistivity 10^4 ohmmeter, which is much lower than Germanium (0.6 ohmmeter) and hence it is wrong to consider the semiconductor as a resistance material. It is important to mention here that resistivity does not alone decide whether a substance is semiconductor or not. For example, an alloy may be prepared whose resistivity falls within the range of semiconductor but this cannot be regarded as a semiconductor. In fact, there are a number of other properties, which distinguish semiconductors from conductors, insulators and resistance materials .

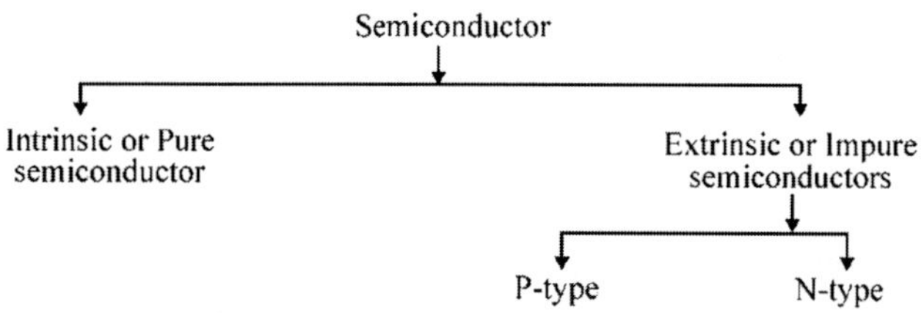

Semiconductors have the following properties:

(i) They have resistively less than insulators and more than conductors.

(ii) The resistance of semiconductor decreases with the increase in temperature coefficient of resistance. For example, germanium is an insulator at low temperatures while it becomes a good conductor at higher temperatures.

(iii) When suitable metallic impurity like arsenic, gallium etc. is added to a semiconductor, its current conducting properties change appreciably. This is the most important property.

Semiconductors are extensively used in electronic circuits. For example, transistora semiconductor device has replaced bulky vacuum tubes in almost all applications. Transistors are only one of the family of semiconductor devices, many other semiconductor devices are now becoming popular .

A *semiconductor in an extremely pure from is known as intrinsic semiconductor or a semiconductor* in which electrons and holes are solely created by thermal excitation is called pure or intrinsic semiconductor. For example, pure crystals (like germanium, silicon), which provide electronhole pairs, are called intrinsic semiconductors. The electrons reaching the conduction band due to thermal excitation leave equal number of vacancies or holes in valence band *i.e. in intrinsic semiconductor the number of free electrons is always equal to the number of holes.*

When an external electric field is applied across an intrinsic semiconductor, the conduction through the semiconductor is by both free electrons and holes. The free electrons in the conduction band move towards the positive terminal of the battery while the holes in the valence band move towards the negative terminal of the battery i.e. the electrons and holes move in the opposite directions. The total current inside the semiconductor is thus the sum of currents due to free electrons and holes. It may by noted that the current in the external wires is only due to electrons as shown is Fig. 8.1. As the holes reach the negative terminal B, the electrons reaching here combine with the holes, the holes are thus destroyed. At the same time, the loosely held electrons near the positive terminal A are attracted away from their atoms into the positive terminal. Now new holes are created which again drift towards the negative terminal B.

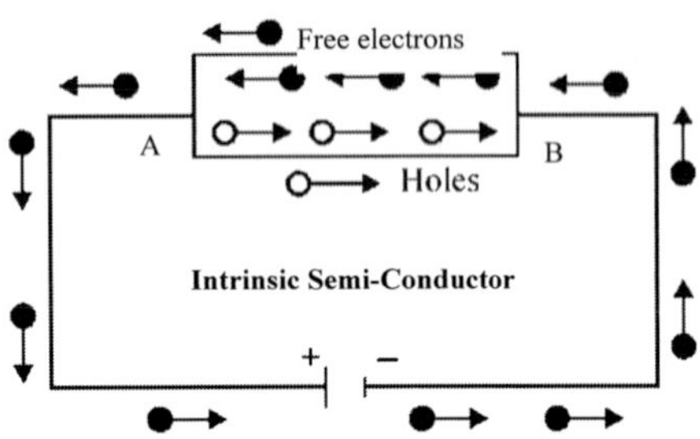

Fig. 8.1: Current Flow in Intrinsic Semiconductor

At room temperature, the intrinsic semiconductor has little current conduction capability. In order to use the semiconductor in electronic devices, its conduction properties should be increased. The electrical conductivity of intrinsic semiconductor can be increased by adding some impurity in the process of crystallizaiton. The added impurity is very small of the order of one atom per million atoms of the pure semiconductor. Such semiconductor is called *impurity or extrinsic semi conductor*. The process of adding impurity to a semiconductor is known as doping.

Usually, the doping material is either *pentavalent atoms* (bismith, arsenic, phosphorus which have five valence electorns) or *trivalent atoms* gallium, indium, aluminium, boron which have three valence electrons). The pentavalent doping atom is known as *donor atom* because it donates one electron to the conduction band of pure semiconductor. The trivalent atom, on the other hand, is called as *acceptor atom* because it accepts one electron from semi conductor atom. The doping materials are called impurities because they alter the structure of pure semiconductor crystals. Depending on the type of impurity added, the extrinsic semiconductors can be divided into two classes :

(i) Ntype semiconductor, and

(ii) Ptype semiconductor.

(A) NType Extrinsic Semiconductor

When a small amount of pentavalent impurity is added to a pure semiconductor crystal during the crystal growth, the resulting crystal is called as Ntype extrinsic semiconductor. Let us consider the case when pentavalent arsenic is added to pure germanium crystal. As shown in Fig 8.2 the arsenic atom fits in the germanium crystal is such a way that its four valence electrons from covalent bond with the four germainum atom.

Fig.8.2: Crystallettice with one Germanium Atom displaced by Arsenic Atom

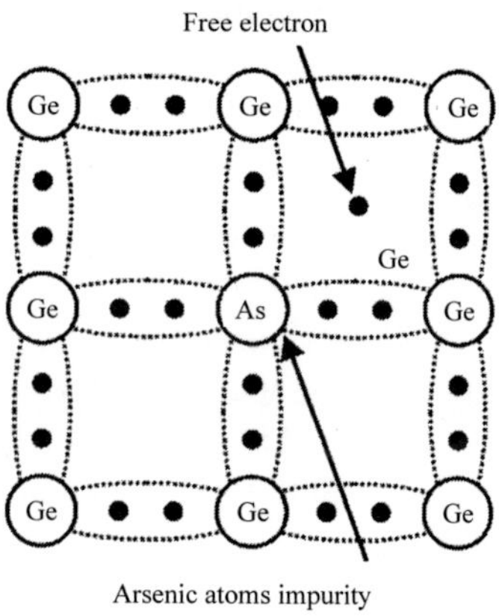

The fifth electron of arsenic atom is not covalently bonded but it is loosely bound to the parent arsenic atom. This amount of energy needed to detach this fifth valence electron form impurity atom. This electron is available as a carrier of current. The amount of energy needed to detach this fifth valence electron from impurity atom is of the order of the only 0.01 eV for Ge and 0.05 eV for Si using arsenic impurity. This energy is very small and may be provided with thermal agitation at room temperature. Such a liberated valence electron is then free to move in the crystal lattice in the same way as free electrons in an intrinsic semiconductor. Although each arsenic atom provides only one free electron yet an extremely small amount of arsenic impurity provides enough atoms to supply millions of free electrons. The energy band description is shown in Fig. 8.3. As seen, in addition to the electrons and holes available in pure germanium, the addition of arsenic greatly increases the number of conduction electrons. Thus, the concentration of electrons in conduction band is increased and exceeds the concentration of holes in the valence band. Here it should be noted that after giving away its one electron, the arsenic atom becomes positively charged ion.

Fig. 8.3: Energy Band Description of Ntype Semiconductor

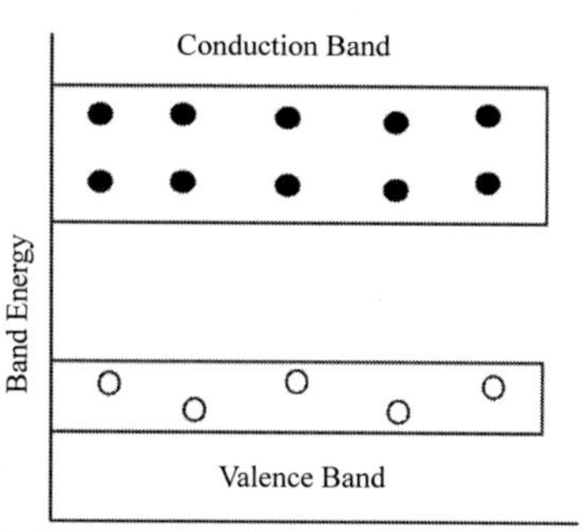

It can not take part in conduction because it is firmly fixed into crystal lattice.

In case of Ntype semiconductor, the following points should be remembered:

(i) In Ntype semiconductor material, there is an excess of electrons. Here the number of holes is small in comparison to parent intrinsic semiconductor because the large number of electrons available fills up the vacancies. Thus, in Ntype semiconductor the electrons are the majority carriers while positive holes are minority carriers.

(ii) Although Ntype semiconductor has excess of electrons but it is *electrically neutral*. This is due to the fact that electrons are created by the addition of neutral pentavalent impurity atoms to the semiconductor i.e. there is no addition of either negative charges or positive charges.

(iii) When Ntype semiconductor is placed between two electrodes and an electric field is applied (Fig 8.4), the excess electrons donated by impurity atoms will travel towards the positive electrode. This constitutes the electric current. This type of conductivity is called negative or Ntype conductivity because the current flows through the crystal is due to free electrons (negatively charged particles). It may be noted that this conduction is just as in ordinary metals like copper.

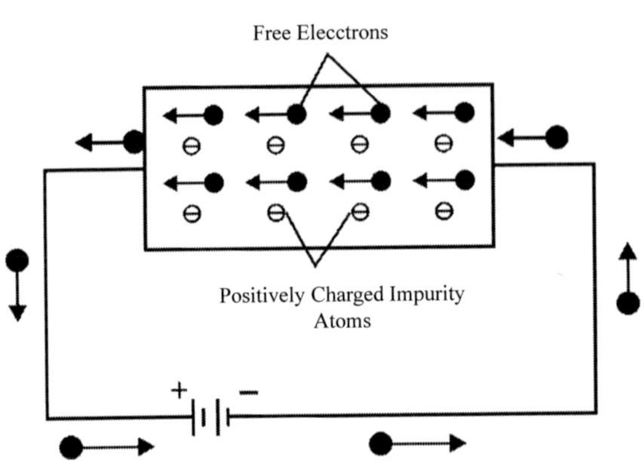

Fig. 8.4: N-type Conductivity

(B) P Type Extrinsic Semiconductor

When a small amount of trivalent impurity is added to a crystal growth, the resulting crystal is called a Ptype extrinsic semiconductor. Let us consider the case when trivalent boron is added to pure germanium crystal. As shown in Fig 8.5 each atom of boron fits into the germanium crystal with only three covalent bonds. This is because the three valence electrons of boron atom form covalent bonds with the valence electrons of germanium atom.

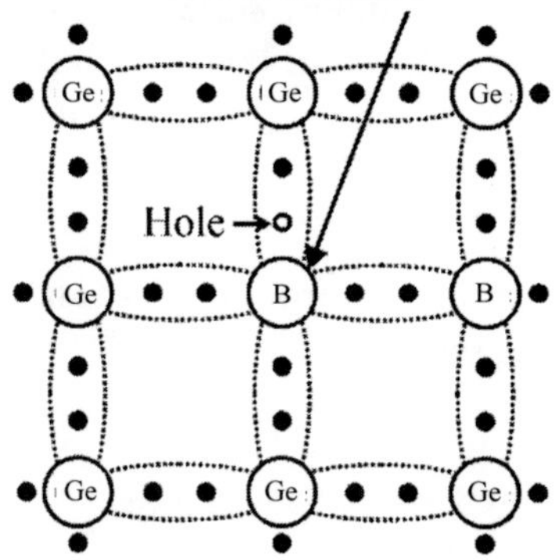

Fig. 8.5: Crystal, Lattice with one Germanium Atom Displaced by Trivalent Impurity Atom (Boron)

In the fourth covalent bond, only germanium atom contributes one valence electron and there is deficiency of one electron, which is called a hole. In other words, we can say that the fourth bond is incomplete, being short of one electron. Therefore, when one boron atom added, one hole is created. A small amount of boron provides millions of holes. The energy band description of Ptype semiconductor is shown in Fig. 8.6. The addition of trivalent impurity produces a large number of holes in the valence band. However, there are few conduction band electrons due to thermal energy associated with room temperature.

Fig. 8.6: Energy Band Description of P-type Semi-Conductor

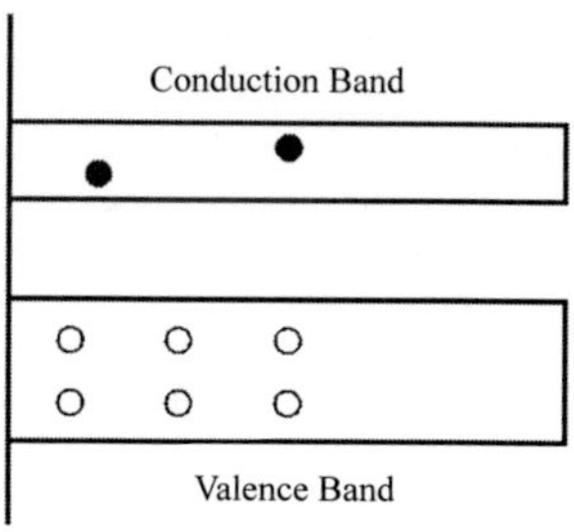

In the incomplete covalent bond, the remaining fourth electron of germanium atom also tries to form a covalent bond. Actually, it does so by taking the advantage of the thermal motion. It steals an electron from the neighboring germanium atom. Due to this stealing action, a hole is created in the adjacent atom. This process continues and the hole moves about in a random way due to thermal effects. Impurity atoms that contribute holes in this manner are termed as acceptors because they accept electrons from germanium atoms. Since current carriers are positively charged particles (holes) this type of semiconductor is called a Ptype semiconductor.

In case of Ptype semiconductor, the following points should be remembered:

(i) In Ptype semiconductor materials, the majority carriers are positive holes while minority carriers are the electrons.

(ii) The Ptype semiconductor remains electrically neutral as the number of mobile holes under all conditions remain equal to the number or acceptors.

(iii) When an electric field is applied across a Ptype semiconductor (Fig. 8.7) the current conduction is predominantly by holes. Here the holes are shifted from one covalent bond to another covalent bond. As holes are positively charged, they are directed towards the negative terminal and constitute the hole current. The hole current flows more slowly than electron current in Ntype semiconductor.

(iv) In Ptype semiconductor, the valence electron moves from one covalent bond to another covalent bond unlike the Ntype where current conduction is by free electrons.

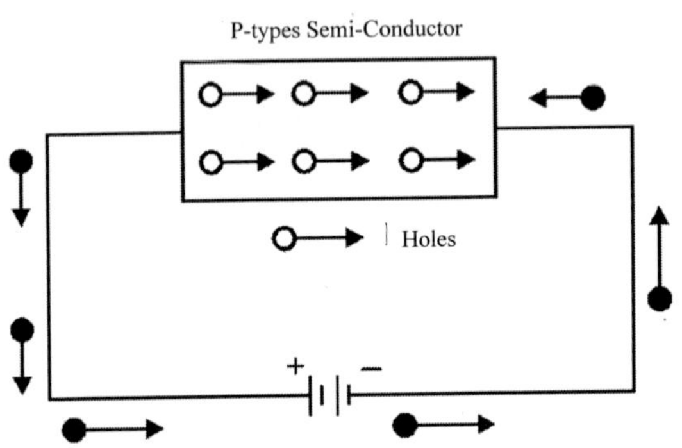

Fig. 8.7: P-type Conductivity

8.4 PN Junction Diode

This is two terminal devices consisting of a PN junction formed either in Ge or Si crystal, when a Ptype material is intimately joined to Ntype, a PN junction is formed. In fact, merelyjoining the two pieces, a PN junction cannot be formed because the surface films and other irregularities produce major discontinuities in the crystal structure. Therefore, a PN junction is formed from a piece of semiconductor (say germanium) by diffusing Ptype material to one half side and Ntype material to other half side. The plane dividing the two zones is known as junction.

A PN junction is illustrated in Fig. 8.8 shows Ptype and Ntype semiconductor pieces before they are joined. In this figure, Ptype semiconductor has negative acceptor ions (shown by encircled minus sign) and positively charged free holes, which moves about on P side. Similarly, Ntype semiconductor has positively donor ions (shown by encircled positive sign) and negatively charged free electrons, which move about N side.

Now let us consider that the two pieces are joined together as shown in Fig. 8.9. As Ptype material has a high concentration of holes and Ntype material has high concentration of free electrons and hence there is tendency of holes to diffuse over to Nside and electrons to Pside. The process is known as diffusion. So due to diffusion, some of the holes form Pside cross over to Nside where they combine with electrons and become neutral. Similarly, some of the electrons from Nside cross over to Pside where they combine with holes and become neutral. Thus, a region is formed which is known as depletion layer or charged free region or space charge region because there is no charge available for conduction. The diffusion of holes and electrons continues till a potential barrier developed in space charge region or charged free region, which prevents further diffusion or neutralization. The potential barrier can be increased or decreased by applying an external voltage.

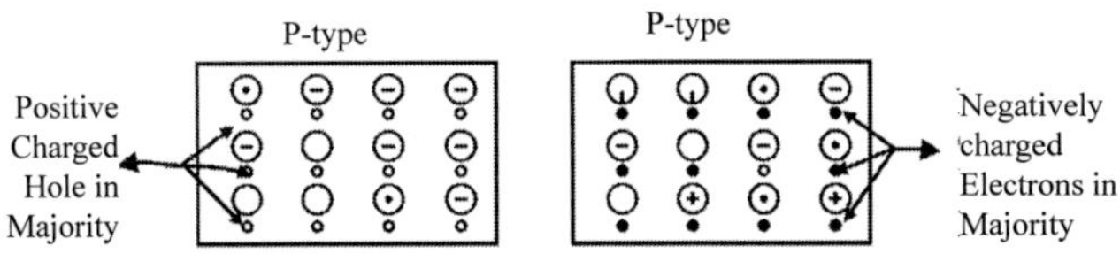

Fig. 8.8: P-type and N-type Semiconductors (not-joined)

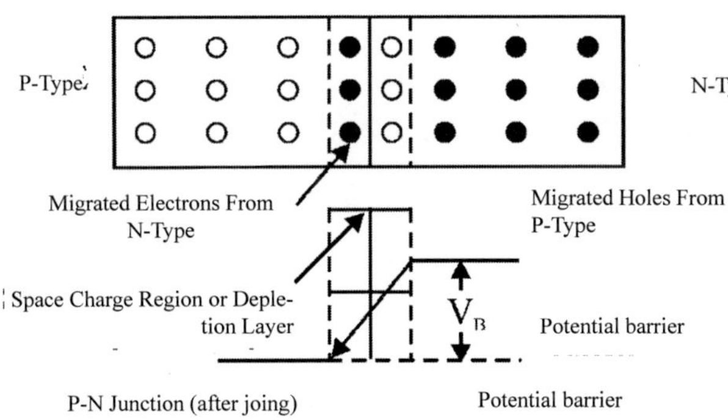

Fig. 8.9: Formation of p-n Homojunction in a Solar Cell

8.5 Solar Cell

Generally, the solar cells are made of silicon. In an intrinsic silicon semi-conductor each atom shares its electrons in the outer most (valence) band with four atoms of silicon (Fig 8.10a), now consider an atom of boron or phosphorous replacing one atom of silicon (Fig 8.10b,c). The phosphorous atoms have one more electron than the number of electrons in the outermost orbit of silicon atom whereas the boron atoms have one electron less.

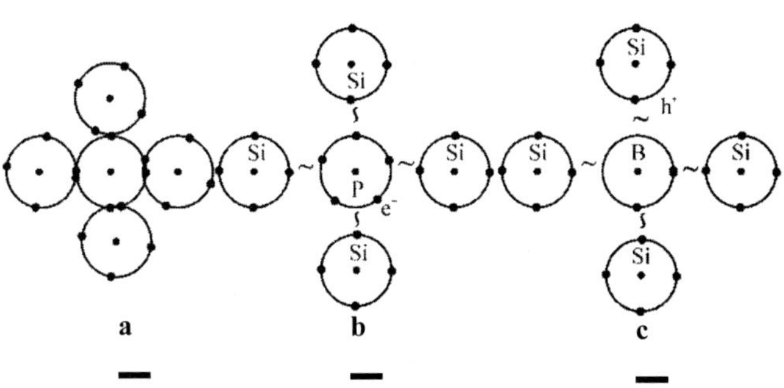

Fig. 8.10: Configuration of Silicon Solar Cell
(a) Sharing of electron in Silicon, **(b)** Phosphorous in Silicon lattice,
(c) Boron in Silicon lattice

Therefore, in presence of phosphorous impurity the silicon material has excess electrons and in known as "donor", whereas presence of Boron impurity makes the silicon material excess number of holes and is known as "acceptor" material these are denoted by ntype and ptype semiconductors respectively. Now consider a situation where both materials are present in the same crystal in a continuous manner. The region where the pand n region met in called the pn junction. This junction is formed when the excess donor electrons from the ntype material cross to the ptype, and vice versa for holes. Due to the movement of electrons and holes excess negative charge on the pside and positive charge on the ntype is available. However, the material as a whole is electrically natural. Here conduction and valence band each have a step at the junction. This is also known as depletion region or zone.

Solar cells involve a pn junction in a semiconductor between a positive layer, which contains movable positive charge or holes and an ntype, which contains movable negative electron. When sunlight enters the crystal, electrons are released and they flow to an electrode and through a wire to the other electrode, where they combine with the positive holes. A barrier at the pn junction prevents the instant recombination of electron and positive holes and causes the electrons to go through the wire, generates current. Here in this process, no material is consumed, no chemical reaction occur and the operative of the cell can continue indefinitely.

Fundamentally, a solar cell is a PN junction diode in which one of the P or N regions is made very thin, so that the light energy is not greatly absorbed before reaching the junction, is used in converting light energy to electrical energy. Such PN junction diodes are called Solar Cells. In the solar cell, the thin region is called the emitter and the other region is base. By shining the light on the emitter, we can get a current in the resistance R_L (Fig. 8.11). The magnitude of current depends on intensity of light.

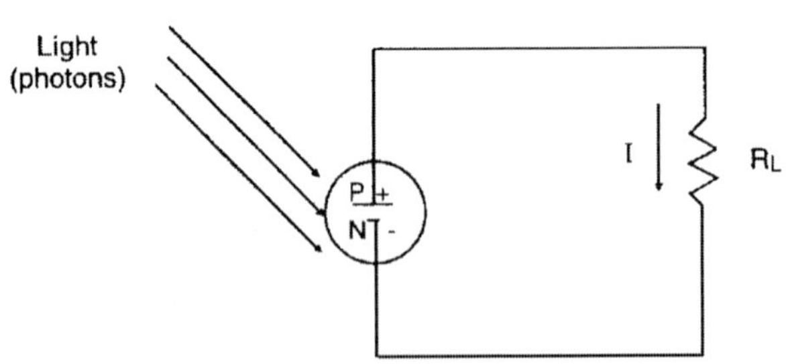

Fig. 8.11: Solar Cell

In general, the solar photovoltaic cell is constructed from a wafer 250-400 mm thick cut from a single crystal. The silicon is doped with boron or phosphorous diffused in from the top surface to give an electrical contact. The second contact is formed by a metal layer on the bottom surface. The top surface has an antireflective coating to reduce loss of radiation by reflection. The top surface will be negative electrically, with respect to the bottom surface. Hence, a typical crystal able silicon cell is constructed. Solar cells operate according to what is called the photovoltaic effect (photolight, voltaicelectricity). In photovoltaic effect, "bullets" of sunlight photonstriking the surface of semiconductor material such as silicon, liberate electrons from the material's atoms. Certain chemicals added to the material's composition help establish a path for the free electrons. This creates an electrical current. Through the solar photovoltaic effect, a typical four-inch silicon solar cell produces about one watt of direct electricity. For efficiency and practically, multiple cells are wired together in series / parallel fashion and placed in a glass-covered housing called a module. The modules themselves can then be wired together into arrays. Solar photovoltaic arrays can produce as much direct current electricity as desired through the addition of more modules. In general, SPV lighting, SPV pumping, SPV refrigerator, SPV telecommunication & SPV radio and television are in use throughout the world .

8.6 Historical Perspective of Solar Photovoltaic Technology

The photovoltaic effect was first observed by French physicist Edmund Becquerel in 1839, when he noticed that illuminating one of two identical electrodes in a weak conducting solution would produce voltage. In the 1870s, the photovoltaic effect was studied in solids such as Selenium. This led to Selenium photovoltaic cells with conversion efficiencies of 12 percent by the 1880s. Although too inefficient for electricity generation, these cells found use as lightmeasuring devices, such, as the "light meters" used in photography.

The modern era of photovoltaic technology for electricity generation began in the mid 1950s. The semiconductor era brought with it the availability of highpurity singlecrystal silicon. In 1954, Bell Laboratories used this technology to produce the first modern cell. This silicon cell had a 4 percent conversion efficiency, which was soon improved to 6 percent. Bell Laboratories fielded the first module in 1956 to power a telephone line. The potentialfree source of electricity was recognized almost immediately. Unfortunately, the cells were very expensive, and this cost and the lack of a market almost neglected the photovoltaic device.

The beginning of the Space Age in the late 1950s provided a strong impetus to extensive research and development into photovoltaic. Space satellites needed a lightweight, longlasting energy source, and photovoltaic was an ideal match to this need depending only on the sun for fuel. The age of the photovoltaic power source in space began with the launch of the Vanguard satellite, and photovoltaic quickly proved be an ideal power source of nearsun orbit applications. Research improved the performance, and by the early 1970s large arrays were being deployed, such as those powering the Skylab.

The tremendous progress made between 1954 and 1973 brought photovoltaic out of the laboratory and made them an important element of the space programme. In addition, even though the modules were not yet economical, this progress established their credibility for terrestrial applications. With the oil embargo of 1973, the current programme to develop photovoltaic as terrestrial energy source began in earnest. Since then, SPV manufacturing industries in many countries have pursued R&D and production of lowcost, high performance SPV systems.

Although still expensive, photovoltaic were used to power remote telecommunications repeater stations in the mid1970s. Much like the space programme, the need was for a reliable power source requiring almost no maintenance and one that could be placed in remote locations. It was the savings in maintenance and fuel supply costs, not the cost of energy that made these photovoltaic systems competitive. As collector prices dropped from hundreds of dollars per watt in 1973 to $10 per watt in 1980, the use of photovoltaic in remote power applications increased.

8.7 Materials for Photovoltaic Applications

There are various types of photovoltaic materials, which can be used to fabricate solar cells for converting sunlight into electricity. Few of these materials are:

(a) **Homo junction solar cells**: Solar cells made of from the same base material across the pn junction. Such as silicon solar cells.

(b) **Hetrojunction solar cells**: Solar cells having two base materials are known as hetrojunction solar cells. Such as SnO_2 on ntype silicon.

(c) **Schottkg Barrier Ms and MIS solar cells**: In such solarcells, the pn junction is formed at a metal semiconductor (ms) interface. This can be prepared by depositing thin films of metals on base materials.

(d) **Polycrystalline solar cells**: The base materials of these cells are always crystalline. Polycrystalline materials can be used for making solar cells using these film techniques.

(e) **Amorphous solar cells**: Materials like glasses are called `Amorphous' materials, they have a short range order. The silicon can be grown in the amorphous state having photovoltaic properties. Such materials can be converted into pn junction cells.

(f) **Direct and indirect band gap solar cells:** Materials like silicon are categorized under this class.

(g) **Liquid interface solar cells**: Such solar cells are formed by using top surface of a liquid electrolyte.

(h) Organic materials: Carbon based organic materials have also semi-conducting or photo conducting properties, those materials can also be used for making solar cells.

(i) Intermediate Transition Based solar cells: n these cells, front surface is coated by phosphors /fluorescent materials which may absorb high-energy radiation to avoid heat generation.

8.8 Present Status of Solar Photovoltaic Technology

(A) Silicon Technology

Being the best studied and understood both as a material and its related devices, Si based PV technologies are the natural choice. The high efficiency, reliability and long life (>> 25 yrs) of crystalline Si cells make it an ideal choice for strategic applications.

A block diagram of the various steps for SCSi PV technology is shown in Fig 8.12. The price of the processed cell depends on the price of solar grads SGSi, the related materials and the energy consumed during these processes. Some processes being energy intensive, and these being upto 50% loss of good material during processing. Numerous innovations have helped to bring down the price in the past. These include:

(i) Improved crystal growth techniques

(ii) Increased (upto 150 cm diameter wafer size)

(iii) More effective wafering techniques to minimize loss of material

(iv) Smaller thickness, down to 300 mm of the wafer

(v) Effective surface AR coatings

(vi) Back surface electric field

(vii) Photon trapping surface morphologies

(viii) Better grid design

(ix) Innovations in the design of the cell active region

(x) Novel surface passivation techniques, etc.

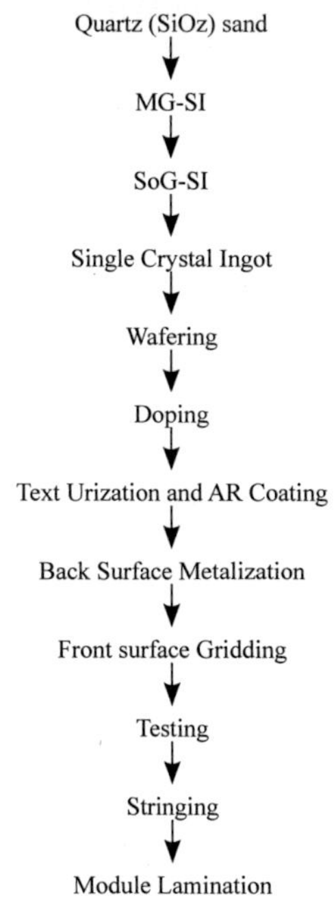

Fig. 8.12: Silicon Solar Cell Technology

As a result: (1) The efficiency of the cells has improved considerably, (2) The production cost of cells has gone down remarkably, (3) The reliability and life of the cell has increased upto 30 yrs, and (4) The energypayback period (EPBP) has decreased from about 4 to 2 years.

Under very special production conditions with precisely structured surface morphology, large area ScSi cells of 18% efficiency are being produced on a limited scale by a couple of companies for special applications. Clearly, with increased number of production steps and precise surface machining by lithographic techniques, such a technology is expensive and not suited at present for largescale manufacturing.

Efforts to cut down the energy budget as also the production cost have been made by experimenting with lower purify grade (solar grade) silicon in the form of cast multi crystalline (mc) sheets, ribbons or melt spun sheets, etc. Performance of mcSi calls has been improved by passivating grain boundaries and surfaces.

Fig 8.13 shows how Sibased PV production has increased and the PV system cost have decreased over the last three decades. The cost of a module (cells connected in series/parallel combination to yield desired power ratings) at present is about double that of a cell and the cost of a complete PV system (which includes such balanceofsystem (BoS) components as battery, invertors, power conditioning unit, etc.) is about double that of the module. It should be noted that Fig. 8.13 includes cSi, mcSi and aSi: H production. The cost of all types of Si based cells is comparable at present.

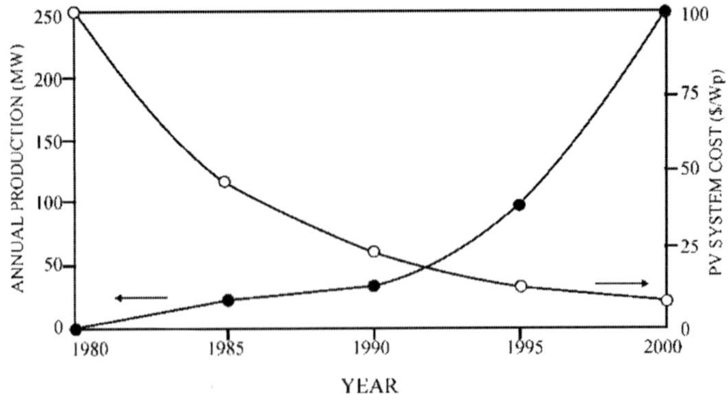

Fig. 8.13: Si-based PV Production and Cost of PV Systems

(B) Thin Film Photovoltaic

There is a general recognition that the only way to cut down PV cost below $2/ Watt is to develop thin film PV systems. Thin film of suitable semiconductors with high optical absorption in the visible offer numerous advantages such as:

(1) Tailor ability of a number of relevant optical, electrical and photoelectronic properties (2) Small thickness (typicallyfew micrometers) and thus very little semi conducting material needed (3) Choice of a number of PVD, CVD and EVD deposition techniques (4) Very large surface area by a batch or continuous process feasible (5) Choice of a variety of thin, cheap and flexible substrates (6) Monolithic integration of cells to form modules during processing (7) Large scale monolithic manufacturing with substrate in, encapsulated moduleout (8) Very high specific power per unit weight on substrate of desired shape and geometry .

The predicted cost benefits of thin film PV attracted massive R&D efforts first on aSi:H and later on Copper Indium Selenide (CIS) and CdTe cells. Only recently, polycrystalline (pc) Si films have been considered seriously. The manufacturing unit process for thinfilm PV, irrespective of the cell material, consists of the following steps:

(i) Preparation of a suitable substrate of appropriate dimensions

(ii) Cleaning

(iii) Deposition of conducting (in some cases transparent and conducting) contacts onto the substrate

(iv) Deposition of the junction forming semiconductor film of desired thickness by an appropriation technique

(v) Deposition of conducting contacts

(vi) Mechanical/laser scribing at different stages of the manufacturing process to achieve monolithic integration

(vii) Encapsulation and finally

(viii) External electrical connections.

Although metallic foils, ceramics and plastics (e.g. Kapton) have been used for some thinfilm PV, glass substrate coated with a transparent conducting oxide film in the preferred choice due to its process ability, ready availability, and reasonable cost. The major material and energy costs for manufacturing are determined primarily be the substrate and one of the expensive materials (e.g. Te, In, Ga, Ag) used in the cell/module. The energy cost and the material use efficiency do vary considerably for different deposition processes. However, considering all parameters of production, calculations show that for large scale (>> 10 MW) manufacturing, the EPBP in all cases varies from a few months to one year and the cell cost ranges form $0.5 to $1.0 per Watt .

Being a new, unusual and matastable material, a lot of R/D efforts were focused on understanding of aSi:H deposition by glow discharge (GD) deposition process, the physics of the material and its junction devices. The cell efficiency reached a level of 9% for single junction and over 12% for triple junction cells after over two decades of work. Several Megawatt PV production plants have been setup in different countries. The low rate of deposition of device quality a Si:H films and thus a rather low throughout and the intrinsic SW degradation effect continue to be serious drawbacks. Serious efforts to overcome both the problems continue to be made with limited success. By using ultrathin intrinsic layers of aSi:H and creating double/triple pin junctions, stability of performance has been improved considerably but not completely. As a result of various innovations in designing the pin junctions, the best cell efficiency has reached 12% level. Stabilized efficiencies of single and multiple junction cells in production are typically 5 and 7% respectively.

A comparative analysis of strengths and weaknesses of presently PV technologies is shown in Table 8.1.

8.9 Efficiency of Solar Cells

Efficiency of a solar cell is defined as the ratio of energy output to the energy input from the sun. The energy output (watt hour or W_h) indicates the amount of energy produced during the day. Most of the sun's energy reaching a solar cell is lost before it can be converted into useful electricity. The minimum amount of energy necessary to free an electron from its band varies with different semiconductor materials. Since solar cells are unable to respond to sunlight's entire spectrum, the solar cells cannot be 100 per cent efficient. Photon whose energy is less than that of the materials band gap, are not absorbed, resulting in a waste of about 25 per cent of the incoming energy. The energy content of photons above the band gap will be wasted surplus, reemitted as heat or light.

Another factor that limits the cell efficiency is the inadvertent recombination of electrons and holes before they can contribute to an electric current. The natural resistance to electron flow in a cell also decreases cell efficiency. Such losses occur in three places; in the bulk of the base material, in the arrow tip surface and at the interface between the cell and the electric contacts leading to an external circuit. Cell efficiency is also affected by temperature. Solar cells work best at low temperature as determined by their material properties. All cell materials give less efficiency as the operating temperature rises. One way to increase cell efficiency is to minimize the amount of light, which is reflected away from the cell's surface .

Thus, the efficiency of a solar cell is affected by several losses. Some of these losses are avoidable and other cannot be avoided under normal conditions of production and utilization. Some of these losses are given in Table 8.2.

Table 8.1: Comparative Analysis of various Solar Cells

Cell / Active Layer Thickness (mm)	Best Cell η (%) Approx.	Production Module η (%)	Life / Stability	Production Technology / Maturity	Remarks
C-Si (SC. Poly)(300)	24	12-15	> 20 yrs. Excellent	Crystal Growth Mature	• Reliable • Cost Limitation
C-Si (Sheet / Film) (20-50)	12	10 (Pilot)	Long Good	PVD & ECD Under Development	• Promising and Viable • Needs further Development
a-Si:H	12	7-8 (triple Jet)	Variable uptp 5 years	GD Mature	• Long Range Commer cial Viability Questionable • Instability and Poor output
CdTe	15.8	9 (Pilot)	Fair Under study	PVD & ECD Under Development	• Simple production process • Problems with Device and Process
Cu-In-Ga-Se	18.8	12 (Pilot)	Several Years Good	PVD & ECD Under Development	• Complex Production Process • Viability Questionable
Go-AS	25	12 (Pilot)	Several Years Good	MOCVD MBE & LPE Mature	• Expensive • Good for space Application

There are few precautions required when working with the SPV system, which are listed below:

(a) The best safety system is an alert mind, a skeptical nature and a slow hand.

(b) Never work on a PV system alone.

(c) Know the system configuration before you start to work on it

(d) Check test equipment before going to the system site.

(e) Wear appropriate clothing and an approved electrical safety hat.

(f) Wear eye protection glasses particularly if working on batteries.

(g) Remove any jewellery.

(h) Wear dry leather gloves to reduce the probability of getting shock.

(i) Measure voltage from all conductors (on PV output circuit) to ground.

(j) Work with one hand whenever possible.

(k) Be skeptical, do not assume that configuration agrees with the electrical diagrams, that is current is not flowing in ground circuit, etc.

Table 8.2: Losses for Retarding Efficiency of Solar Cell

S. No.	Nature of loss	Amount of contribution in percentage
1.	Top surface constant losses	3
2.	Losses due to reflection at the top surface	1
3.	Photons not utilized due to their lower energy content than the band gap	23
4.	Excess energy of photons lost as heat	33
5.	Quantum efficiency losses if thickness of the cell is less than the minimum required thickness	0.4
6.	Collection losses; to minimize the collection loss better design is to be developed	Varies
7.	Voltage factor loss	20
8.	Curve factor loss	4
9.	Series resistance loss	0.3
10.	Shunt resistance loss	0.1
	Power left to be delivered	20 Approx.

8.10. Design of Solar Photovoltaic System

In order to design a solar photovoltaic system, it is necessary to collect necessary information and data related to size of system to be adopted for a particular job. The basic objective of the design should be to develop a reliable, cost-effective system for a particular application. Further optimization of the system for customer, simulation of the system, convincing the customer that it is a reliable system, which performs according to the requirements. The selection of individual components in a PV system also needs careful consideration. Because important matter to be seen for a perfect SPV system are expected amount of sunlight, tilt optimization and correction, energy demand, daily load pattern and required reliability and properties of solar modules, batteries and environmental conditions. In general, design of system size procedures is as given:

Requirement of the customer → Gathering of the input data → Determination of system reliability → Tilt optimization → Trade off calculations → Decision of system size → Simulation of selected system → Detailed engineering of the system.

The design procedure includes following points.

1. Estimate total energy by adding total power required for all the applications. Consider their total rated wattage and operational hours per day, whether they are ac or dc appliances. Arrange these usage as per following table.

S. No.	Appliance/ load	Voltage (AC/DC)	Rated Amphere or Wattage	Daily use hours	Wh/day (AC/DC)

SubTotal AC = …… (Wh/day); DC = ……… (Wh/day)

Adjust AC loads for inverter losses = AC load/ηdc-ac (Wh/day)

The efficiency of DC to AC inverter ranges from 85 to 98%.

2. Total Daily load; DC load + AC loads = Total daily load (Wh/day)
3. Determine system characteristics for sizing a PV system namely hours of peak sunlight for a particular place, battery efficiency, battery regulator efficiency etc.

$$\text{Array load} = \frac{\text{Detailed Energy Consumption}}{\text{Battery efficiency} \times \text{Charge regulatore efficiency}}$$

4. Calculate PV array size and battery capacity by using following

$$\text{PV Array size(Peak W or } W_p = \frac{\text{Array load}}{\text{Insolation} \times \text{Mismatch factor(0.85)}}$$

$$= \frac{\text{Total daily load(Wh / day)}}{\text{Peak sunlight hours} \times \eta_{bat} \times \eta_{reg} \times \text{Mismatch}}$$

where

η_{bat} = Battery efficiency (typically 7590 per cent)

η_{reg} = Regulator efficiency 85 per cent includes losses due to dirt and cables etc.)

Similarly, Battery Capacity (A_b)

$$= \frac{\text{Total daily load((Wh / day)} \times \text{Storage period in days}}{\text{Battery voltage} \times \eta_{bat} \times \text{DoD}}$$

where DoD = Battery maximum depth of discharge (from 20 to 80 percent)

Also, battery capacity can be calculated as

Daily energy consumption

$$\text{Battery capacity} = \frac{\text{Daily energy consumption}}{\text{Maximum allowable discharge} \times \text{Nominal voltage}}$$

8.11 SPV System and Components

Schematic of a simple photovoltaic system and its basic components shown in the Fig. 8.14.The load for most of the small stand-alone systems utilizes power in direct current (dc) form as generated by SPV modules. In addition to the SPV modules, other components are storage batteries; blocking diode to prevent loss of battery charge through the cell overnight, and a regulator to prevent overcharge of the battery in period of high insolation levels, the connecting wires, load, supporting structure, etc. Depending on the type of applications, some other system components may be needed. In case higher power values are required, several modules must be connected together to form an array. This array can be of any size and be mounted on any appropriate support structure, depending on the intended use. With the help of storage batteries, which can be sized according to the amount of back up needed, the fluctuations between day and night and cloudy periods can be evened out without any major difficulty.

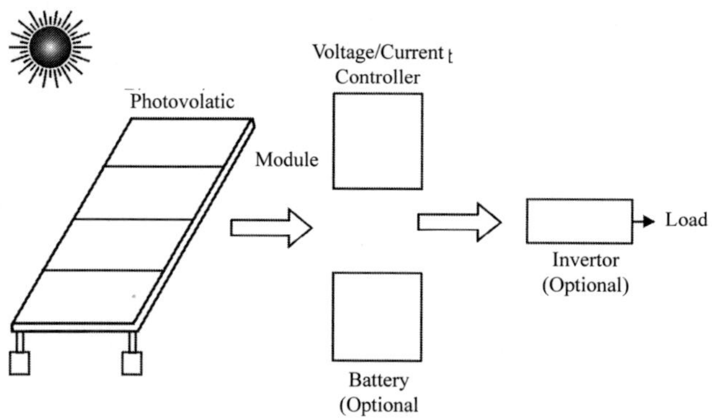

Fig. 8.14: SVP System

Three important considerations are the optimal combinations of the individual components, proper sizing and siting. Another important point is to raise the total system efficiency value by matching these individual components and to achieve a long lifetime for the system as a whole, while keeping the maintenance requirements as low as possible.

The PV systems are grouped in several ways. Some common classifications are:

(a) **Stands alone system:** A PV system that is not connected to the utility grid. Most stands alone system use batteries to store the energy produced during day light hours for use at night or on cloudy days.

(b) **Grid connected system:** A PV system tied to the electric utility's power distribution grid. The power not used by load is transferred to the grid, the load can receive the power from the grid when the PV system is not generating enough to satisfy demand.

(c) **Flat plate system:** A PV system comprised of modules that are flat in geometry and use natural (unconcentrated) solar irradiant to produce electricity, and some power is generated even on the cloudy days. The sum of the direct and diffuse solar irradiance is called total global irradiances.

(d) **Concentration system:** A PV system comprised of modules that have concentrating optics as a part of their structure. They use only the direct beam solar irradiance focused by lenses or mirrors, a concentrating system will not generate power on cloudy days. The intensities produced by these modules also produce intense heating, that must be dissipated by active or passive cooling mechanisms.

(e) **Fixed tilt system:** It is PV array with modules at fixed tilt angle and orientation. The array may be mounted on a rooftop, on a pole, or on the ground. These systems use flat plate modules only.

(f) **Tracking system:** A PV system with modules mounted on a tracking unit that follows the sun. Single axis trackers follows the sun daily from east to west & two axis trackers include elevation control to correct for seasonal NorthSouth sun movement. Tracking system are more expensive than fixed tilt system but also produce more electrical energy per unit area because they follow the sun and collect the maximum available irradiation at all times. Tracking system may employ either flat plate or concentrator modules.

(g) **Hybrid system:** Any system, which makes use of more than one powerhouse.

Example 8.1: Assuming that the average daily solar radiation at a particular place is 600 W/m^2. Calculate the size of array with 30% efficient solar cells to produce 2 kW power.

Solution

Total energy required to produce 2 kW power is

$$\text{Cell efficiency} = \frac{\text{Energy Output}}{\text{Energy Input}}$$

$$\text{Energy Input} = \frac{\text{Energy Output}}{\text{Cell efficiency}}$$

$$\text{Energy Input} = \frac{2000}{0.30}$$

$$= 6666.67 \text{ W}$$

With the given solar radiation, we can estimate the size of the array for the solar cells as,

$$A = \frac{\text{Energy Input}}{\text{Solar Radiation}}$$

$$A = \frac{6666.67}{600}$$

$$A = 11.11 \text{ m}^2$$

Example 8.2: A 100 kW peak system that was installed on a University building cost a total of Rs. 85 lakhs after rebates from the government. Expected life of solar panel is about 25 years. What is the cost per kWh of electricity generated over this lifetime? Assume the solar cell produces at peak power for 6 hours in the day, throughout the year.

Solution:

Given

Capacity of the system is: C=100 kW

Cost of the system: Cs= Rs. 85 lakhs

Lifetime of peak system: n=25 years=25×365=9125 Days

Value of electricity generated in kilo watt per day.

P = Installed capacity (kW) × Peak hour per day

=100 × 6 =600 kWh per day

Power generation during lifetime:

= Power generation per day × lifetime in days

= 600 × 9125

P = 54,75,000 kWh

Cost per kWh of electricity is

$$Cost / kWh = \frac{Cs}{P}$$

$$= \frac{85,00,000}{54,75,000}$$

= 1.55 Rs.

Example 8.3: Determine the peak power of a PV system which provide an average electric power of 15 kW. The average power generation of a solar panel is 18% of the peak power generation. Determine Peak power

Given

Average electric power provided by system (Pavc)=15 kW

Efficiency of power generation of a solar panel at peak = 18%

We can determine the peak power generation using following equation

$$P_{peak} = \frac{P_{ave}}{\text{Peak Efficiency}}$$

$$P_{peak} = \frac{15}{0.18} = 83.33 \text{ kW}$$

8.12. Applications of Solar Photovoltaic

The electricity generated by solar photovoltaic system can be used for the following applications: Telecommunication, Railway signaling, Cathodic protection, Navigational aids, Traffic warning light, Remote instrumentation, Crop spraying, Water pumping, Vaccine refrigeration, Lighting system, Battery charging, Consumer products, Educational kits, Entertainment, Offshore application, Space applications, Rural electrification, Grid connected houses, Hydrogen production and alarm systems, etc.Solar photovoltaic system are used for meeting out various essential requirement, which are given in Fig. 8.15

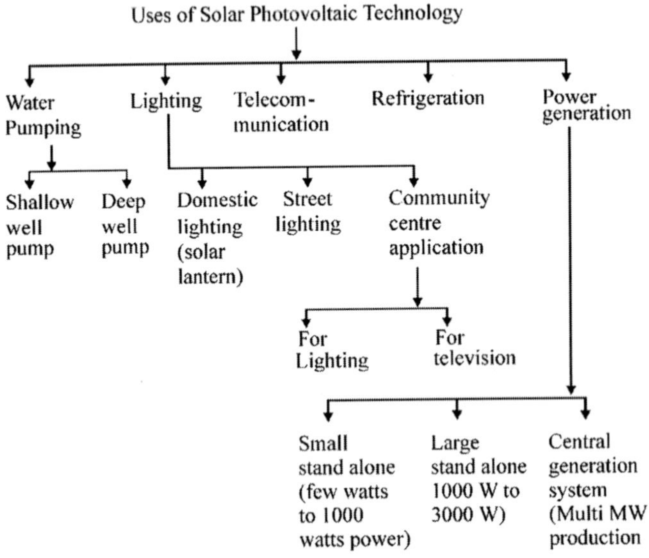

Fig. 8.15: Application of Solar Photovoltaic System

8.13. Present Status of Application of PV in India

Considerable experience has been generated in the country in the development and use of solar PV systems for several applications. The technology for the manufacture of solar cells, modules and systems has been developed indigenously. Systems for such applications as street lighting, domestic lighting, community lighting, water pumping for drinking water supply and irrigation, vaccine refrigeration, communication, etc. has been demonstrated and field tested in the country. About 10,000 villages and hamlets have already been provided with amenities such as street lighting, lights in homes, community centers, and night schools/adult education centers and community television, etc. About 1200 solar PV water-pumping systems have also been supplied to individual and institutional users as part of the SPV demonstration programmes for meeting small irrigation and drinking water requirements. Some deep well water pumping systems has also been installed. About 80 solar PV centralized systems of capacity between 1 and 25 kw (peak) have been set up in different location in the country. The SPV systems have created a significant impact and are meeting the basic energy requirements of the people living in rural, remote and un electrified areas.

Currently, there are 16 manufacturers of crystalline silicon solar cells and nearly 38 manufacturers of PV models, including three public sectors companies among which Bharat Heavy Electricals Limited is the major production unit for Solar cell, PV modules and systems. The major areas where SPV systems are applicable are as follows:

1. Railway

The CEL has produced entire power requirement for signalling utilised at Railway stations in Northern Railways. They has also developed solar powered level crossing warning systems for Indian Railway.

2. Oil Sectors

SPV based cathodic protection system for oil pipeline is also available. CEL has designed, developed, installed and commissioned all the cathodic protection system in western region of ONGC and low cost distributed cathodic protection system for Indian Oil Corporation.

3. Village electrification

Village electrification through solar PV system is very attractive option for the villages, which are not grid connected.

Apart from above specialized items, various commercialized gadgets are also available in market based on solar photovoltaic technology. These commercial products are given as below

(A) Solar Lantern

A typical solar lantern consists of a small photovoltaic module, a light source, a high frequency invertor/ballast, battery, charge controller and appropriate unit. During the day hours, the module-facing south is placed in the sun and it converts the solar radiation into electricity and charges the battery, which is connected, to the lantern through a cable. In the evening, the lantern with the charged battery is disconnected from the module and is available for indoor or outdoor use.

In another model, the module is integrated with the rest of the components in a flat house roof. In either case, there is no need for a mounting structure. However, a small and simple stand can be used to support the module. There is also a provision of central charging station in a village, where a sizable array are used to charge individual lanterns brought to the station daily by the villagers.

The solar lantern is a very simple unit, which can be used for lighting. It has no operating and maintenance cost except the change of battery after about four to five years of use.

(B) Street Lighting System

The solar PV based street lighting System has following three models, namely.

a) Single module, three hours operation, and fluorescent lamp with timer controls.

b) Two module, six hour operation, fluorescent lamp with timer control and

c) Two module, all night operation, with compact fluorescent lamp (CFL) and day/night sensor.

The governments have many schemes for the installation of the streetlights in rural areas and other remotely located places. As there is no installation cost or cost of setting up power lines such units are very cheap in far flung areas having no access to electricity.

(C) Domestic Lighting System

The following three types of domestic lighting systems are available in our country.

a) Fixed domestic lighting units (one module, 2 CFL and 7 W each with sealed battery and controls).

b) Portable lighting unitslanterns (10 W module, 57 watt CFL and sealed battery)

c) Service connections from photovoltaic power plants.

Each state government has set up a network for the purchase of the domestic units in rural and far-flung areas.

(D) Community PV Lighting System and PV Power Plants

For village level use, two types of systems are available:

a) A 300 Wp PV installation for the community centers in villages, largely comprising lights.

b) Village based PV power plants of 2 to 10 KW capacity or even higher capacity plants.

A typical village-based PV power plant consists of an array, a battery bank housed in a small shed, a power conditioning unit, meters & controls and distributing system. The connected load may be for lighting, TV, radio and drinking water supply pumps. The cost of installation of such systems is shared by the state government, central government and villagers.

(E) Solar Water Pumping

A solar water pumping system extensively seen for irrigation of agricultural land. The conventional energy cost can be eliminated through solar photovoltaic panels. The solar water pumping system significantly accelerate the development of agriculture in India. The solar energy might be the easiest way for the farmers to generate electricity specially for those farmers living off the electricity grid with poor infrastructure around their residence. Solar pumps are clean, simple and energy-efficient alternative to grid electric and fuel-driven pump sets. Approximately 40 % of the world population depends on agriculture and its main source of income, access to water remains an ongoing struggle for many people.

A solar photovoltaic (SPV) water pumping system consists of a PV array, a DC/AC surface mounted/ submersible/ floating motor pump set, electronics, if any, interconnect cables and an "On-Off" switch. PV Array is mounted on a suitable structure with a provision of tracking.

The solar PV pumps are reliable in regional and remote areas and easy to transport and relocate as per the needs of the farm. The initial cost of system is high and suitable in regions with a lot of sunlight.

The system operates on power generated using solar PV (photovoltaic) system. The photovoltaic array converts the solar energy into electricity, which is used for running the motor pump. The pump set may be submersible type or and surface type as shown in Fig 8.16 (a & b). The system requires a shadow-free area for installation of the Solar panel. Different pump capacity for surface / shallow water sources are presented in Table 8.3 and for submersible / borewell, tube well is presented in Table 8.4.

Table 8.3: Different Pump Capacity for Surface / Shallow water sources

Particulars	DC Pump			AC Pump			
	1 HP	2 HP	3 HP	1 HP	2 HP	3 HP	5 HP
PV Arry size (Wp)	900	1800	2700	900	1800	2700	4800
Total Dynamic Head (m)	10	10	20	10	10	20	20
Shut off Dynamic Head (m)	12	12	25	12	15	25	30
Water Discharge (LDP) at 7.15 kWh/m^2 Solar radiation on PV array	90,000 @ 10 m head	1,80,000 @ 10 m head	1,30,000 @ 20 m head	80,000 @10 m head	1,50,000 @ 10 m head	1,25,000 @ 20 m head	2,10,000 @ 20 m head

Table 8.4: Different Pump Capacity for Submersible / Borewell, Tube well etc.

Particulars	DC Pump				AC Pump			
	1 HP	2 HP	3 HP	5 HP	1 HP	2 HP	3 HP	5 HP
PV Arry size (Wp)	1200	1800	3000	4800	1200	1800	3000	4800
Total Dynamic Head (m)	30	30	50	50	30	30	50	50
Shut off Dynamic Head (m)	45	45	75	70	45	45	75	70
Water Discharge (LDP) at 7.15 kWh/m^2 Solar radiation on PV array	42,000 @ 30 m head	60,000 @ 30 m head	60,000 @ 50 m head	1,00,000 @ 50 m head	40,000 @30 m head	55,000 @ 30 m head	55,000 @ 50 m head	90,000 @ 50 m head

Fig. 8.16a&b: Types of Solar Water Pumps

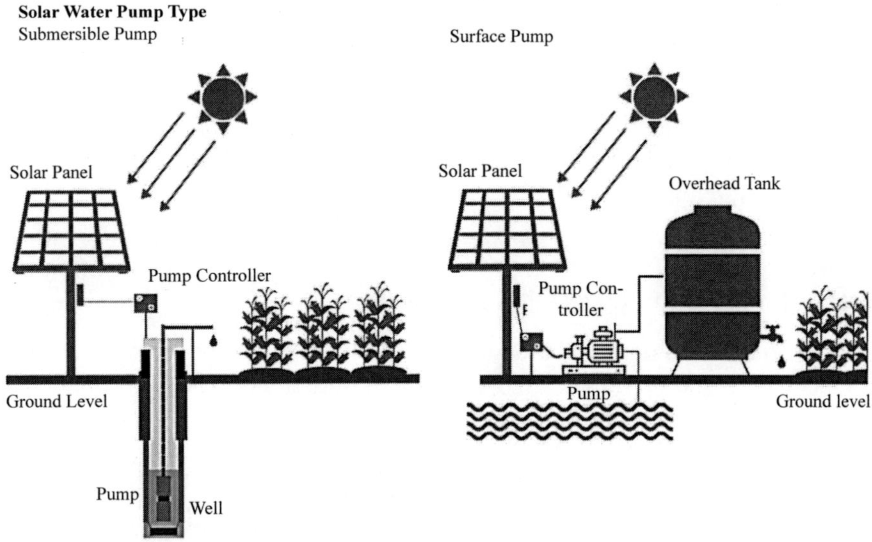

(F) Solar Photovoltaic Deep Well Pump

The solar photovoltaic water pumping system commonly used for deep well pumping comprised of a solar photovoltaic array having 42 panels, a DC - AC inverter, a submersible pump and electric cables. Each panel of SPV array is having 36 multicrystalline solar cells interconnected in series with 1.8 KW peak capacity. The size of one panel is 1.0 m x 0.41 m. The array is having an area of 17.22 m^2 of seven subarrays of 6 panels, each connected in series and these sub arrays are further connected in parallel to get a maximum of 150 V open circuit voltage and 10A short circuit current. The power produced during sunshine hours is transmitted through electric cables to a submersible pump (5 cm. Diameter) via a DC - AC inverter. Anticipated life of the array is around 2025 years, during which only their surface need to be cleaned occasionally. To attain an optimum DC output of the solar array, the array are to install on the roof at a tilt angle of the place with the horizontal plane. The array has south facing orientation to get maximum absorption of sunlight by the solar cells.

Presently available the solar photovoltaic deep well pumping system consists of a "GRUNDFOS SOLARTRONIC SA1500" DC - AC inverter used to convert DC power produced by the solar array into AC power. This inverter is operated at a nominal voltage of 105 V (maximum 1700 W and 160 V). The solartronic inverter is installed vertically as it is cooled by convection. The inverter is placed in a box near well to avoid direct sunlight, which may cause overheating and subsequently may hinder the operation of the system.

The solartronic system is capable of measuring DC current, DC voltage and internal temperature. In case of faults in any of the supervised or associated parameters, the solartronic system stopped and showed the fault at the specific diodes on the front cover. The essential tool of SPV deep well pumping system is a submersible pump, which is also of GRUNDFOS make. The schematic representation of solar photovoltaic submersible pump installed in the bore well is illustrated in Fig.8.17.

Testing of Solar deep well pump consists of following measurements

(i) Development of Power Characteristics

In order to develop power characteristics of the array, the voltage and current produced by it should be known. For this a multimeter (0-250 V) is connected in parallel combination to the array.

No load Testing

In no load testing, the pump is to disconnected and the voltage produced by the array be measured by multimeter at every half an hour interval from 0900 to 1700 hrs. on a normal sunny day. The corresponding solar intensity (w/m^2) is also measured from the pyranometer.

Fig. 8.17: Solar Photovoltaic Deep Well Pump

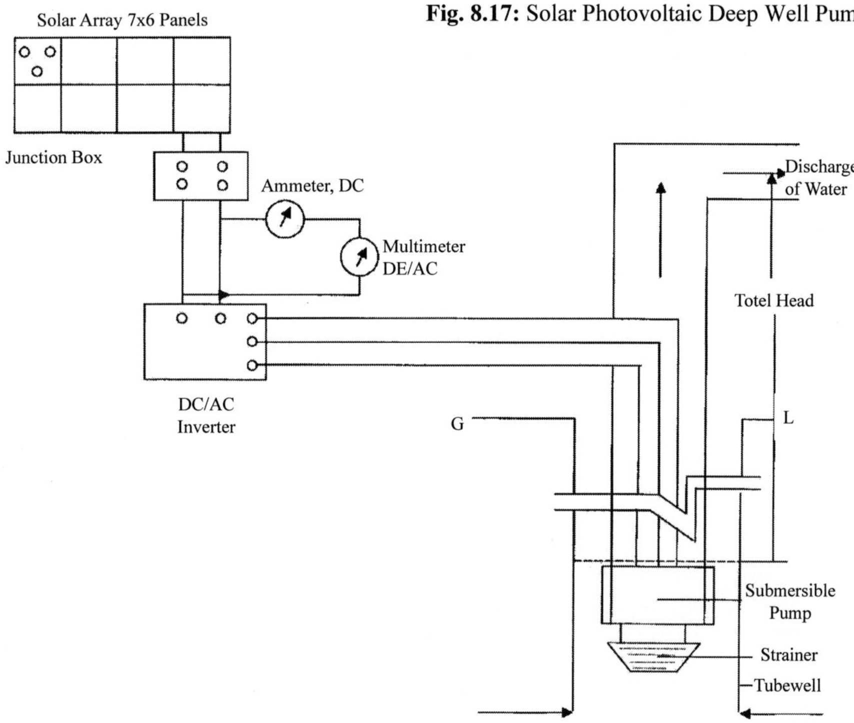

On load Testing

During on load testing, initially the pump at certain pumping head is connected to the solar array and voltage and current development be measured. The discharge through delivery pipe is also measured by volumetric method using a graduated bucket and stopwatch for every half an hour interval during sunshine hours. Corresponding value of solar intensity (w/m^2) is also measured from pyranometer. The performance of the system is evaluated in terms of PV efficiency, pumping efficiency and system efficiency over different values of solar intensity in a day. The PV efficiency, pumping efficiency and system efficiency are expressed by the following equations:

$$\text{PV efficiency } \% = \frac{\text{Output Power}}{\text{Input Power}} \times 100$$

$$= \frac{\text{Voltage produced} \times \text{Current developed}}{\text{Corresponding solar intensity} \times \text{Area of the array}} \times 100$$

$$\text{Pumping efficiency } \% = \frac{\text{Power needed to lift water(W)}}{\text{Power supplied by the array}} \times 100$$

$$= \frac{\frac{Q \times H}{76} \times 746}{\text{Voltage produced} \times \text{current developed}} \times 100$$

where,

Q = Discharge of water through delivery pipe (lit/sec)

H = Pumping head (meter)

And thus,

$$\text{System efficiency \%} = \left(\frac{\text{PV efficiency(\%)}}{100} \times \frac{\text{Pump efficiency(\%)}}{100} \right) \times 100$$

For determination of power developed by the array at any value of solar intensity between the observed limits, the power characteristics of the system are to be developed. The net power developed at observed values of solar intensity is the product of corresponding voltage and current (For D.C. supply).

8.14 Design of Solar PV Technology for Water Pumping

(A) Daily Insolation levels

The power output from the PV array will depend upon the insolation and availability of sun per day. The insolation varies from one location to another, month to month because of seasonal and climatic changes. If water requirements are in the same range of the whole year, solar design calculations should be based on the month with the lowest insolation of the year. This will ensure adequate water supply through out the year. For irrigation water, the months with the lowest insolation often correspond to those in which crop demand for water is lowest. If water consumption varies round the year, the system design should be based on the ratio of water required to the insolation available. The month in which this ratio is largest will determine the PV array size.

(B) System sizing

The size of a PV water pumping system will depend on the water requirement, the total head and the solar insolation. The pattern of water use should also be considered in relation of system design and storage requirements enough to provide for daily water requirements and short periods of cloudy weather.

We can calculate the size of PV array by using the equation

$$E = \frac{\rho \times g \times h \times v}{3.6 \times 10^6}$$

where

E = hydraulic energy required (kWh/day)

ρ = density of water (1000 kg/m^3)

g = gravitational acceleration (9.81 m/sec^2)

h = total hydraulic head (m)

v = Volume of water required (m^3/day)

By putting these values, the above equation reduces to

E = 0.002725 hv (kWh/day)

The required PV array size can be calculated from equation

A = E / eFI

where

I = average daily solar insolation (kWh/m^2)

F = array mismatch factor (%)

e = daily subsystem efficiency (%)

A = Cell / array area (m^2)

(C) Orientation and Direction of the PV Array

The PV array should be positioned in such a way that the sunlight is utilized to its maximum that is true south (in the Northern Hemisphere). The local declination which depends on the location and changes with the times should, however, be taken into account. When the direction north south has been found, the lines a and b are marked at right angles to the northsouth line to indicate the position of the PV array. The PV array should not be shaded by obstructions like nearby building or trees.

(D) Determination of Tilt Angles

Module surface tilted at a right angle to the sun's rays catch the most sunshine per unit area. An angle equal to the local latitude is the closest approximation to that tilt or slope on annual basis. The tilt angle should be selected in accordance with the latitude in which the solar water pumping system is to be installed. If the water requirements do not remain uniform throughout the year, a higher or lower PV array tilt angle might be advantageous and lead to better system performance.

(E) Determining Peak Water Flow

The maximum required water flow in liters/hour would be approximately the system's requirement divided by the number of sunhours. Dividing this figure by 360 second/hour gives the maximum expected water flow in liters/second. For example, to meet drinking water requirements of a typical village, 8000 liters of water requirements, 8 sunhours per day of insolation is available, the peak flow rate from the system will be 8000 litres/5 sun hour = 1600 litres/hour = 0.4 litres/second.

(F) Water Production

In designing a PV water pumping system, it is essential to know the amount of water a well can produce. A correctly operating pumping system should not exceed the well's water production. To determine the amount of water produced by a well, a portable pump is needed which is capable of pumping water at a rate at least as high as the peak-required rate.

First, measure the depthtowater in the well. Install the pump and let it pump water until the water level in the well stabilizes. Now with the pump still operating, measure the depthtowater again. Repeat the measurement at several time intervals to ensure that the water level has stabilized. Now measure the water flow rate by filling a container of known volume and measuring the time required to do so the flow rate should be as high as the peak flow rate required. If the water level drops to the bottom of the well, the well does not produce enough water for the village needs. So, either the well should be deepened or there has to be another well to meet the water requirements of the village.

(H) Wells

Wells are normally constructed to the depth of the local water level plus and additional depth to account for pumping draw down. Well must be designed to provide uncontaminated water free from bacteria and abrasive that wear out pump components. Usually, wells are lined with RCC, plastic or steel wall casing extending to the aquifer.

Well yield is a crucial design parameter that must be considered relative to the maximum water-pumping rate. Water extraction rate should not exceed the well yield or the water table may fall below the suction of the pump. Many pumping system failures are directly attributable to well design and yield problems.

(I) System Configuration

Water pumping systems can be configured for three general application ranges: These are shallow water table, lowflow intermediate and deep-water table application ranges of commercially available motor pump sets.

(i) Shallow well application (7 meters water level). The photovoltaic water pumping system for shallow well applications will consist of a well with a surface mounted singlestage centrifugal pump directly coupled to a DC motor (Fig.8.17).

(ii) Intermediate(Fig.8.18)/Deep well Application (2040 meters water depth) (Fig.8.19).

Fig. 8.18: Shallow Water Table Applications

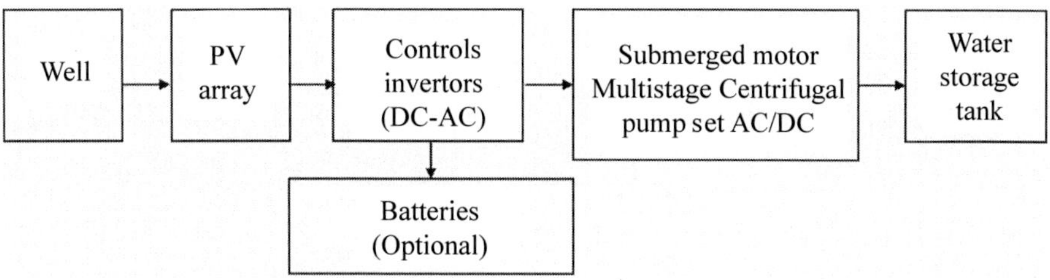

Low flow (30,000 liters/day or less)

The surfacemounted centrifugal pumps are suited for locations where the water levels do not exceed 7 meters. Beyond 7 meters, the centrifugal pump will lose suction and will not be able to lift water from the well.

Fig. 8.19: Intermediate Water Table Applications

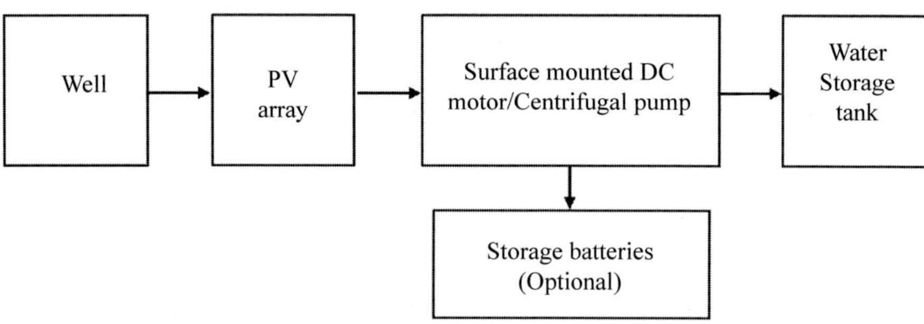

For intermediate and deep well applications, the choice is a jack pump or submersible motor pump set. At water depths of 20 meters, jack pumps are good choice for water if demand is below 30,000 liters per day. In this, mounted DC motor drives a jack pump. The system comprises power-conditioning equipment to match the cyclic continuous photovoltaic power output.

High flow (over 30,000 liters per day)

To meet water requirements of more than 30,000 liters per day, a submersible motor pump set is recommended. The system will consist of a well, PV array, controls and invertors and AC motor pump set in case of an AC system. A DC motor pump set is used to PV array capacity of 1 kWp. No invertors are required for a DC system. This type of system uses a multistage centrifugal pump with a submersible motor. For peak power demands beyond 1 kWp and AC system energizing AC motor, which operates, multistage centrifugal pump is assumed.

Fig. 8.19: Deep Water Table Applications

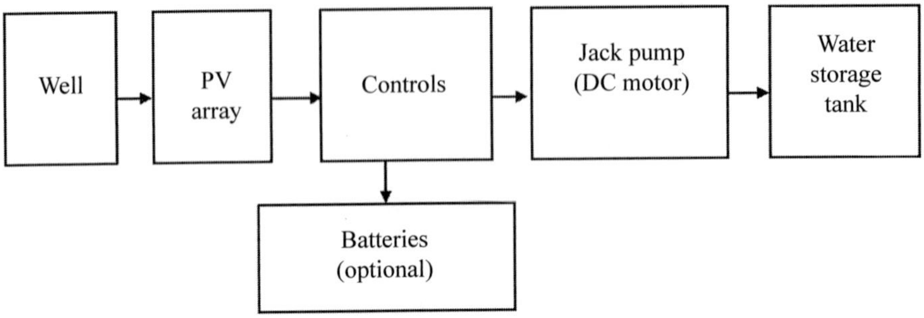

(i) Operation and maintenance

Operation and maintenance requirements for a solar photovoltaic water pumping system are minimal. Maintenance is required only of the motor/pump set such as the lubrication of parts and replacement of brushes. The cleaning of the top glass surface of the photovoltaic modules with a duster is required regularly, in some cases once a week or once in 15 days. Cleaning of the glass surfaces with water may be done, if required. Batteries can be considered only in the case of motors requiring high starting current. If the batteries are of stationary, tubular type not much maintenance is needed and only addition of distilled water in required.

9

Solar Heating and Cooling

9.1 Introduction

Basically, for heating or cooling the space using solar energy, the solar energy is to be collected, stored and distributed properly in the space to be heated or removed from the space to be cooled. There are two basic approaches for solar houses, there are:

(a) Active solar house technology

(b) Passive solar house technology

In active approach, the solar energy is collected and stored in some separate solar energy collector and then energy is distributed in the space where heating is to be done using electrically operated pumps and fans coupled with radiator etc. The storage of solar energy in active solar house may be in the form of storage as sensible heat or storage as latent heat or chemical heat storage. As per requirement of heating the size of solar collector and storage system be decided. Generally, the active concepts of solar houses are used for heating purpose. Active concept can also be used for space cooling applications. Solar vapor compression and absorption system can be used for this purpose, essentially, this includes incorporation of solar heat collectors, thermal storage system, auxiliary heat supply system (to compensate intermittent nature of sun energy) and control systems. Further, there are three ways of solar space heating i.e. solar air systems, solar liquid systems and solar heat pump systems.

The passive solar house technology may be used for space heating as well as for cooling. In this approach, orientation and layout of the buildings is so arranged, that three function of solar houses i.e. solar energy collections, storage and distributions is made by natural means and no electrical, mechanical or electronics controls are used. In the passive house, all the building components such as walls, roofs, windows, partitions, ventilation's walls etc. are so selected and arranged architecturally, so that solar collection, storage and distribution is managed automatically. In this approach, solar energy oriented parameters such as solar radiation's, outside air and internal metabolism, air flow rate are used for heating purposes, whereas sky & space temperature, outside wind and wet surface are used for cooling purposes. The building materials (stone, bricks, cement etc.) and the various thermal processes (thermal radiation, natural & forced convection, conduction, air stratification, evaporation, thermosyphoning etc.) are so integrated along with building components and thereby efficient & convenient atmosphere is attended with the house. Depending upon site, climate & activity and particular needs, a proper balance is needed among building components, building elements and thermal processes. The passive heating & cooling process follows a sequence of several phases. Although the overall process might be described in different ways, the major steps are:

(a) Identifying suitable building components for desired cooling and heating.

(b) Identifying and establishing the criterion of acceptance of different building elements (stone, bricks etc.)

(c) Incorporating suitable thermal process

(d) Integrating optimally step (a), (b) & (c)

9.2 Active Solar House Technology

In the active solar houses, separate solar collectors are needed for heating the space as per requirement. The heating of house in winter is one of the major requirements for all types of houses. The technique of heating house with the help of some gadgets based on solar energy is an ancient technology. Basically, the active solar houses consisted of following major parts (Fig. 9.1).

(a) Solar Collectors

Different types of solar collectors are used for heating a fluid. These solar collectors may be air collectors, liquid collectors or heat pump type. In the air collectors, the air is heated and after words the heat is transferred to living space from storage unit by means of air or liquid whereas in liquid collectors, any suitable liquid preferably water is used for heating purpose.

(b) Storage System

The collected solar energy in turn is stored in suitable structure either based on concept of sensible heat storage or latent heat storage. The chemical storage systems are also employed depending upon its availability and cost involvement. In fact, storage devices are used to store heat for use at night and on intermittent days.

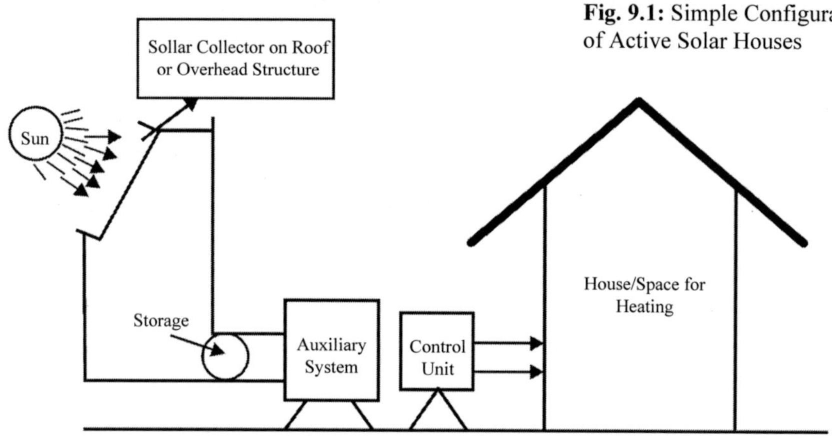

Fig. 9.1: Simple Configuration of Active Solar Houses

(c) Auxiliary Heating System

Viewing to intermittent nature of sunlight, it is always advisable to have auxiliary system for night or for seasons when solar energy is not available. Basically auxiliaryheating system is provided to supply heat when required or to compensate performance of solar heating devices. However, such system generally turns out to be costly.

(d) Controls

Some sort of controls either electrical or mechanical or electronically operated are used for supplying stored energy into space for heating purpose.

Basically, solar houses are meant to get more solar energy in winter and less in summer, so that this reduces the heating and cooling load for the building. Therefore, active solar houses, apart from its usual components, must have following prerequisition:

It must be adequately insulated, so that optimum resistance to flow of heat is maintained.

- It must be properly oriented to take advantage of nature gifts.
- It must be optimally glazed, so that heat acquisition is maintained naturally as well as heat radiation should not go through openings.
- It should be adequately sealed against air leakages, so that there would be minimum heat losses.
- It must be made up of appropriate materials & combinations of materials, which should have adequate properties of absorption, admittance, storage, etc.

The addition of above mentioned component in the active houses boosted the overall performance of the system and thus reduces cost of cooling load in summer. Therefore, in the designing of active houses, proper planning is made in construction of house apart from its general parts.

Apart from above essential components of active solar heating systems, there are many other additional components that are required as per climatic and operating conditions, these are:

- A heat exchanger is required in the collector and storage unit, this required when the fluid in collector unit and storage unit is different.
- Draindown type of collector, essentially for liquid/water collector in the places where chances of water freezing in there in the night.
- Multilayered and multiplex Storage System, Storing the heat at different temperature for different applications and for different periods.
- A device for exhausting the surplus heat for preventing damage against boiling of water.

9.3 Types of Active Solar Houses

Basically, there are three type of active solar house, which are as follows:

- Active Solar House equipped with Air Collector.
- Active Solar House equipped with Liquid Collector.
- Active Solar House with Heat Pumps Mechanism.

All these three technologies are widely tried and are in use for space heating. However, every system has got its own advantage and disadvantage, therefore, depending upon its technical feasibility, economically viability and efficiency, one has to adopt an appropriate type of system. The right choice of an active system depends in general, on the following factors:-

(1) The expected time of heating and expected time of solar energy availability on the place.

(2) The geographical locations of the place.

(3) The nature of the load to be connected i.e. complete house heating, partial house heating and single room heating.

(4) The degree of reliability needed for the process.

(5) Availability of scope of auxiliary space heater & the manner in which it would supplied heat.

(6) An economic analysis that determines how much of the total usually, annual loads should be carried be solar and how much by auxiliary energy heater.

(7) The temperature requirement in the space i.e. temperature at which heat would be available and used for space heating.

(8) Temperature stratification, if any

(9) The means of controlling & their costs.

9.4 Solar Air System for Active Approach

The solar air system for active heating of building essentially consisted of a solar air collector (for heating air), a storage unit (generally rock bed type), auxiliary system (conventional energy based air heater), automatic air dampers (for directing air flow), air handling duct, blowers and pumps, control unit with necessary sensors, etc. The working of this system includes heating of air through solar collectors and storing hot air in the rock bed type storage and then incorporating a suitable heat distribution system for heating the space as per requirement in following modes:

a) Heating the space directly from the hot air supplied through solar air collectors during day hours.

b) Heating the space from the storage unit during late hours or when sunlight is not available.

c) Heating the space partly through direct contact of solar air collector and rest of heat through auxiliary air heater during day hours.

d) Heating the space partly through storage heat and rest from the auxiliary air heater, when sunlight is not available.

e) Heating the space from the complete stand by mechanisms operated through auxiliary air heater only.

The airbased system offers many advantages as compared to liquid based system there are summarized as follows:

a) Only one medium (preferably air) may be used for collecting, storing and transferring the heat into space, this would yield higher overall heating efficiency.

b) Air systems are more durable, convenient and reliable. The efficiency of air system is also better as compared to other working medium.

c) The combination of thermal & operating parameters of air like specific heat, flow rates, density, low inlet temperature can yield higher collector outlet temperature.

d) The problem of corrosion is eliminated in air heating system.

e) Air collectors are more economic and air leakage is not serious.

f) The mechanism involved in control unit is simple and it is readily available.

g) There are no freezing and boiling problems like waterheating system.

Though most of space heating systems uses hot air for operation, but along with its merit, the air heating system involves few disadvantages, which are as follows:

a) It involves relatively high pumping cost, because large volume of air is required in heating.

b) Since large volume of storage system is required, therefore relatively large volume of storage system is required. Hence more cost involvement as compared to other system of heating.

c) The combining of solar air heating system with auxiliary air heater (preferably air conditioning system) is quite difficult.

Various types of solar air heater are used in active heating system of building, few commonly used are as follows:

- Simple conventional air heater.
- Vcorrugated type air heater
- Finned type air heater
- Overlapped glass plate air heater
- Matrix type air heater
- Porous bed air heater.

9.5 Solar Liquid System for Active Approach

The solar liquid system consists of heating the liquid through solar collectors. The mode of operation of solar liquid system is similar to solar air system. The liquid heating system consists of liquid flat plate collectors, storage systems, and auxiliary system, radiant or connective panels, heat exchangers & other control units. As compared to air heating system, this liquid heating system is having atleast three heat exchangers, one in the collection unit, second in the storage unit and third in room, where heating is done. The water heating system has few advantages as compared to air heating system, such as:

- The water heating systems are widely used and can readily supplied combined space heating and cooling mechanism.
- As an added advantage, this system can supply hot water for domestic applications.
- The liquid collectors operate at higher efficiencies at same collector inlet temperature and can provide domestic hot water efficiently compared to air collectors.
- It requires less space as compared to air collectors.
- The disadvantages in liquid system are:
- Freezing & boiling problem in collector water.
- Corrosion problem because of water circulation.
- Hazards due to leakage.
- Higher cost due to additional heat exchangers, leak proof joints & corrosion resistant metals etc.
- Law overall efficiency because of more number of heat exchangers.
- Comparatively low durability.

The general operation, mode of action, size of storage unit and heat distribution system is more or less similar to air heating system and it depends on the system heating mechanism and on the system heating load and geographical locations. There is various types of liquid collectors, some of them are:

- Corrugated sheet type flatplate collector
- Rollbond type of flatplate collector.
- Tubein plate type of flatplate collector
- Water trickle type collector.
- Evacuated tube type collector.

9.6 Solar Heat Pump Type Active Houses

The heat pump technique is based on mechanical engineering gadget, which provides heating or cooling by using a reversible refrigeration cycle. Essentially a simple heat pump consists of following components (Fig. 9.2)

(1) Compressor

(2) Condenser

(3) Evaporator

(4) Working fluid (Freon)

(5) Expander

Basically, the heat pump in heating mode attracts heat at low temperature from outside air and rejects heat at higher temperature to the room air. Where as in cooling mode, a reversing value reverse the roles of evaporator and condenser thus heat is extracted from indoor air providing cooling and rejected to the outside air. In case of solar assisted heat pump, the coefficient of performance increases because the source temperature increases.

Fig. 9.2: Operation of Heat Pump

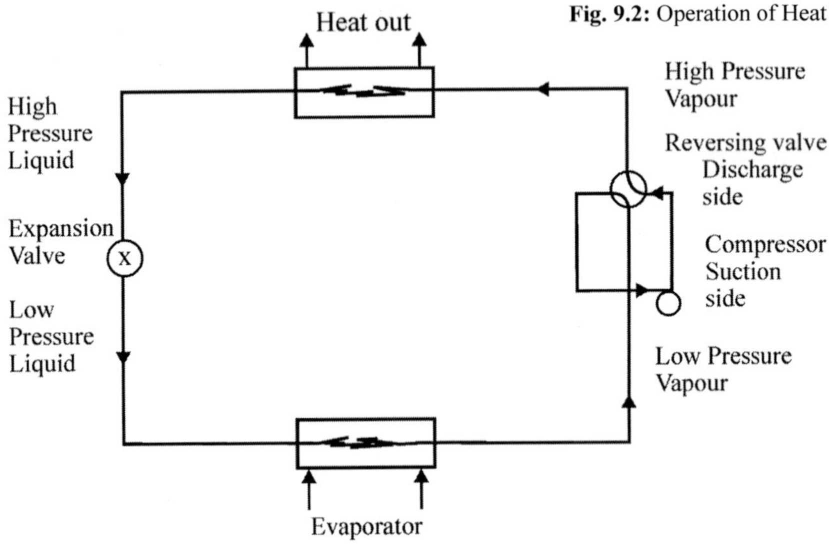

In the solar assisted heat pump the working fluid is in the gaseous state, which is compressed and send to the condenser, where it gets cooled to liquify & releases large amount of heat to the surrounding. Then, through expansion value, this liquified working fluid gets vaporized. The evaporation in the evaporator requires large amount of heat, which is drawn from the surroundings. This large amount of low temperature heat is supplied to the evaporator by solar heated air or water. In this system by using appropriate value and controls the heating in winter and cooling in summer can be provided by the same unit. The solar assisted heat pump can be of following types:

- Air to Air type.
- Air to Water type
- Air to Earth type
- Placement of Evaporator & Condenser respectively.

9.7 Solar Active Cooling Techniques

The "active" features are usually incorporated, since they make possible to control the internal climate & heat distribution more precisely. The active solar systems use solar panels for heat collection and electrically driven pumps or fans to transport heat/cold to the living area or to the storage. Electronic devices regulate the collection, storage and distribution of heat within the system.

Active solar cooling is defined as utilization of sun's energy to help offset net cooling load of space conditioning and refrigeration of a building. Many options are available in this direction, which include absorption cycle, desiccant and solar driven "heat engine" cooling etc. However, some mechanical assistance is also required for operating the system.

The first attempt for coupling solar energy to the building was made by Abel Pifre in 1872. He has used steam from a solar heated boiler to operate an absorption cooler. However, in 1950 the technology was perfected to certain extent and wide spread interest in this concept began. One of the earliest solar regenerated desiccant systems was built, tested and reported by Lofin in 1955. The important feature of this concept was the storage of energy in the form of chemical energy using a concentrated solution thereby reducing or eliminating the need for conventional thermal storage. In the 1960's, solar powered cooling system for space conditioning and refrigeration purpose was successfully demonstrated.

There is also wide spread interest in desiccant cooling because desiccants can be combined with direct or indirect evaporative coolers to eliminate all refrigerants in the cooling cycle. As in the present context, everyone is aware of global warming and chlorofluorocarbon issue, therefore it is believe that desiccants will become more popular, as regeneration of the desiccants can be accomplished with typical flat-plate collector and neither desiccant nor absorption cooling systems use CFCS. It also eliminates chances of carbon dioxide emission, which contribute to global warming.

Since cooling is required in the summer when the sun is shining, or when heat is available in the atmosphere. Because of this advantage, the solar refrigeration (preservation of products) and air conditioning (cooling the space) required a significant position in present context.

The Refrigeration and Air conditioning using solar energy can be achieved by one of the following means:

- Vapour compression systems using heat engines or photovoltaic system driven electric motors.
- Vapour absorption system using liquid absorbents such as LiBa B H_2O, H_2O, NH_3, LiCl - H_2O, NH_3 - $LiNo_3$, R22 - DMF, NH_3 - NaSCN, or using solid absorbents such or $CaCl_2$ - NH_3, silicagel H_2O, Zeolites-H_2O.

- Vapour Jet System
- Evaporative cooling system
- Thermoelectric cooling system.

A solar air conditioning system consists of following components:

- Field of solar collectors (simple flat plate collector or evacuated rube collector or concentrating collector) depending on the temperature requirement.
- A heat storage device, which gets heat from solar collectors and is used to operate the cooling device.
- A solar cooling device (based on Absorption or Rankine Cycle).
- A cold storage device.
- A heat rejection device.
- Air handling system.

Operation wise, the field of solar collector heat the heat transfer fluid, which is used to operate the cooling device. A part of the heat can be stored in the storage unit. The heat collected from the building is rejected to the atmosphere using a tower or any other suitable heatrejecting device. If air is cooled by the cooling device then it is directly supplied to the building to be cooled or if chilled water is produced then it is circulated through fan coil units and a part of a chilled water is stored for use when the cooling device is not in operation. The schematic of solar cooling system is shown in Fig. 9.3.

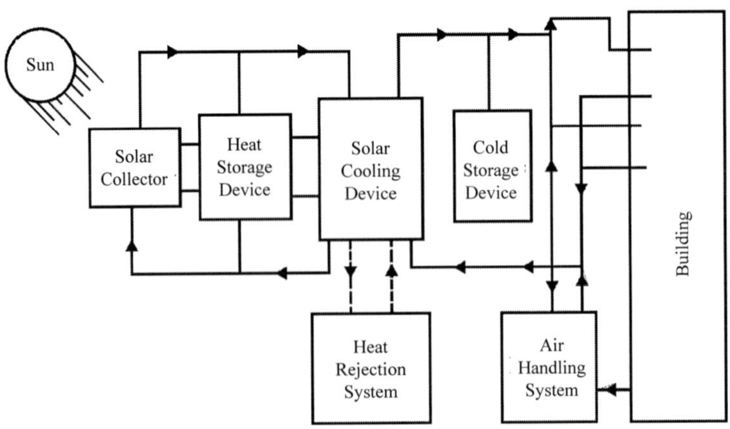

Fig. 9.3: Solar Cooling System

9.8 Evaporative Cooling

Evaporative cooling is a reliable method to provide cooler air during hot weather. It is a process that uses the effect of evaporation as a natural heat sink. Sensible heat from the air is absorbed to be used as latent heat necessary to evaporate water. The amount of sensible heat absorbed depends on the amount of water that can be evaporated. Evaporative cooling can be direct or indirect; passive or hybrid.

In direct evaporative cooling, the water content of the cooled air increases because air is in contact with the evaporated water. In indirect evaporative cooling, evaporation occurs inside a heat exchanger and the water content of the cooled air remains unchanged. Since high evaporation rates might increase relative humidity and create discomfort, direct evaporative cooling can be applied only in places where relative humidity is very low. Whereas, when it occurs naturally it is called passive evaporation. A space can be cooled by passive evaporation where there are surfaces of still or flowing water, such as basins or fountains. Where evaporation has to be controlled by means of some mechanical device, the system is called a hybrid evaporative system.

Example 9.1: Determine the degree of cooling if 1% water is being evaporated in evaporating cooling system.

Solution

Assuming that initial mass of water is – M_w

Amount of water evaporated – M_{we}

Specific heat of water – C_p

Degree of cooling - Δt

Latent heat of water – 2260 kJ/kg

Energy Balance

$(M_w - M_{we})\, C_p\, \Delta t = \lambda M_{we}$

$$\Delta t = \left(\frac{M_{we}}{M_w - M_{we}}\right) \times \frac{\lambda}{C_p}$$

$$\Delta t = \left(\frac{1}{100}\right) \times \frac{2260}{4.18}$$

$\Delta t = 5.4\ ^{o}C$

9.9 Solar Driven Vapour Compression System

The conventional Vapour Compression System (VCS) can by converting solar energy into mechanical energy through one of the following means.

9.9.1 Heat Engines

It convert heat energy into mechanical energy or work of air conditioning. The operation of heat engine can be governed through various thermodynamic cycles such as Rankine cycle, Brayton cycle and Stirling cycle. In order to get low temperature Rankine engine is most appropriate. The working of Rankine engine is similar to the steam engine or engine using organic vapour. The use of water vapour (steam) driven Rankine cycle based heat engine is used for high temperature range above 250°C, whereas some type of Ranking cycle based heat engine consistent with flat plate or line focussing collectors of low concentration ratios is suitable for low to medium temperature applications (between 90° to 150°C). The Stirling cycle and Brayton cycle based heat engine driven vapour compression type of air conditioning unit is suitable for high temperature application. The operation of heat engine based on different thermodynamic system is shown in Fig. 9.4.

Fig. 9.4: Solar Driven Vapour Compression System

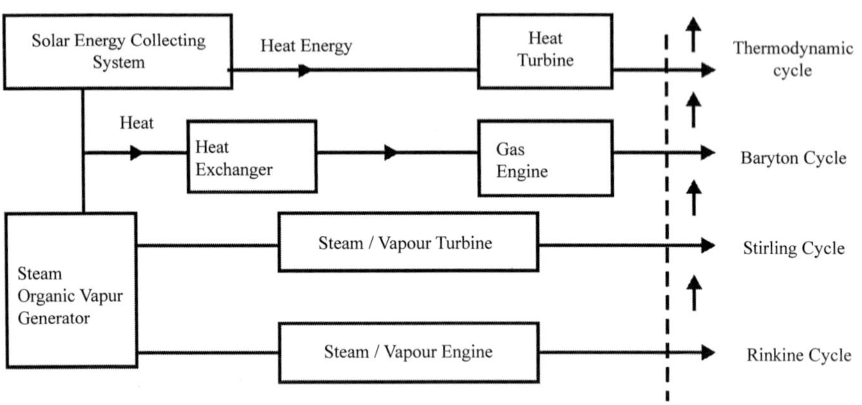

The Solar Rankine vapour compression system based air conditioning systems are available with different capacities. The overall coefficient of performance (which is the ration of the amount of cooling to the energy input) for a Rankine cycle operated solar cooling system is of about 0.3 to 0.4. The COP of Rankine cycle depends on the efficiency of solar collecting field. The solar Rankine vapor compression cooling process can be used in the heat pump mode also, and for electricity generation as well when cooling is not required. However, controls of the system need attention during variable solar insolation.

9.9.2 Solar photovoltaic based electric motor operated vapour compression system

The vapour compression cooling process operated by photovoltaic panels gives a coefficient of performance in the range of 0.25 to 0.35. It may be because solar cells have lower efficiency. This system can also be used in the heat pump mode and for electricity generation when cooling is not required. However, it is very costly system, it is due to high cost of solar cell.

9.9.3 Solar Operated Vapour Absorption System (VAS)

This type of system requires very little quantity of solar energy as compared to vapour compression system. Therefore, it is a prospective device for exploiting solar thermal energy for producing cold. Essentially it includes two working fluids i.e. refrigerant and absorbent. The selection of absorbent and refrigerant is so made that the absorbent has a high affinity for the refrigerant. Essentially absorbent B refrigerant solution is liquid. In this type of system, solar energy is used for heating a strong solution, which is rich in refrigerant. The heating is through solar collector and such that the vapour pressure equals the saturation pressure in the condenser. The refrigerant in vapour form goes to the condenser, while the weak solution returns to the absorber through a recuperator and a throttling value. In the condenser the refrigerant gets condensed rejecting heat and comes in the liquid form at high pressure. The refrigerant now passes through the expansion value and evaporates in the evaporator, thereby taking heat from the surrounding. Thus the cold air goes out from the evaporator and cools the space. After that the low vapour pressure refrigerant from the evaporator goes to the absorber where it is reabsorbed with the liberation of heat. The low pressure, rich reabsorbed refrigerant solution is now pumped to the generator at highpressure to complete the cycle. In this system the evaporator and absorber of the system are in the lowpressure side and the generator and condenser are in the highpressure side of the system. A recuperator heat exchanger is also used to transfer heat between the solutions passing between the absorber and the generator.

The operating principle of absorption cooling in shown in Fig. 9.5.

The vapour absorption system suitable for solar energy operation can be classified as follows:

- Intermittent Cycle
- Continuous Cycle - Closed and Open type

The coefficient of performance of the cooling system based on absorption closed cycle is 0.1 to 0.2 depending on the collector efficiency.

Intermittent vapour system especially developed for operation with solar energy take into account the intermittence due to daily variation of solar radiation. In this system, ammonia as refrigerant and water or NaSCN as absorbent and Freon 22 as refrigerant and DMF as absorbent can be used.

This system yields low coefficient of performance because high heat losses are associated with alternately cooling and heating of the unit. As a result of it large physical size of the system is required for a given capacity. Further, as the nature of this system concerned, it is not capable for producing cooling on continuous basis.

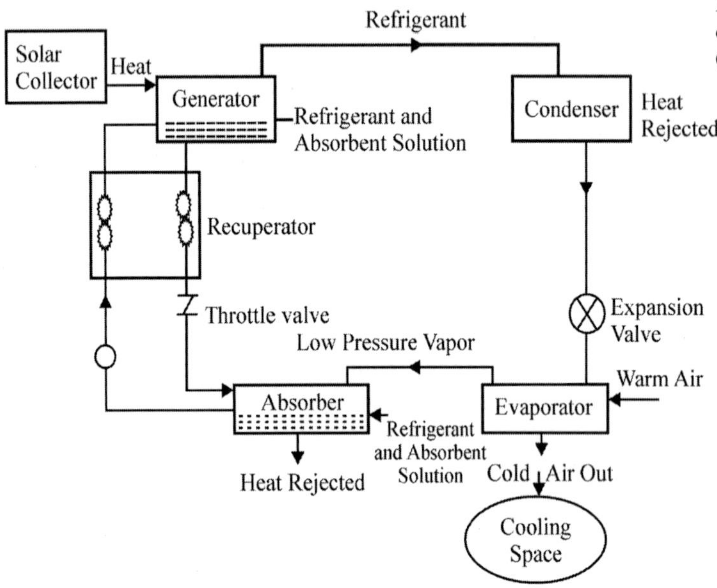

Fig. 9.5: Principle of Absorption Cooling System

However, the intermittent VAS is simple in design and construction and it does not require refrigerant pump. The operation of this system is shown in Fig. 9.6.

In a continuous vapour absorption system, heat can be recovered continuously by using a regenerative heat exchanger to transfer heat from the hot weak solution to the cold strong solution. In case of closed cycle continuous vapour absorption system, method of supplying heat of generator. In such integrated collector generator systems, it will be necessary to have storage reservoirs for weak solution, strong solution & the liquid refrigerant.

The open cycle continuous VAS is similar in operation to the closed cycle except that the weak solution is regenerated by losing the refrigerant to the atmosphere in place of the refrigerant being recovered in a condenser.

Fig. 9.6: Principle of Intermittent VAS

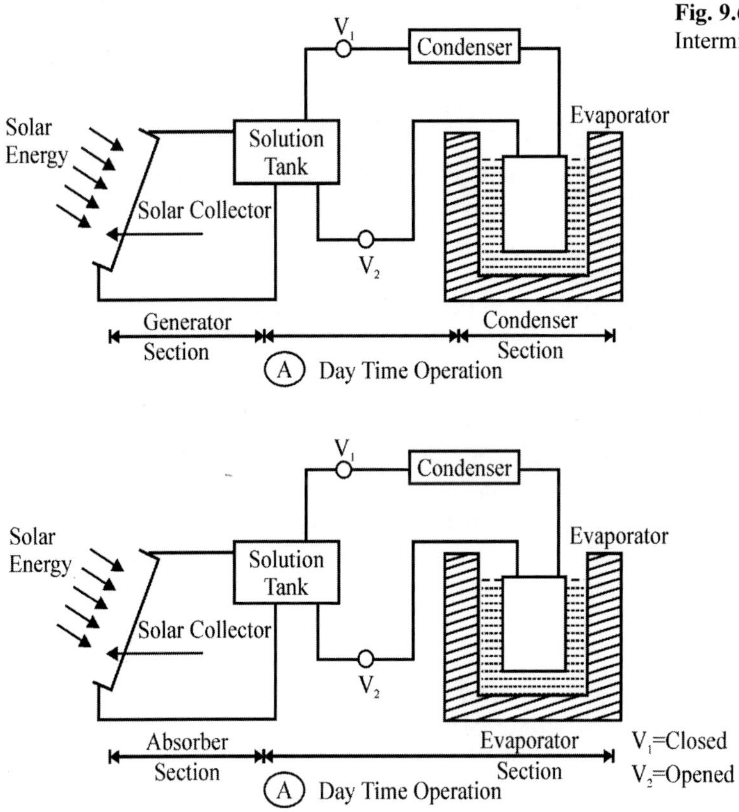

9.9.4 Desiccant Cooling

Dehumidification is also a method of cooling. In fact, dehumidification is a method of removal of moisture from the room air. This can be carried by using absorbent followed by evaporative cooling of air. This is very promising way of active cooling for hot and humid climate. There are few liquid & solid materials, which have property of attracting and holding water vapour known as desiccants. These materials can be used for dehumidify the air and can thus be used for cooling the building. These desiccant may be absorbents or adsorbents type. The desirable characteristics of desiccant are:

- They should be thermally and chemically stable.
- They should be nontoxic, nonflammable, odorless & noncorrosive.
- Their cost should be minimum, as far as possible also they should be readily available.

Both liquid and solid desiccant can be used in open as more as in closed cycle.

9.10 Passive Solar Architecture Heating

In a passive solar heating system, the solar energy collection, storage and distribution is done by natural means and for this no external means either electrical or mechanical means are used. In designing and constructing buildings in order to take advantage of the Sun's energy and the subsounding environment. The solar passive building itself collects stores and uses solar energy and is designed to take advantage of site specific features for natural heating, cooling and lighting.

In this system, various building elements such as walls, roof, windows, partition etc. are so architecturally selected and arranged that they contribute effectively in solar energy collection, storage, transportation and distribution of thermal energy as per requirement. In passive solar house, heating is done by natural onsite energy sources such as solar radiation, outside air and internal metabolism, while cooling is meet out through sky and space, outside air and wet surface. Both direct as well as diffuse solar radiation is used for heating the house. Apart from this, building elements such as construction materials like stone, bricks, concrete, water insulation, glazing, shading, reflectors etc. and various thermal processes such as thermal radiation, natural & forced convection, conduction, air stratification, evaporation, themosyphoning etc. are combined in various ways to get desired heating or cooling.

The passive house concept was use around 2500 years ago by the ancient Greeks. They were using this concept to reduce the requirement of firewood for space heating in winter and to cool their houses on hot summer days. The buildings of that time were having open, southfacing porches, which permitted winter sunshine into the main living rooms. Further, they were also using shade during the hottest parts of the summer. They were using dark stone floors and thick masonry, which absorbed the solar heat within the building and released gradually in the evening when outside temperature reduces. Heavy walls at the rear sheltered buildings from cold northern winds and low walls at the front and sides cut down draughts. Greeks built high-density houses and big cities on solar heating principles, typical examples of these cities are Olynthus and Priene.

The Romans also used glazing (glass surface) to retain the heat for longer duration and to warm the houses. One of typical example was the Pueblo Indian

City of Acoma in North America, which had three extended terraces that ran east to west and were built in tiers to make the most of the winter sun. In order to protect against full blaze of the summer sun, roof of each tier was layered with and other material to insulate the rooms.

In ancient Colorado, the Mesa Verde people have also used solar housing heating and cooling concept. They built their houses under an overhanging cliff so that during winter, when sun is low in the sky, sunlight could enter the house and heating takes place. Further, in summer, the buildings were shaded by the overhang and so stayed cool.

In India, old forts built by various kings reflect in built solar passive architectural concepts. The Fatehpur Sikri and the Red Fort are excellent example for this. In the Deegh Palace near Bharatpur in Rajasthan, summer coolness was achieved by sprinkling water on the walls and winter warmness was achieved by covering the walls with quills. Our other historical buildings are standing example of solar houses, which made use of architectural elements like Screens, Water sprinklers, Ventilators, Skylights, Chowks, Verandas, Windows, Judicious use of Space, building design & materials for optimum temperature comfort levels etc. Different climates have produced traditional architectural styles, which are suitable to local climatic conditions. For examples, Compact massiveness for hot arid regions with cold harsh winters and double roofs in desert areas that have dependable sunshine plus wind for convective cooling.

9.11 Solar Passive Architecture - Conceptual Details

Solar passive architecture incorporates several features such as shape and orientation of building, shading devices and use of appropriate building materials for conserving energy used in heating, cooling & interior lighting of building considering prevailing solar radiation conditions in the area.

There are few important points that are kept in view before drawing up building plans. Orientation of building in right direction to get the maximum benefit of natural ventilation and lighting as it reduces demand for artificial lighting and also provides natural cooling & heating is one of important consideration for proper building plans. The proper orientation of building must include following points:

- The living room, the bedrooms and the kitchen must be windy and naturally ventilated.
- The wind must blow into the living room from the south/southwest direction through the entrance door, if possible.
- A French door (high door) may be oriented in the east and designed to allow sunshine into the room in winter but not in the summer. Thus the

design and position of door must takes into account the movement of the sun in all seasons.

- Care must be taken to see that neither the living room nor any of the bedrooms have a wall that directly faces the west.
- The roof ceiling is also at a sufficient height in the living room so there is enough space for the hit air to rise and the living area near the floor remains comparatively cooler.
- A light well may be incorporated in the house, which keeps the home fresh and cool and it is a source of natural light that keeps the inside of the building bright throughout the day. If possible, it should be provided right at the building bright throughout the day. If possible, it should be provided right at the centre of the house so that when doors of the living room are opened, cool air from the outside rushes in, driving out the hot air within the house though this light well, which is provided with lowers at the top.
- It is not enough to improve the insulating value of the house. This saves thermal energy but will increase the electrical energy load for artificial light, if it involves an inadequate area of glazing. A proper balance is to made between the benefit of natural lighting and the loss of heat through windows.
- Consideration must be given to the profile of the building, the location of the windows and the depth of the usable floor area in relation to the line of the windows, so that the utmost benefit can be obtained from natural lighting.
- The building envelope must be tight, so that the leakage of unwanted air from outside can be eliminated. The windows are placed in such a manner that they must be open able in order to give relief to the occupants by the provision of adequate natural ventilation.
- Building mass should be properly deployed so that it can be useful for storing unwanted solar heat during peak hours for use at a later time when they can offset a heat loss.
- The location of thermal insulation in a building is quite important when it is placed on the inside of the house it minimizes the effect of the building mass and when placed on the exterior it increases the effect.
- If possible, a basement is provided under the living room. Three fourths of the basement are underground and one side should be partially exposed to get fresh air. This keep the basement cools and ventilated. As there are enough natural lights and the earth embankment on the three

sides keeps it cool, the basement is a favourite place to work or relax on hot summer afternoons.

- External shading should be arranged to minimize unwanted solar gains in summer.
- The material chosen for building construction should be reviewed in relation to the energy used in their manufacture. As an example, reinforced concrete beams appear to need less energy for their manufacture, transport & erection on site than do steel beams. Similarly, brick seems to be more economical than concrete, in terms of energy & timber windows frames better than aluminum.
- Growing ample greenery around a concrete structure (house) is also a passive solar feature. Green plants absorb heat and offer shade to the walls and the roof from the direct assault of the sunrays. Plants and trees, because of their dark colour, large structure area and evaporative cooling, don't reflect heat towards a building nor store heat for later reradiation to the building.

Thus a complete passive solar heating system includes collection of energy through south facing glass, storage of energy through the use of thermal mass (usually in the form of concrete, brick, water or phase change material), regulation of energy through provision of overhangs, shades or other insulating material for windows or providing a door to close off a sun space at night and distribution of energy, which usually consists of vents, dampers, duct work and small fans. Therefore, the design of houses according to passive approach requires a detailed understanding of the complex interrelationship between architectural textures, human behaviour and climatic factors. By arranging local features such as vegetation, topography & solar exposure, the climate in the house can be favourably altered to improve thermal comfort. Apart from heating and cooling, there are many other factors, which need to be considered for one comfort levels These are relative humidity, air flow and mean radiant temperature (temperature radiated from objects in the living area). If the humidity in room is low, the air movement high and the walls & floors of the room become cold in nature & as a result of it we feel cool inside the room even when actually the temperature is high. But if the humidity is high, air movement low and radiant temperatures high and as a result of it, inside of room become warmer, even if the temperature is low.

9.12 Types of Passive solar heating systems

The passive solar house are classified as:

- Direct Gain Type
- Indirect Gain Type (Thermal Storage wall and thermal Storage roof)
- Isolated type Gain System (Attached Sunspace type & convective loop type).

9.12.1 Direct Gain Type

This is one of the most common type approaches, which include capturing the sun's rays directly by using large south facing window and glazed surfaces on the walls and on the roof.

The placement of glazing over the window facing south or the entire southfacing wall permits solar radiation entry inside the room. Almost all the solar radiation enters in the room is converted into heat energy (Fig. 9.7).

Fig. 9.7: Direct Gain Type

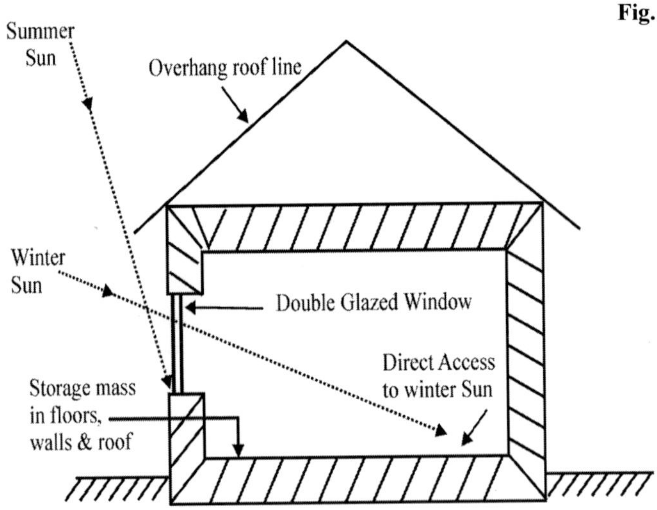

The heat loss from the room is reduced by using a double glazed window. An appropriate overhang above the window is provide which only permit winter sun to enter, whereas summer sun is not allowed to enter inside the room. The seasonal variation reflects the position of sun, during summer the elevation of the sun is high (Fig.9.8).

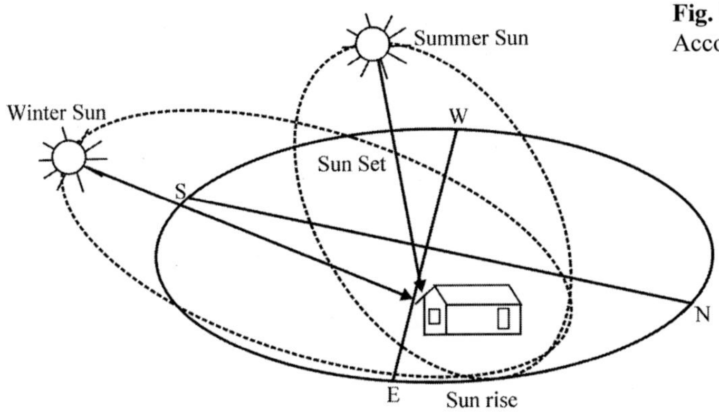

Fig. 9.8: Position of Sun According to Season

A sufficient thermal mass in the form of massive floor area and wall is required, which absorb sunlight during day hours and reflect heat during night hours. Sometimes, insulation, over the window is also required which is used during night, when heat loss is more than heat gain through windows.

9.12.2 Indirect gain type

Idea behind this type of approach is similar to that direct gain type solar houses, except that heat is collected in some special architectural feature such as thermal mass is placed between the glazed solar collection area and the interior space and thus solar radiation is blocked from entering directly into the room to be heated. Here the heat is transferred indirectly from the storage mass by natural convection and radiation. The heat transfers from the storage mass either wall or roof is continuing as long as its temperature is above that of the inside temperature of the rooms.

Indirect gain type solar houses provide effective way of heating the houses and reducing the large oscillations in the room air temperature. Essentially thermal storage wall of indirect gain type solar house may be constructed of masonry or concrete materials or formed by the use of water filled containers. The masonry or concrete wall is known as Trombe wall whereas water filled containers type of wall is known as water wall. The schematic of Trombe wall and water wall is given in Fig. 9.9.

Operation wise, when solar radiation passes through the glazing and strikes the wall and as a result of it heat is absorbed by the dark exterior of the wall. After certain duration, temperature of wall rise and than heat begins to flow into the interior of the building. This time period is called time lag of the wall. Which depends on the thickness and the material of the thermal storage wall. Also, there is air gap between the wall and glazing surface. In this air gap, hot air moves from bottom to top generally due to natural convection. This heated air goes in the room through the upper vent while the cool air from room through the bottom sent enters in the gap. This circulation is continuing till the wall goes on heating the air. In case of peak summer, when the temperature of air inside the air gap increases to large extent, in this case upper dampers provided at the top of glazing surface, opened out ward and thereby excess heat can be removed.

Such a thermal storage wall made of masonry and concrete is known as Trombe wall, while the thermal storage wall made up of drums or barrels of water is known as water wall. The water wall, which consists of barrels of water, which, stacked over each other, collects, stores and distributes the heat into the room. An insulating sheet is also provided on the glazing surface, which can be opened and closed as per requirement of the heat entry inside the room. During sunshine hour, the insulating layer is removed, which allows solar energy to enter inside the room, while during off sunshine hours, insulating layer is placed over the glazing to prevent back flow of heat from the room to outside.

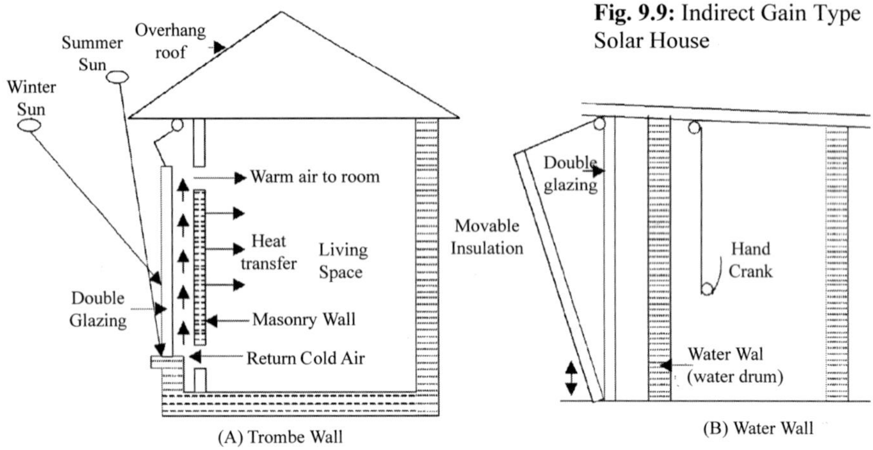

Fig. 9.9: Indirect Gain Type Solar House

Thermal storage roof is also reflecting indirect gain type of system. The basic difference between thermal storage wall and thermal storage roof is that in the storage roof case the interposed thermal storage mass is on the building roof instead of a wall generally, a metal roof is used for conducting heat effectively. Over the roof, water bags made of transparent or black plastic sheet filled with water or any other massive materials are placed, which used to store heat during day hours and heats the room below it during day hours as well as during night hours or when ever heat is required. A provision of movable insulating shutters over the water bags is also provided, which direct heat supply to room, whenever it is required.

9.12.3 Isolated gain type solar house

This is a system of passive house in which direct as well as indirect gain concept is integrated to get full advantage of passive technology. It may be attached greenhouse type or convective loop type. In case of attached greenhouse type a additional sunspace is provided between the interior space (where heating is done) and the solar heating glazing surface. This sunspace is on the extreme south side, which can be used as green house for raising vegetables or flowers. It has been observed that this attached sunspace has large air temperature swing.

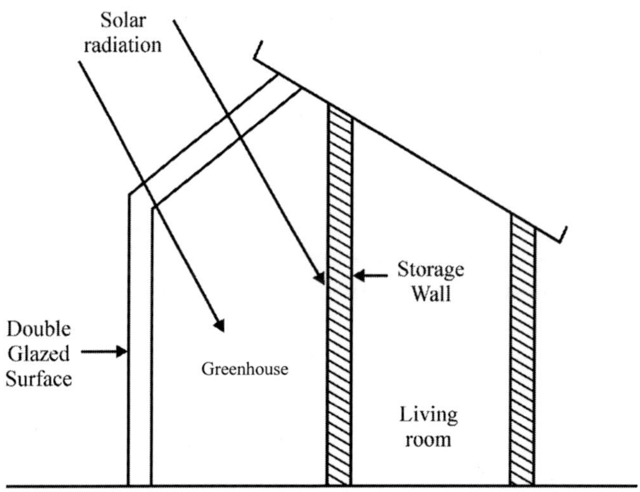

Fig. 9.10: Isolated Type Solar Passive House

The temperature is quite high during summer. The heat gain in the green house is imparted to the thermal storage wall, which is in between room (where heating is required) and green house. The heating is conveyed through radiation and convection made inside the living room (Fig. 9.10).

Convective loop isolated type solar house is similar to the general characteristics of active system, in which a separate solar collector & thermal storage bed is provided. In this system, air after getting heated through solar air collector goes directly into the living room for heating or through a rock bed storage unit for heating the storage unit when heat is required vents are opened to the thermal storage to allow warm air to flow into the room by convection (Fig. 9.11).

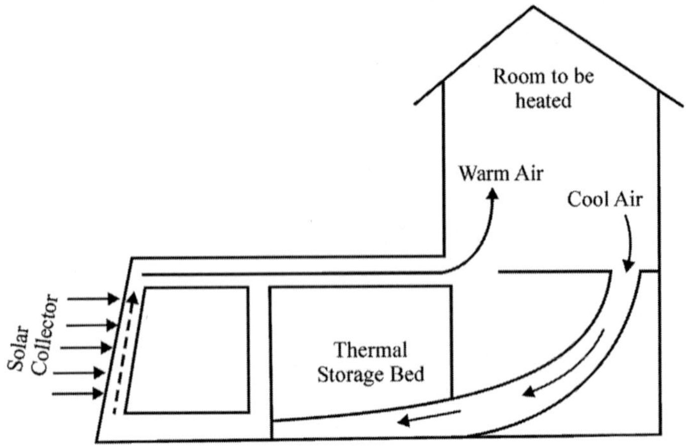

Fig. 9.11: Convective Type Solar Passive House

9.13 Natural lighting

Natural lighting is nothing but it uses sunlight and diffuse radiation from the sky to provide light inside buildings. It is most important for building. It works when the power grid goes down. Interest in natural light has significantly increased in the last few years as the many benefits of natural light have become clearer. Several studies have proved that natural light can improve moods, spirit, work performance, and human health etc.

9.13.1 Visible Light

Normally, human eye perceives a portion of the electromagnetic spectrum from violet to red light from about 380 to 750 nm. For the most part, natural lighting is concerned with the visible spectrum.

9.13.2 Healthy Light

Natural light is necessary for humans. In higher latitudes, short winter days are known to cause seasonal affective disorder (SAD). Seasonal affective disorder (SAD) is a type of depression that's related to changes in seasons — SAD begins and ends at about the same times every year. SAD can be treated with early-morning exposure to light levels of 2,000 to 12,000 lux, depending on the severity of symptoms. Buildings with poor natural light can cause SAD even in the summer. Therefore, desirable sunlight to be allow in the building which essential for human body to produce vitamin D.

9.13.3 Sources of Natural Light

The sun is the ultimate source of all natural light. However, only direct sunlight is used for natural lighting. The proportion of direct sunlight depends on building orientation and design, variation in ground surface, and climatic conditions.

9.14 Human Thermal Comfort

Energy balance on human body

Buildings account for nearly 40% of global energy consumption and about 40% and 15% of that are consumed in HVAC and lighting systems respectively. Comfort Indoor thermal environment is essential for human being as 90% of their time spend inside the building. Such comfort improves the overall productivity, and. satisfaction level. Indoor thermal comfort usually refers to occupants' feelings of room comfort (Zhao et al. 2017). To ensure the thermal comfort of human body, the heat produced in the body should be equal to heat rejected by the body to its environment. Fanger (1977) developed a model to establish a steady state energy balance on the human body and it is as follow:

$$Q_{meta} = \pm Q_{skin} \pm Q_{resp}$$

$$= (\pm Q_C \pm Q_R \pm Q_E)_{skin} + (\pm Q_C \pm Q_E)_{resp}$$

$$Q_{meta} = [h_c A_{skin} (t_{skin} - t_{amb}) + h_{rad} A_{rad} (t_{skin} - t_{amb}) + h_{evp} A_{wet} (\varphi_{skin,Sat} - \varphi_{air,sat}) h_{fg,skin}]_{Skin} + [m_{resp} C_{pa} (t_{ex} - t_{amb}) + m_{resp} h_{fg,b} (\varphi_{ex,Sat} - \varphi_{air,sat})_{resp}]$$

Where

Q_{meta} is the net rate of heat production through metabolism

Q_{skin} is the heat loss (+) or gain (-) through skin

Q_{resp} is the heat loss (+) or gain (-) through due to respiration

t is temperature

φ is hunidity ratio

A is area

Subscript C designates convection, R radiation, and E evaporation, ex – exhaled air

$$Q_{sen} = Q_C + Q_R$$

$$= h_C \left(t_{cloth} - t_{amb}\right) + h_R \left(t_{colth} - \overline{t}_r\right) A_{cloth}$$

Combining heat transfer coefficient

$$h = h_C + h_R$$

considering operating temperature t_o

$$t_o = \frac{h_C t + h_R \overline{t}_r}{h}$$

Considering convective and radiative heat transfer coefficients are equal

$$Q_{sen} = h(t_{cloth} - t_o)\, A_{cloth}$$

Heat transfer equation by eliminating cloth surface temperature

$$Q_{sen} = \frac{A_{skin}\left(t_{skin} - t_o\right)}{R_{cloth} + \dfrac{A_{skin}}{hA_{cloth}}}$$

Similarly, total heat transfer from the skin

$$Q_{lat} = \frac{A_{wet}\left(\varphi_{skin,\,Sat} - \varphi_{air,sat}\right)h_{fg,skin}}{R_{evp,\,cloth} + \dfrac{A_{wet}}{h_{evp}A_{cloth,wet}}}$$

$$Q_{meta} = \frac{A_{skin}\left(t_{skin} - t_o\right)}{R_{cloth} + \dfrac{A_{skin}}{hA_{cloth}}} + \frac{A_{wet}\left(\varphi_{skin,\,Sat} - \varphi_{air,sat}\right)h_{fg,skin}}{R_{evp,\,cloth} + \dfrac{A_{wet}}{h_{evp}A_{cloth,wet}}}$$

$$+m_{resp}\left[C_{pa}\left(t_{ex} - t_{amb}\right) + h_{fg,b}\left(\varphi_{ex,\,Sat} - \varphi_{air,sat}\right)\right]$$

Average convective coefficient for seated people in still air Mitchell (1974) is

$h_C = 3.1$ (W)/(m^2 °C) $(0 \le V \le 0.2)$

$h_C = 8.3\, V^{0.6}$ W/m^2 °C) $(0 \le V \le 0.2)$

Where V is velocity of air in m/s

Average convective coefficient for active people in still air (Gaggec1974) is

$h_C = 5.7(M-0.85)^{0.39}$ W/m^2 °C) $(1.1 \le M \le 3.0)$

Where M is the normalized metabolic rate in unit of met

The amount of metabolic heat dissipated during different human activity are listed in Table 9.1. The metabolic heat generation rate divided by skin surface area of an adult male i.e. DuBois surface area and it is equal to 1.8 m^2.

Table 9.1: Typical Metabolic Rates for Some Common Activities

Activity	Met - Metabolic Rate	
	W/m^2	Met Unit 1 Met = 58 W/m^2
Reclining, Sleeping	46	0.8
Seated relaxed	58	1.0
Standing at rest	70	1.2
Sedentary activity (office, dwelling, school, laboratory)	70	1.2
Car driving	80	1.4
Graphic profession - Book Binder	85	1.5
Standing, light activity (shopping, laboratory, light industry)	93	1.6
Domestic work -shaving, washing and dressing	100	1.7
Walking on the level, 2 km/h	110	1.9
Standing, medium activity (shop assistant, domestic work)	116	2.0
Building industry - Brick laying (Block of 15.3 kg)	125	2.2
Washing dishes standing	145	2.5
Domestic work - washing by hand and ironing (120-220 W)	170	2.9
Building industry - forming the mold	180	3.1
Walking on the level, 5 km/h	200	3.4
Forestry - cutting across the grain with a one-man power saw	205	3.5
Volleyball, Bicycling (15 km/h)	232	4.0
Agriculture - digging with a spade (24 lifts/min.)	380	6.5
Running 12 min/mile, Forestry - working with an axe (weight 2 kg. 33 blows/min.)	500	8.5
Sports - Running in 15 km/h	550	9.5

10

Wind Energy

10.1 Introduction

Winds are the motion of air around the Earth. This movement in air is caused by the uneven heating of the planet's surface by the sun. The idea of using wind as a form of power is not new. The traditional applications of wind were primarily as sources of kinetic energy for rural, agricultural and a limited number of industrial uses such as pumping water and grinding grain.

Presently wind energy can be used for two major applications, such as wind mills for pumping water for drinking as well as for irrigation purposes and second application is as aero-generator for electricity generation for domestic and industrial uses. In addition to this, presently wind energy battery charger are also available, which can store energy for lights, radio communication, hospital equipment and to power various emergencyrelated equipment.

The wind energy has an enormous resource, but the problems of utilizing the winds are many and varied. Wind energy is very diffuse in nature and local topographical features significantly alter the prevailing winds, thus leading to the extremely site specific nature of wind energy.

Scientists have estimated that as much as 10% of the world's electricity could be prepared by wind generators by the middle of 21 century. India has estimated wind power potential of 40,000 MW. India now, ranks fourth in the world in wind power generation.

The world's largest wind farms are in California (USA), where wind turbines can generate power up to about 1120 MW. The first wind mill used as source of electric power was built in Denmark in 1890.

10.2 Beaufort Wind Scale

The Beaufort Wind Scale is a scale that is used by seamen and coastal observers to estimate wind speed. The scale was created by British Rear Admiral Sir Francis Beaufort in the year 1805, and it was derived from his observations of sea conditions. The relation between wind speed and Beaufort scale is presented in Table 10.1

Table 10.1: Beaufort Wind Scale

Beaufort Number	Wind Speed at 10 m Height (m/s)	Description	Wind Turbine Effects	Land Effects	Sea Effects
0	0.0 – 0.4	Calm	-	Smoke rises vertically	Mirror smooth
1	0.4 – 1.8	Light	-	Smoke drifts, vanes unaffected	Small ripples
2	1.8 – 3.6	Light	-	Tree leaves move slightly	Definite waves
3	3.6 – 5.8	Light	Small size turbines start	Tree leaves in motion, flags extend	Occasional breaking crest
4	5.8 – 8.5	Moderate	Start up of electrical generation	Small braches moves	Larger wave, white crests common
5	8.5 -11.0	Fresh	Useful power generation at 1/3 of capacity	Small trees sway	Extensive white crests
6	11.0 - 14.0	Strong	Rated power range	Large braches move	Larger waves, foaming crests
7	14.0 – 17.0	Strong	Full capacity	Trees in motion	Foam breaks from crests
8	17.0 – 21.0	Gale	Shut down initiated	Walking difficult	Blown foam
9	21.0 – 25.0	Gale	All wind machines shut down	Slight structural damage	Extensive blown foam
10	25.0 – 29.0	Strong gale	Design criterion against damage	Trees uprooted; much structural damage	Large waves with long breaking crests
11	29.0 – 34.0	Strong gale	-	Widespread damage	-
12	>34.0	Hurricane	Serious damage	Disaster conditions	Ship hidden in wave trough

10.3 Wind Energy and Its Characteristics

The energy in the wind is the kinetic energy due to stream of air particles of density (ρ), moving with a velocity (V) through an area (A) swept by the rotor of diameter (D) (Fig. 10.1).

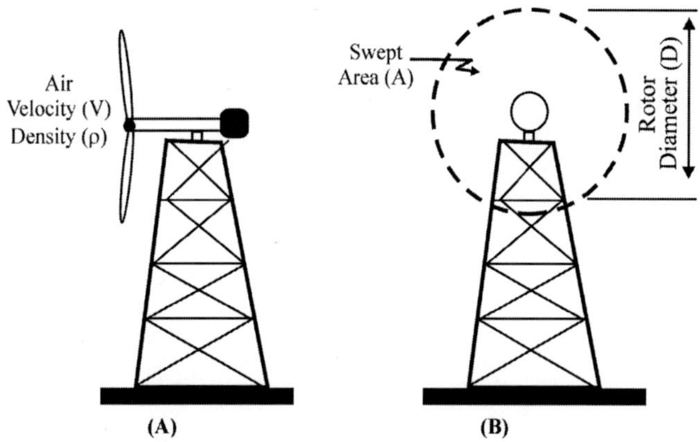

Fig. 10.1: Energy in the wind Due to **(a)** Air velocity and air density **(b)** Swept area

Harnessing the Power from Wind

Kinetic energy available in wind

$$KE = \frac{1}{2} MV^2 \qquad 10.1$$

Mass flux across the turbine

$$M = \rho\, AV \qquad 10.2$$

Hence kinetic energy

$$KE = \frac{1}{2} \rho A V^2 \qquad 10.3$$

Where ρ is the density of air (kg/m^3), A is the turbine swept area (m^2), and V is wind velocity (m/s).

To understand the power generation from a wind turbine, Betz model is taken into account (Fig. 10.2)

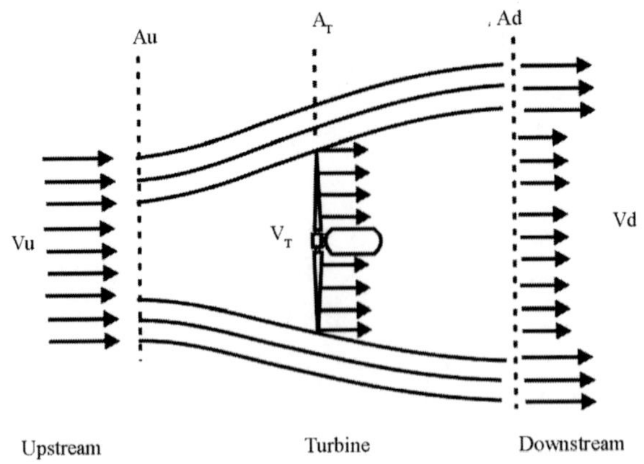

Fig.10.2: Betz Model of Airstream on Turbine

The thrust on the turbine which reduce the momentum per unit time

$$F = MV_u - MV_d \qquad 10.4$$

Force applied by airstream on turbine, power extracted by turbine

$$P_T = FV_T$$

$$P_T = M\,(V_u - V_d)V_T \qquad 10.5$$

The losses in energy per unit time

$$P_L = \frac{1}{2}\dot{M}\left(V_u^2 - V_d^2\right) \qquad 10.6$$

Equating eq. 10.5 and 10.6

$$\dot{M}\left(V_u - V_d\right)V_T = \frac{1}{2}\dot{M}\left(V_u^2 - V_d^2\right) \qquad 10.7$$

$$V_T = \frac{1}{2}\left(V_u + V_d\right)$$

The power output from turbine can be expressed as

$$P_T = \frac{1}{2}\rho A_T V_T\left(V_u^2 - V_d^2\right) \qquad 10.8$$

Substituting value of V_T in eq. 10.8

$$P_T = \rho A_T\left(\frac{V_u + V_d}{2}\right)\left(\frac{V_u^2 - V_d^2}{2}\right)$$

$$= \frac{1}{4}\rho A_T\left(V_u + V_d\right)\left(V_u^2 - V_d^2\right) \qquad 10.9$$

For maximum power generation from turbine Eq. 10.9 differentiating with respect to V_d and equating derivative to zero, i.e.

$$\frac{dP_T}{dV_d} = 3V_d^2 + 2V_uV_d - V_u^2 \qquad 10.10$$

This quadratic equation has two solution i.e. $V_d = \frac{1}{3}V_u$ and $V_d = V_u$

In actual practice $V_d \prec V_u$; hence $V_d = \frac{1}{3}V_u$

$$P_{T\max} = \frac{8}{27}\rho A_T V_u^3 \qquad 10.11$$

$$= \frac{16}{27}\left(\frac{1}{2}\rho A_T V_u^3\right)$$

$$P_{T\max} = 0.593 P_{total} \qquad 10.12$$

It is clear from eq. 10.12 that maximum power available through turbine is only 0.593 time of total available wind power. The factor 0.593 is known as **Betz limit**.

Thrust on Rotor

$$F = \frac{1}{2}\dot{M}\left(V_u^2 - V_d^2\right)$$

$$= \frac{1}{2}\rho A_T\left(V_u^2 - V_d^2\right)$$

$$= \frac{1}{2}\rho\frac{\pi}{4}D^2\left(V_u^2 - V_d^2\right)$$

$$= \frac{\pi}{8}\rho D^2\left(V_u^2 - V_d^2\right)$$

As earlier we find out the condition for maximum power output from turbine is

$$V_d = \frac{1}{3}V_u$$

Substituting the value of Vd in eq. 10.13 we have

$$F_{\max} = \frac{\pi}{9}\rho D^2 V_u^2 \qquad 10.14$$

Torque

The maximum conceivable torque (T_{max}) on turbine rotor would be occurring when the maximum thrust can be applied. For a propeller turbine of radius R

$$T_{max} = F_{max} R$$

Thrust, F become maximum when $V_d = 0$

$$F_{\max} = \frac{1}{2}\rho A_T V_u^2$$

$$T_{\max} = \frac{1}{2}\rho A_T V_u^2 R \qquad 10.15$$

For a working wind machine producing shaft torque T, the torque coefficient C_T is defined as

$$T = C_T T_{max} \tag{10.16}$$

The tip speed ratio 'λ' is defined as the ratio of outer blade tip speed V_{tip} to the upstream wind speed V_u

$$\lambda = \frac{V_{tip}}{V_u} = \frac{\omega R}{V_u} \tag{10.17}$$

Where ω is angular velocity of rotor and R is the outer blade radius

$$R = \frac{\lambda V_u}{\omega} \tag{10.18}$$

Substituting the value of R in eq. 10.17 we have

$$T_{max} = \frac{1}{2}\rho A_T V_u^2 \left(\frac{\lambda V_u}{\omega}\right)$$

$$= \frac{1}{2}\rho A_T V_u^3 \left(\frac{\lambda}{\omega}\right)$$

$$= P_{Total}\left(\frac{\lambda}{\omega}\right) \tag{10.19}$$

The maximum shaft power P_{max} derived from the turbine can be expressed as

$$P_{max} = T_{\omega}$$

$$= C_T T_{max} \omega$$

$$C_p P_{Total} = C_T P_{Total} \frac{\lambda}{\omega} \omega$$

$$C_P = C_T \lambda$$

$$C_T = \frac{0.593}{\lambda} \tag{10.20}$$

Solidity

The solidity can be defined in two way i.e. (i) Blade solidity and (ii) Chrod solidity

Blade solidity (S_b) is defined as the ratio of total blade area to the swept area of turbine

$$S_b = \frac{NbR}{\pi R^2} = \frac{NB}{\pi R} \tag{10.21}$$

Chord solidity (S_c) is defined as the total blade at a given radius by the circumferential length at that radius

$$S_c = \frac{Nb}{2\pi R} \tag{10.22}$$

Where N is number of blades, b is the blade width R is blade radius

In the initial inception a high solidity machine required high torque, but shortly generates optimum power even at low rotational speed. Therefore, high solidity machine prefers for water pumping and low solidity machine used for electricity generation.

The theoretical maximum power that may be captured by a wind machine was shown by Betz in 1920 to 16/27 or 59.3% of the incoming energy in the wind.

The major factor influencing energy in the wind is the wind velocity. As shown above, the power is a factor of the cube of the velocity, so doubling the wind speed will increase the available power eight times. However, wind energy is very diffuse in nature and also the velocities are extremely variable.

The practical output in terms of electrical energy or amount of water lifted can be calculated by using following equations

Power = $0.5\ A\ \rho\ V^3.C_p$ watts

where

V = Average wind speeds m/s

A = Swept area of rotor, m^2

C_p = Ratio of available power extracted to the theoretical power available

In case of wind mill $C_p = 0.4$ and generating efficiency = 0.8

Thus, effective power = $0.19\ A\ V^3$ watts

In case of Aerogenerator, Effective power = $0.2\ AV^3$ watts.

The energy output is found multiplying with the duration of the period concerned. This is calculated for a year showing the annual output as a function of the average wind speed (V).

The efficiency of wind machine depends on wind speed, with type of turbine & with the nature of load. However, the wind speed is playing an important role. As the speed increases from a low value, the turbine is able to overcome all mechanical and electrical losses and start delivering electrical power to the load at cut in speed V_c. The rated power output of the generator is reached at rated wind speed V_R. Above that speed, some wind power is spoiled to maintain constant power output. At the furling speed V_F, the machine is shut down to protect it from high winds. V_R can be increased by using a larger generator for a given wind turbine. This increases the static friction so that V_c will also increase. The cut in speed will typically be about one half the rated wind speed, which corresponds to cut in power of oneeight the rated power. Below V_c most of power in the wind is used to overcome mechanical and electrical losses.

Wind is highly variable source and there are several methods of characterizing this variability. Most common is the power duration curve to select V_c and V_R for a given wind site. Seasonal and diurnal variation has significant effect on wind. Load duration data are required to judge the appropriate height. Diurnal variation is less with increase in heights. Due to seasonal variation average power may typically vary from 65 to 125%. This factor must be suitably considered in the design parameters.

Wind speed increases with the height because of friction at earth surface. The rate of increase is given by

$$V/V_0 = (H/H_0)^{1/7}$$

Where V is the predicted wind speed at height H & V_0 is the wind speed at height H_0. This translates into substantial increase in power at greater heights.

Example 10.1: Calculate the power converted from a wind machine at wind speed 12 m/s, air density 1.23 kg/m³, and power coefficient is 0.4. The blade length of wind machine is about 52 m.

Solution

Blade length = radius = 52 m

$$A = \pi r^2$$

$$= \pi \times 52^2$$

$$= 8495 \text{ m}^2$$

Available power

$$P = \frac{1}{2}\rho A V^3 C_p$$

$$= \frac{1}{2} 1.23 \times 8495 \times 12^3 \times 0.4$$

$$= 3.6 \text{ MW}$$

Example 10.2: A wind turbine travels with the speed is 10 m/s and has a blade length of 20 m. Determine wind power.

Solution

Given:

Wind speed v =10 m/s,

Blade length l = 20 m,

air density ρ = 1.23 kg/m^3,

area A = πr^2

$= \pi \times 400$

$= 1256$

The wind power formula is given as,

$$P = \frac{1}{2}\rho A V^3$$

$= 0.5 \times 1.23 \times 1256 \times 1000$

P = 772440 W.

10.4 Design of Wind Energy System

The primary step of the design of wind energy system includes data collection and its analysis.

10.4.1 Data collection

As air moves across the surface of the earth, its speed and direction are changed by local topography as well as by local heating and cooling. At any particular site, trees, buildings, terrain, proximity to large bodies of water, or other small scale influences can disturb the flow of air. The combined effects can lead to highly variable winds and results in the good wind energy sites.

Topographic surface roughness increases the friction on the boundary layers of air at the surface and it may result in shearing of the boundary layers. Fig. 10.3 indicates this effect, and the need to site windmills and the measuring equipment above the affected regions.

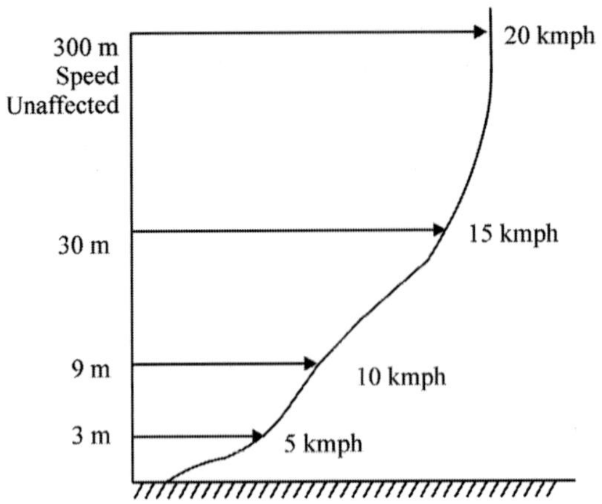

Fig. 10.3: Effect of Surface Roughness on Wind Velocity

In areas of hilly or mountainous terrains, wind velocities and directions will be seriously altered by local topography. Such features affect airflows on much larger scale than surface roughness but their combined effects should always be taken into account. Nonflat terrain can be divided into two categories:

(i) Elevated terrain

(ii) Depressions.

(i) Elevated Terrain: The main topographical features associated with elevated terrain are ridges, hills and cliffs. Their effects are generally two fold, namely:

1. The elevated terrain can act as huge towers, thereby exposing a windmill to winds of high velocities.
2. Airflow can be accelerated when moving around elevated terrain, thus enhancing the wind velocities.

However, elevated terrain can also have adverse effects, as slowing of the winds, turbulence and wind shear can occur, but proper site analysis can eliminate these hazards.

(ii) Depressions: Terrain features such as valleys, basins, gorges and passes are associated with depressions. Generally, depressions will lead to areas which are sheltered from the prevailing winds and are most likely to be affected by the diurnal changes in wind patterns. While some valleys may have a funneling effect on the winds and increase their velocities, depressions do not usually offer good sites for wind energy.

10.4.2 Data analysis

The most commonly applied technique for assessing wind energy is to first tabulate the available data, and then calculate monthly, daily and, in some cases, hourly average wind speeds. These mean wind speeds are plotted on a graph of "Mean Wind Speed VS Time", or alternatively the wind speeds can be adjusted to provide mean power densities (i.e. power per unit area) and the results then analysed. Integrating these figures over the particular time period gives the power densities available over that time period.

However, when considering a wind energy conversion system (WECS) will have a cutin speed, rated speed and cutout speed. Analysing the data along with the specifications of different WECS is much more useful and the actual energy output from the WECS may be calculated for the time period over which the data was taken.

There are at least two major constraints on wind turbine design one is to maximize the average power output. The other is to meet the necessary load factor (which is the ratio of average electrical power to the rated electrical power) requirement of the load. Load factor is not of major concern if the wind electric generator is acting as a fuel saver on the electric network. But if the generator in pumping irrigation water in a synchronous mode, for example, load factor is very important.

Design for wind turbines have been driven by three basic concepts for handling wind load i.e. (1) Withstanding the load (2) Shedding or avoiding of loads (3) Managing load mechanically and/or electrically. Important characteristics of first design concepts are optimization for reliability, high solidity but nonoptimization blade pitch, low tipspeed ratio (TSR), three or more blades. Turbines based on second concept have design criteria like optimization for performance, low solidity, optimum blade pitch, high TSR/ optimum blade pitch etc. where as in third concept design considerations like optimization for control, two or three blades, moderate TSR, mechanical & electrical innovations (flapping or hinged blades, variable speed/low speed generators).

10.5 Wind Measurement Equipment

The type of equipment to be used for estimating the power in the winds depends largely on the type of data required. If the WECS is to be used for estimating annual power output, mean wind speeds and wind speed distributions for one year are required. If the WECS is to be used for providing power to match peak power periods, more detailed records of the diurnal changes are required. For small power applications (e.g. pumping of water) mean annual or monthly wind speeds are sufficient.

Wind measurement apparatus consists of wind sensors and displays or recorders. The wind sensors are cup or propeller anemometers which sense wind speed and wind vanes for sensing wind direction.

The wind sensors usually produce an electrical signal, but in some cases it is mechanical, which is monitored by a recorder. These recorders may be counters, strip chart recorders or magnetic tape recorders and provide a record of the wind velocities which can be analysed. The simplest recorder is the windown anemometer which has a mechanical counter measuring the amount of wind passing the sensor. For obtaining average wind speeds this system is adequate, but if a more detailed record is required, either the strip chart or magnetic tape recorders must be used. However, strip chart and magnetic recorders are more expensive and delicate instruments.

Strip chart recorders require manual tabulation of the recorded data and this can be a tedious and inaccurate process. Magnetic tape recorders, on the other hand, are easily processed with computers, but such facilities must be available.

In selection of the wind measurement sites, attention should be given to the various topographical factors that affect wind speeds. The anemometers should be placed at height where they will be least affected by the local environment, or at the height where they will be least affected by the WECS, if a commercial system is being considered.

Where ever possible, wind data available from a local weather station should be consulted and some kind of correlation should be made between the existing data. If a good correlation exists, then some indication of how typical the data is for that period will be obtained and greater assurance in the accuracy of the data will then exist.

10.6 Wind Energy Conversion Systems

The use of crude windmills to pump water and to power simple machines dates back a long time. A picture of windmill in a Chinese vase of about the third millennium B.C. is the earliest, but the Persians were using windmills for grinding grain by seventh century A.D. Until the early twentieth century, windmills were in common use for providing electricity, on a small scale, and for pumping water in rural areas. In addition, the wind was the motive power for the European colonization of most of the world in the sixteenth and seventeenth centuries. In more recent times, the windmill is being assessed as a potential source of power generation both on large and small scale. In general, WECS are divided into two types, horizontal axis machines and vertical axis machines, and some typical examples of both types are shown in Fig. 10.4. The horizontal axis machines are those in which the axis of rotation is parallel to the direction of the wind. These WECS suffer the disadvantage of having to be oriented into the wind when the wind direction changes, and in small machines this is usually accomplished by a simple tail vane held by spring tension. In large horizontal axis WECS, sophisticated electronic sensors and motors may be required to achieve the desired orientation.

Verticalaxis machines have their axis of rotation at right angle to both the Earth's surface and the direction of the wind, and they have the advantage that do not have to be repositioned in to the direction of the wind. The simplest vertical axis WECS is the Savoniustype rotor, which is high torque and low speed machine that can be constructed very cheaply from old drums (Fig. 10.4). This machine is vary suitable for pumping water and generating small quantities of electricity (kW). The efficiency, however, vary at about 1015% depending on the wind speed, so far larger requirements, the Savonius is not recommended. Larger vertical axis rotors, such as the Darrieus turbines are being developed extensively for large power generating systems, but their applicability may be hindered by the need for high average wind speeds (38 kmph) and an inherent efficiency which is lower than for horizontalaxis propeller machines.

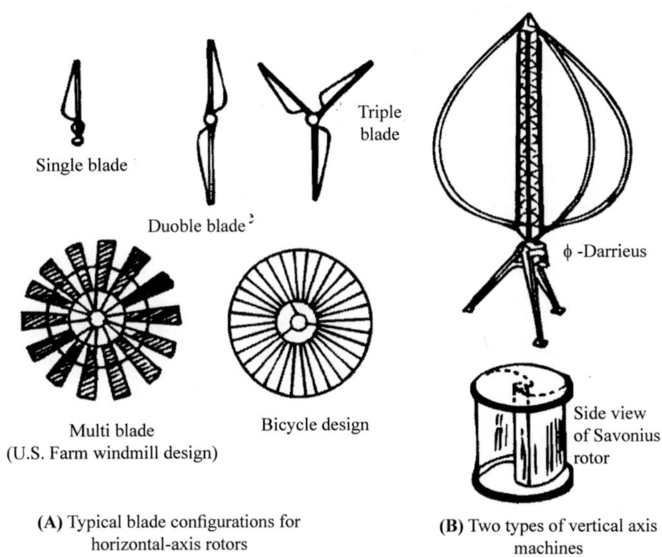

Fig. 10.4: Wind Energy Conversion System

The advantages and disadvantages of horizontal axis and vertical axis machines are given in Table 10.2. The earliest wind mills were of the fixed blade variety, but to improve efficiency, particularly of large WECS, pitch modulations mechanisms are now being utilized. These mechanisms change the angle of attack of the blades according to changes in wind speed. This results in additional expense to the system, but the increased efficiency at low wind speeds usually offsets this cost disadvantage.

The Fig. 10.5 shows some performance curves for different WECS. The power coefficient(C_p) is a measure of the efficiency of the wind turbine, and it is seen that A, the high speed double blade type horizontal axis rotor, is the most efficient. For large power generating systems, this is the design that is now achieving the most interest. The simple Savonius rotors and US multiblade windmill are low efficiency machines, but for meeting small power requirements they are very well suited as they are the most inexpensive machines and may frequently be fabricated with a minimum of sophisticated equipment.

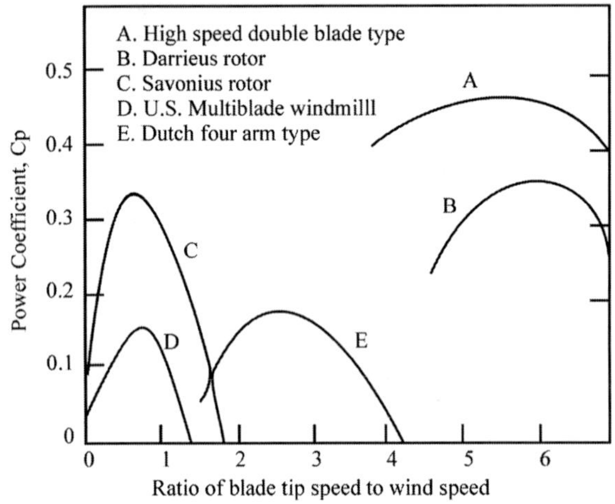

Fig, 10.5: Typical Performance Curves for Various Wind Turbines

Table 10.2: Comparison of Horizontal & Vertical Axis Machine

S.No.	Type of Machine	Advantage	Disadvantage
1.	Horizontal axis Machine	Rotor weight is less	Pitch changing mechanism for speed control requires maintenance
		High speed are possible	Tower required for mounting
		Solidity ratio are less	Up wind rotors should have enough clearance from the tower to avoid striking
		Large experience available	Downwind rotor required for large size
		Available in sub kW to MW range	Directional control is necessary
2.	Vertical axis machine	Less land for structure	Solidity ratio is high
		Directional control not required	Rotor is heavy
		Surrounding of blades can be used against gusts	Suitable for small power generation
		High rpm is possible	High starting torque required

10.7 Wind Energy Devices

The wind energy is successfully used for water pumping, grinding of grains and for electricity generation. The details of devices available in the field in India are as follows:

10.7.1 Wind Pump

The size of wind pumps currently available range from 1 to 8 m rotor diameter. The pumping height and average wind speed influence the average power output which can range from a few watts to about 1 kW. For higher power demand, wind electric pumping systems (WEPS) can be applied, incorporating a wind generator (available in larger diameter) driving an electronic motor pump combination. At higher power output levels (10 kW or more), the pumping system could be integrated with a small electric grid supplying electricity for other purposes besides water pumping. For average rural Indian applications, the daily output ranges from 1.5 to 500 m^3 at a head of 20 m or roughly from 0.1 to 3 kWh/day. The reasonable costs of pumping water run upto Rs. 20 per m^3. The pumps are available which can be operational at wind speeds of 56 km per hour or more.

10.7.2 Aero-generator

The country ranks fourth in the world in wind power, with 37.65 GW installed capacity.

Wind farms, which constitute a cluster of gridconnected wind electric generators of 250 - 500 kW aero generators, have proved to be a feasible method of power generation on a large scale. It is well suited in locations where the annual average wind speed is at least 18 kmph or the energy content is above 2000 kWh/m^2/year. As wind farms need large open spaces (810 hectares/MW), the availability of land area in windy region is extremely important. But within a wind farm, only 5 to 10 per cent of the land area is actually required for installation of the wind electric generators and roads, etc. the remaining land can be put to productive use such as crop growing, cattle grazing etc.

The average capital cost of a wind farm project works out to Rs. 4.50 to 6.90 crore per MW which is fairly comparable with cost of setting up a conventional power plants. Based on this investment and normal methods of cost calculation, the average cost of electricity generation in wind power projects is in the range of Rs. 2.40 to 2.80 per unit, depending upon the site.

10.7.3 Wind Battery Charger

Wind energy generators can also be used to charge batteries, which can store energy for lighting, radio communication, hospital equipment and to power various emergency related equipment. These wind energy generators are known as Wind Battery chargers. Battery chargers are wind machines generally with a rotor diameter of 1 to 2.5 m and an output of 50500 W at a 10 m/s wind speed. The components include a wind powered generator, a convertor and a container for the batteries. Presently, battery charger of rating 1.5 kW to 4 kW are available.

10.8 Uses of Wind Energy Systems

Wind energy systems are very reliable and versatile technology which have been used for hundreds of years for different purposes.

(a) **Water Pumping :** The wind has been used as a reliable and inexpensive water pumping power source for generations. Either a mechanical or electric water pumping system could be ideal for rural and remote locations to supply livestock, a household or even a small community (Fig.10.6).

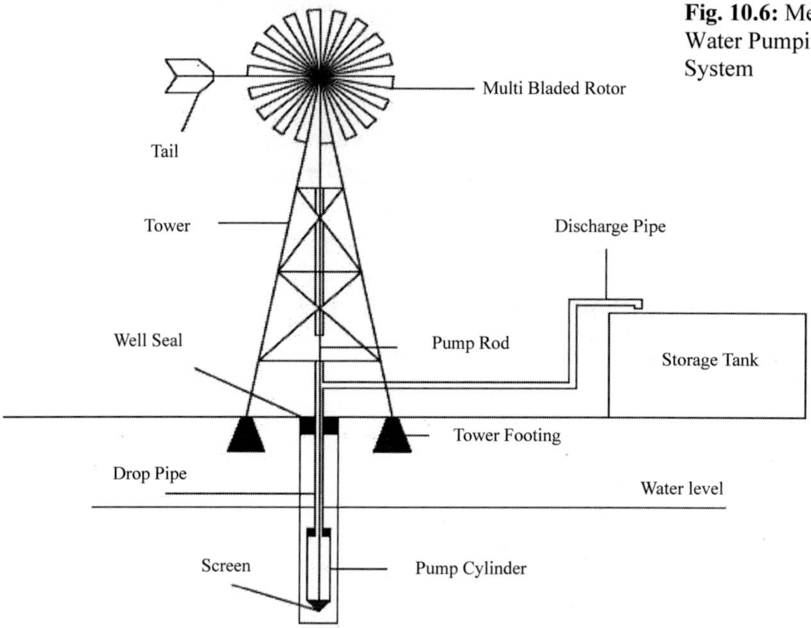

Fig. 10.6: Mechanical Water Pumping System

(b) **Remote Communities :** In remote communities where diesel generators often supply electricity, the use of wind energy not only makes environmental sense but also it makes economic sense. Larger wind energy systems can reduce reliance on expensive and greenhouse gas-producing generators.

(c) **Recreation :** Using the wind as an energy source for cottage or boat could be efficient and inexpensive when compared to fossil fuel generators. An environmentally friendly wind energy system could power lights, radios and small appliances.

(d) **Farm and Ranch :** Today's wind energy systems can do much more for a modern agricultural operation beside pumping water. Because they are ideal where remote, low voltage power is required. Wind energy electrical generators are used for such farm systems as electric fences and yard lights.

(e) **Home Use :** Rural home owners interested to reduce the environmental impact of their energy use can reduce their reliance on grid power with a wind energy system. Even a mini wind energy system saves electricity generated from fossil fuels or other conventional fuel.

(f) **The Right System :** There are several types of wind energy systems. There are stand-alone systems which provide power solely from the wind. A stand-alone system may have a method for storing energy when wind conditions are not good. Usually, batteries are used for storage. There are hybrid systems which use another source of power, perhaps solar panels or a diesel generator, to supplement the energy provided from the wind. Often, a switching mechanism starts the generator remotely when the wind turbine cuts out. There are also mechanical systems which are used to aerate ponds or pump water for livestock, irrigation or household water supplies.

10.9 Economics of Wind Power

The unit cost of electricity generation (U_c) through wind farm can be expressed as:

$$U_c = \frac{IR}{E_{ao}} + M$$

where:

I is the capital cost

R is capital recovery factor

E_{ao} is annual energy output

M is annual operating and maintenance cost

The annual recovery factor can be estimated as follows:

$$R = \frac{i}{1-(1+i)^{-n}}$$

where:

i is the required annual rate of return

n is the number of years over which the investment is to be covered.

An annual energy output, E_{ao} (kWh) can be calculated as follows:

$$E_{ao} = (HP_r F)\, N_t$$

where:

H is number of operating hours

P_r rated power of turbine

F is annual capacity factor of the turbine

N_t is the number of turbines in wind farm.

The operating and maintenance cost, M, per kWh can be expressed as:

$$M = \frac{KI}{E_{ao}}$$

K represents the annual operating cost as a fraction of total capital cost. It is usually taken as 2.5 % of capital cost.

11

Biogas Technology

11.1 General

Biogas is generated through a process of anaerobic digestion of organic materials. The materials normally used are human and animal waste, crop residues, agro-industrial waste and other biomass in a combination with water. Biogas production technology contributes in following ways:

(a) It provides a better and cheaper fuel for cooking, lighting and for power generation.

(b) It produces good quality, enriched manure to improve soil fertility.

(c) It provides and effective and convenient ways for sanitary disposal of human excreta, improving the hygienic conditions.

(d) It generates social benefits such as reducing burden on forest for meeting cooking fuel by cutting of tree for fuelwood, reduction in the drudgery of women and children etc.

(e) As a smokeless domestic fuel, it reduces the incidence of eye and lung disease.

(f) It also helps in generation of productive employment.

The biogas technology is primarily based on fermentation of cellulosic rich organic matter under anaerobic conditions. The methane producing bacteria become more active after the creation of anaerobic conditions, thus, the gas produced become rich in methane & carbon dioxide, beside traces of hydrogen sulphide, ammonia, oxygen, water vapour etc. depending on the feed material and other conditions. For burning of gas, it is necessary that methane contents should be more than 50 per cent. The complete knowledge of process of methane production including its chemistry, physiology, bio-chemistry, micro-biology and bacteriology has been studied over years in many parts of the world. The culture and isolation of microbes under various environmental conditions have been successfully experimented to improve rate of reaction. The conditions, under which cattle dung, organic wastes, bushes, weeds etc. will produce methane, can be understood through chemistry of fermentation.

11.2 Chemistry of Fermentation

The complex organic materials is decomposed by acid producing bacteria which consume oxygen, producing fatty & organic acids and carbo dioxide. The oxygen environment which inhibits the methane bacteria is replaced by CO_2 which helps the growth of methanogen to produce methane from disintegrated cellulose organic material as shown in Fig 11.1. the methane formation process can be completed in two main phases:

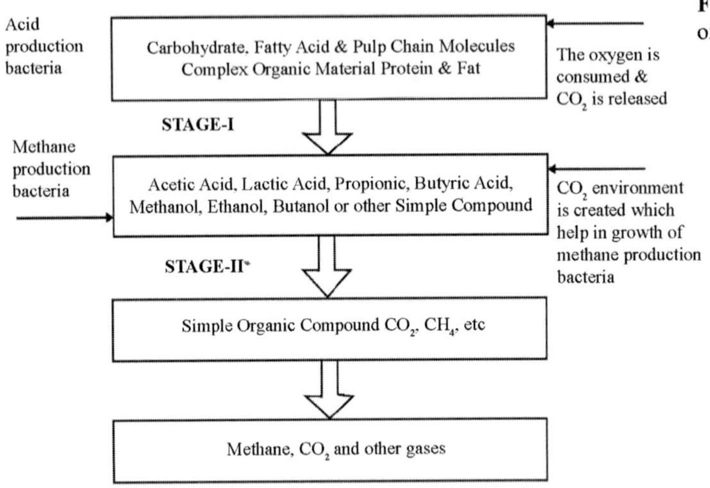

Fig. 11.1: Chemistry of Fermentation

11.2.1 Acid phase including Hydrolysis

The complex organic substance polymers can not be reacted by the bacteria as these substances can not pass through cell remembrances. The enzymes are released by bacteria disintegrating complex substance to make them digestible during hydrolysis. There-after organic substances are oxidized through acid producing bacteria. In simple soluble materials volatile acid formation starts directly during acid phase. The acidogens consume oxygen present in the feed material released carbon dioxide & also fatty & other organic acids such as formic, acetic, propionic, butyric, lactic acid etc.

The acidogens multiply rapidly and can tolerate abrupt change in environment. Their main function is to disintegrate complex organic matter into simple, low molecular weight compounds, excrete enzyme and liquify the organic compounds. A significant quantity of methane upto 70 per cent, is produced during this phase.

11.2.2 Methane Phase

The simple disintegrated compounds, fatty and volatile acids produced during acidogen phase are converted by methanogen into methane & CO_2 in 60 to 70 & 30 to 40 per cent respectively. The methane producing bacteria become more active, and grow rapidly in CO_2 environment under certain specific conditions.

The following reactions take place during this stage.

Individual reaction include:

Acid break down into methane

$$2C_3H_7COOH + 2H_2O \rightarrow 5CH_4 + 3CO_2$$

Oxidation of ethanol by CO_2 to produce methane and acetic acid

$$2CH_3CH_2 + CO_2 \rightarrow 2CH_4COOH + CH_4$$

Reduction of carbon dioxide with hydrogen to produce methane

$$CO_2 + 4H_2 \rightarrow CH_4 + 2H_2O$$

Both the above phases are inter dependent as such a careful balance between the each individual reaction & group of reactions and environmental conditions should be maintained otherwise the methane production will be severely affected. The acid formation phase should be properly balanced by methane phase, otherwise accumulation of acid will inhibit the production & growth of bacteria, lowering pH. Although different kinds of bacteria are active, the by-product of one is consumed as food by other, yet the volatile acids help fast multiplication of methanogens. The carbon chain & nitrogen present in enzymes & protein also support methane production. The methanogens are very sensitive to rapid change in environment, presence of oxygen and fall in temperature. Their activities retard and decrease to multiply when temperature drops below 10 °C as their activities & multiplication becomes optimum at a temperature around 35 – 37 °C.

The ideal conditions which results in optimum producing of methane can be created by balancing several parameters among which pH, C-H ratio, temperature, inorganic/toxic substance are major. This limits the scope of use different raw materials and necessity of pretreatment for use in biogas plants.

11.3 Characteristics of Biogas

(a) Characteristics of methane and carbondioxide

The gas produced during anaerobic digestion of natural organic material consists of a mixture of notably methane (CH_4) and carbondioxide (CO_2) with small amount of other gases, in particular, hydrogen sulphide (H_2S) and hydrogen (H_2), all in very small quantities. Methane is a simplest and most abundant hydrocarbon and the chief constituent of natural gas.

Typically, digester gas, or biogas as it is sometimes called, has between 60 to 70% CH_4 (occasionally more), with 40 to 30% CO_2 the remaining gases are in vary small quantities and are often difficult to detect. The gas is combustible with a calorific value of about 19.73 MJ/m^3.

(b) Hydrogen sulphide

Although the data available show wide variations in the amount of hydrogen sulphide present in digester gas, normally the expected value for a digester that is working well on sewage sludge is between 100 to 3000 mg H_2S per cubic meter.

(c) Hydrogen

Usually no more than 1 or 2% maximum is to be found in digester gas, probably because there are many bacteria present in the sludge which can utilize it readily, particularly the methanogenic bacteria. Although the gas is highly inflammable, the small concentrations present have little effect on the calorific value of digester gas. Hydrogen is not poisonous, although of course if present in large quantities in air may cause asphyxiation due to a reduction in the percentage of oxygen present.

(d) Carbon Monoxide

This is a toxic gas since its affinity for haemoglobin is greater than that of oxygen. The proportion in digester gas is usually well under 1% and often more like 10 to 100 ppm.

(e) Nitrogen

Any appreciable quantity of nitrogen showing up in routine gas analysis is usually indicative of a leakage of air into the system. There will always be a small quantity present since the raw sludge will usually contain some due to previous contact with air. Usually, less than 4% is present in digester gases, and often much smaller concentrations are present.

(f) Oxygen

This is dangerous gas when present in digester because of the risk of explosions when mixed methane leaking into surrounding air, which is more hazardous situation since, with only just over 5% methane in air a mixture occurs in which a flame is selfpropagating.

The gas thus produced by the above process in a biogas plant does not contain pure methane and has several impurities. A typical composition of such gas obtained from the process is as given in Table: 11.1

Table-11.1: Constitutes of Biogas

Methane	60.0%
Carbon-dioxide	38.0%
Nitrogen	0.8%
Hydrogen	0.7%
Carbon-monoxide	0.2%
Oxygen	0.1%
Hydrogen Sulphide	0.2%

The calorific value of methane is 35.16 MJ / m^3 and that of the above mixture is about 19.73 MJ/ m^3. However, the bio-gas gives a useful heat of 12.56 MJ/ m^3.

11.4 Socio-Economic Aspects of Biogas Technology

From the national economic point of view, the biogas yields following economic benefits:

(a) Biogas technology, which is based on recycling of readily available resources in rural areas, gives comparatively cheaper and better fuel for cooking, lighting, and power generation.

(b) An individual can reduce the consumption of commercial energy sources such a firewood, coal, kerosene, LPG, etc. by adopting waste recycling technology which vigorously help in reducing the family fuel budget.

(c) The problem of uncertainty of availability of commercial energy can be resolved by use of biogas technology.

(d) The rural population of the country uses firewood for meeting their cooking requirements, this reduces the national forest wealth. Our forest area can be conserved by using biogas.

(e) The dependency on chemical fertilizer for better agricultural production has increased to a great extent after independence in India. Biogas slurry can be proved a best organic fertilizer which helps in improving soil fertility and crop production.

(f) Biogas technology reduces the import of chemical fertilizer by using home made organic fertilizer and also petro-products.

(g) Biogas technology utilizes effectively, the man power and resources, resulting in self-sufficiency and self-reliance in the society.

In majority of the 5,97,270 (Census 2011) villages in India the fuel is collected & not purchase as such there is no cash benefit derived from the use of biogas technology but still it provides individual benefits, such as:-

(a) In rural areas, the human energy is wasted in collection of firewood, which can be effectively utilized in other useful works on the field, or in other employment thus improving the family living standard.

(b) Fuel and fertilizer are the major item where individual can save the money. This money can be reinvested in irrigation, livestock, creation of farm assets, etc. alternatively the money saved will increase the crop production.

11.5 Social Benefits

Biogas is one of best options in rural areas, which provides self-sufficiency in energy and help in increasing standard of living of rural people and community. The social benefits from biogas utilization are as follows:

(a) Biogas burns giving shootless flame and smokeless cooking, as such it provides cleanliness in the house.

(b) The cooking on biogas is faster and also women is not required to waste her time to collect fuel from forest, as such it reduces the drudgery of women who can use her free time for other developmental activities.

(c) The biogas provides lighting in the rural areas, which are far away from electricity supply lines, thus, it help the children to use their time in study at home.

(d) It helps in generating of employment opportunities to village artisans. This also stop the migration of people from rural to urban areas in search of employment.

(e) Generally, rural women and children, spend their energy and time for collection of fuel, chopping, collecting wood and crop from distant places, energy and time saved due to use of biogas can be used for constructive work or going to school for education.

In one way, in many areas, the setting up of the biogas plant has become a status of symbol and sign of adopting developmental plans, thus pleasing staff involved in Rural Development.

The social and cultural habit of rural masses such as religious beliefs, educational back ground, socio-economic status, settlement pattern of an individual and community are some of the important factors which influenced the acceptability of biogas technology to a great extent.

11.6 Environmental and Health Benefits

Biogas system contributes to maintain clean healthier environmental by processing human and animal wastes. The rapid industrialization, population growth and more transportation means have increased the pressure on environment and hence presently the environment protection is the domain among scientist, researcher, and voluntary organizations. The scientist are worried about environmental protection and looking for appropriate means for the same, which are locally available. In the present circumstances' biogas utilization would be a solution for environmental protection for healthy and prosperous society. Environmental benefits from biogas are enormous, some of them are as under:-

(a) Lungs and eye diseases are very common among village women and children due to smoky kitchen. Biogas utilization reduces the disease spread. Thus reducing health problem, rush in the hospitals and wastes of national wealth.

(b) Reduces the cases of burning women and sanitation problem in villages through systematic collection and through proper processing of animal dung and human excreta.

(c) It helps to prevent deforestation consequently; it control soil erosion and floods.

The technology which is being popularized to be accepted by the society is also facing number of barriers. Benefits and barriers go side by side, one has to accept both the parts of the coin, some time its better and sometime sweet.

11.7 Barriers in the Implementation of Biogas Technology

Biogas technology is having multi-facet usages at village level. However, for effective dissemination of programme at village level certain barriers related to Economic, Social and Health are encountered.

11.7.1 Economic Barriers

The Indian farmers plough back their farm income on social obligation such as marriages and deaths etc. but socially they do not accept any new technology unless and until it is approved by their social organization groups. As such there is no direct and mandatory benefits from utilization of biogas technology, even then people want to invest their additional income on the items which yield direct benefits to them. To accept this technology, following economic barriers come in their way:

(a) The investment on construction of a family sized biogas plant is generally high. It varies between Rs. 26,000 to Rs. 36,000 per plant depending on its capacity. Regular maintenance and repair works are also necessary which add to the cost of biogas generation.

Presently cement and steel are very costly, therefore the cost of plant can be reduced by using low cost locally available materials which are suited to local conditions. The other method to reduce the retention time and hence make the digester small.

(b) The economic return from biogas are generally low. The practical experience explains that fuel represents only 20 per cent possible return on capital investment in the case of small farmers where cost of biogas generation remain always high. In this case biogas slurry is important in many ways. Biogas slurry, which is an enriched fertilizer, can change the condition of the soil. Generally, the indirect benefits from the biogas need to be popularize.

(c) In India, majority of farmers whatever they earn, they spend on their livinghood. The important barrier in acceptance of any new technology in India is shortage of requisite funds. Looking to this problem right from independence, national and state government provide load and subsidies to farmers.

11.7.2 Social Barriers

Biogas technology which is a technically sound option for rural development requires social acceptance and willingness for people's participation, before implementation. Social acceptance of any promising technology depends upon number of factors. Biogas utilization which requires people's participation at every level is affected by individual taste, attitude, belief, habit, behavior and prejudices. The other social barriers are as follows:

(a) People do not want to use human excreta as manure and even do not want to use gas which is produced from human excreta. In this situation technology is rejected by majority of people.

(b) People often find it difficult to change their habits of defecating in the open fields. Even they oppose the construction of latrines at home under Swachh Bharat Mission.

(c) People do not want to attend plant daily and even do not want to prepare slurry because they feel some restriction on their free movcment.

(d) Lighting is not the priority of rural people, without it also they can manage their livelyhood.

(e) In Indian villages the living pattern is generally in cluster or scattered. In biogas utilization both systems have own pros and cons. If people live in cluster, they do not allow to install biogas plant in backyard and if they live scattered then community type biogas plant can not play effective role.

11.7.3 Health Hazards

Waste recycling process requires involvement of the society to a great extent, during the process of production and utilization as well as in the use of the effluent. It requires handling human and animal excreta. Care is required while handling night soil and its sludge as fertilizer. This may cause health problem to operator.

Care must be taken during utilization of biogas. Since it is toxic in nature and direct inhaling of gas for checking its quality is not recommended from the health point.

11.8 Types of Biogas Plants

Number of designs of biogas plants have been developed according to the various parameters as illustrate in Fig. 11.2. All designs have mixing unit, pre digester, digestor, gas storage and drying beds as important parts.

The following most important types of biogas plants are described here:

(i) Fixeddome plants

(ii) Floatingdrum plants

(iii) Balloon plants

(iv) Horizontal plants

(v) Earth pit plants

(vi) Ferrocement plants

(vii) Up flow Anaerobic sludge blanket (UASB)

(viii) Community Biogas plant

(ix) Institutional Biogas plant

(x) Landfill plants.

Of these, the two most familiar types are the fixeddome plants and the floating drum plants.

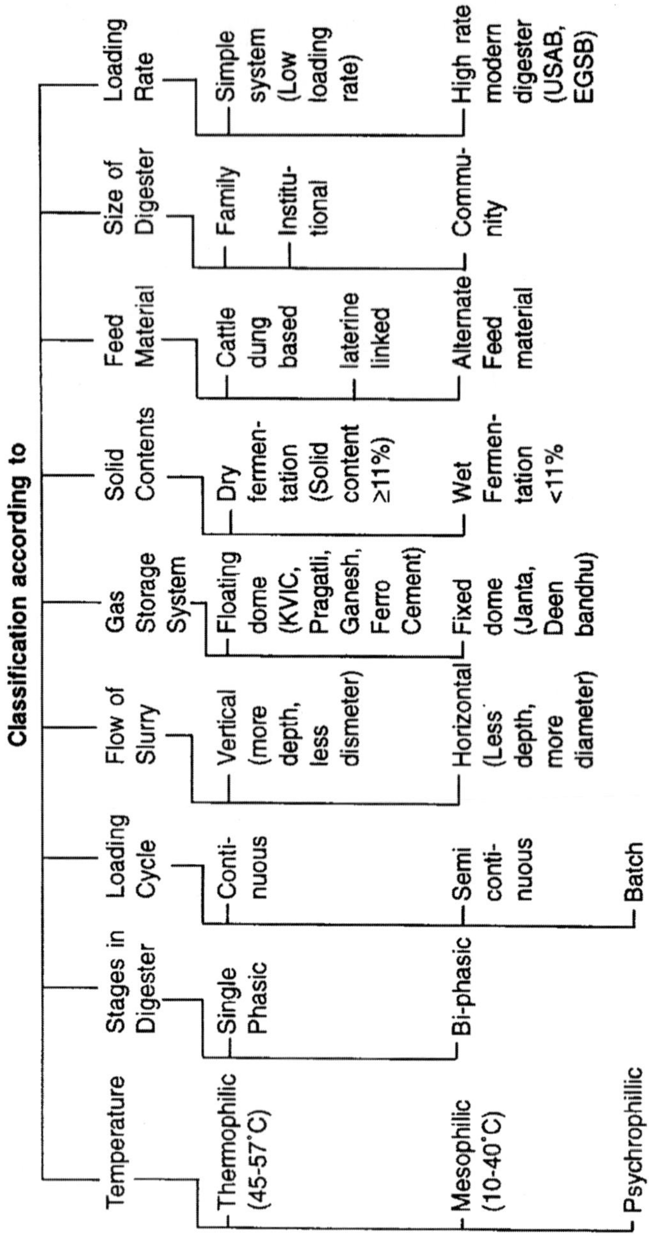

Fig. 11.2: Classification of Biogas Plants

(A) Fixed dome plants

Fixed dome plants are appropriate technology for developing countries, especially for rural areas. They can, nevertheless, in certain dimensions, also be used for agroindustrial and community wastewater treatment.

In the beginning, fixeddome plants were mainly built in the people's Republic of China, but are now spread in several countries.

A fixeddome plant comprises a closed, domeshaped digester with an immovable, rigid gasholder and a displacement pit. The gas is stored in the upper part of the digester. When gas production commences, the slurry is displaced into the compensating tank. Gas pressure increases with the volume of gas stored. If there is little gas in the holder, the gas pressure is low.

The digesters of fixeddome plants are usually made of masonry. They produce just as much gas as floatingdrum plants, but only if they are gas tight. However, utilization of the gas is less effective as the gas pressure fluctuates substantially.

Burners cannot be set optimally. If the gas is required at constant pressure (e.g. for engines), a gas pressure regulator or a floating gasholder is necessary. Engines require a great deal of gas, and hence large gasholders. The gas pressure then becomes too high if there is no floating gasholder.

The top part of a fixed dome plant (the gas space) must be gas tight. Concrete, masonry and cement rendering are not gas tight. The gas space must therefore be painted with a gas tight (for example Latex or synthetic paints) material.

Another possibility to reduce the risk of cracking is the construction of a weakring in the masonry of the digester. This "ring" is a flexible joint between one lower and one upper part of the hemispherical wall.

Fixed dome plants can handle fibrous substance in combination with animal excreta, since the motion of the substrate breaks up the scum each day.

Fixed dome plants must be covered with earth up to the top of the gasfilled space as a precautionary measure (internal pressure up to 0,10,15 bar). The earth cover makes them suitable for colder climates, and they can be heated as necessary. Due to economic parameters, the recommended minimum size of a fixed dome plant is 5 m^3 digester volumes and up to 200 m^3 is known and possible.

Fixed dome plants are characterized by low initial cost and a long useful life (20 years), since no moving or rusting parts are involved. The basic design is compact and wellinsulated. Fixed dome plants create employment locally.

However, during the constructions of fixed dome type plants, masonry is not normally gastight (porosity and cracks) and therefore requires the use of special sealant of the construction of weakring. Cracking often causes irrepairable leaks.

Also, fluctuating gas pressure complicates gas utilization, and plant operation is not readily understandable. Fixed dome plants can be recommended only where construction can be supervised by experienced biogas technicians.

Types of fixed dome plants

There are different models of fixed dome plants:

(i) BORDA Model

The BORDAplant is a combination of the floatingdrum and the fixed dome plant. It combines the advantages of the fixed dome plant with the processstability of the floatingdrum plant, which guarantees a constant gas pressure.

(ii) CAMARTEC Model

The CAMARTEC has a simplified structure of a hemispherical dome shell based on a rigid foundation ring only and a calculated joint of fraction, the so called weak/strong ring, which separates the gas storage area from the lower bottom part and prevents the upper dome from propagating cracks (Fig. 11.3). These can occur at the footing of the dome because of high tensile stresses that masonry cannot bear.

(iii) Deenbandhu Model (meaning "friend of the poor") with a Hemisphere Digester

This model is a lowcost rural household popular biogas plant that has been used in the India since 1986.

Biogas digesters by and large remain beyond the reach of most rural households. There has been an apparent need to reduce the cost of biogas units and bring them within the reach of a larger segment of the population. Plants of both KVIC and JANATA design involve considerable investment. A possible remedy was introduction of the "Deenbadhu biogas plant meaning "friend of the poor" They are built of brick and cement. They are plastered inside to make them tight.

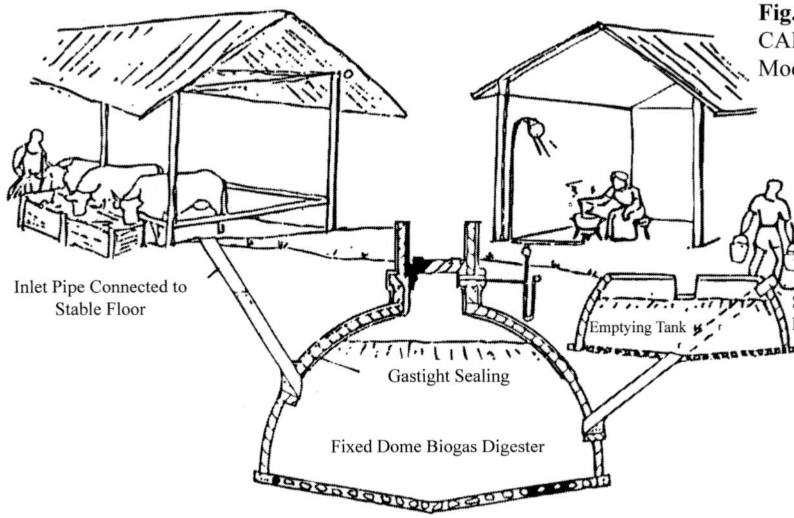

Fig. 11.3 CAMARTEC Model

The Deenbandhu plant consists of one bottom segment and another top hemisphere over it. The bottom segment is for the digester and the top hemisphere is for both digester and gas holder. The mixing tank is connected to the digester by 6" diameter asbestos cement (AC) pipe. Through the outlet hole provided in the digester, the digester slurry is pushed to the outlet tank & it over flows through another hole provided in the outlet tank as shown in Fig. 11.4. The Deenbandhu type plant costs about 20% less than that of Janta model and 40% less than the floating drum type of same size capacities. However, gas pressure is a variable, skilled labour are required for its construction and availability of gas is not visible form outside.

The plant is filled initially with homogenous slurry made from cattle dung mixed with water in the ratio of 1:1 upto the second step of outlet tank. As the gas generates and accumulates in the empty portion of the plant, it presses the slurry in the digester displacing it into the outlet displacement chamber. The slurry level in the digester falls where as in the outlet chamber it starts rising. When the gas value is opened, the gas flows through the outlet pipe to the point of utilisation, as pressure is exerted due to the difference in the levels of slurry in digester and in the outlet tank. As the gas is used, the slurry begins to return into the digester from the outlet tank. This falls and rise is continuous as per the usage of the gas.

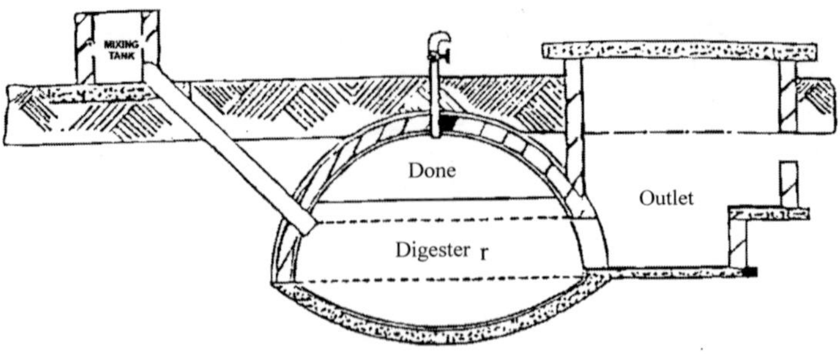

Fig. 11.4 Deenbandhu with Hemisphere Digester

However, a new version has recently been developed by Raymond Myles. An attempt has been made to reduce the cost of construction of Deenbandhu biogas plants by adopting locally available materials like Bamboo instead of steel. The bricks have been replaced with wooden bamboo, which constitute the reinforcement in a woodconcrete construction. The advantages are convincing. The bricks stand for half of the price of the installation. The bricks demand transport and coal for the baking as well. Bamboo on the other hand can be grown locally and with the necessary guidance, woven in the village.

Furthermore, the new construction provides work for even more hands in the community, especially female labour is used for the time consuming weaving work. The woven construction has made it possible to reduce the price giving even more families the opportunity to improve their standard of living and making it attractive for them to stay on the countryside.

With the new technique there is no use any longer for brick and steel. They only need to be delivered a few sacks of cement from outside. Even the tubes are made locally. An ordinary PVCtube serves as model. When the concrete is dry the plastic tube can be drawn out and used again somewhere else.

The local mason work three weeks on the construction of the biogas installation inclusive excavation. A tube is laid out to the kitchen and the gas devices are installed. After this the installation is ready to be used, the local extensionoffice helps the family with the correct use of the biogas installation, as well for security as for the achievement of a good, stable biogas production.

(iv) JANATA Model (meaning, "people") with a Brick Reinforced, Moulded Dome

The JANATA(Fig. 11.5) model is a fixeddome model built from bricks replacing the steel model used before. The feature of this model is that the digester and the gasholder are parts of an integrated brick masonry structure. The digester is made of a shallow well having a dome shaped roof.

The inlet and outlet tanks are connected with the digester through large chutes and these tanks above the level of the junction of the dome and the cylindrical portion are known as inlet and outlet displacement chambers, the gas pipe is fitted on the crown of the dome and there is an opening on the outer wall of the outlet displacement chamber for the discharge of digested slurry. Presently it has completely replaced by Deenbandhu biogas plant.

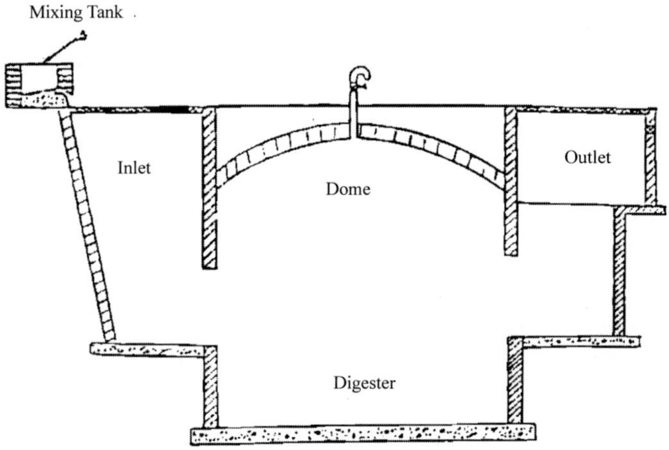

Fig. 11.5: Janata Model with a Brick Reinforced Moulded Dome

(B) Floating Drum Plants

In the past, floating drum plants were mainly built in India. A floatingd rum plant consists of a cylindrical or dome shaped digester and a moving, floating gasholder, or drum. The gas holder floats either directly on the fermentation slurry or in a water jacket of its own.

The drum in which the biogas collects has an internal and external guide frame that provides stability and keeps the drum upright. The gas drum is prevented from tilting. If gas is drawn off, the gasholder falls again (Fig.11.6).

Floating drum plants used chiefly for digesting animal and human excreta on a continuous feed mode of operation, i.e. with daily input. They are used most frequently by small to mid size family farms (digester size: 515 m^3) or institutions and large agroindustrial estates (digester size: 20100 m^3).

Floating drum plants are easy to understand and operate. They provide gas at a constant pressure, and the stored volume is immediately recognizable.

However, disadvantages wise in floating gasholder plant the steel drum is relatively expensive and maintenance intensive due to the necessity of periodic painting and rust removal. Therefore, the lifetime of the drum is short i.e. up to 15 years; in tropical coastal regions about five years. If fibrous substrates are used, the gas holder shows a tendency to get "stuck" in the resultant floating scum.

In spite of these disadvantages, floatingdrum plants are always to be recommended in cases of doubt. Waterjacket plants are universally applicable and especially easy to maintain. This would not stick, even if the substrate has a high solids content.

Waterjacket plants are characterized by a long useful life and more aesthetic appearance (no dirty gasholder). Due to their superior hygiene, they are recommended for use in the fermentation of night soil and for cases involving pronounced scumming, e.g. due to rapid evaporation, since the gasholder cannot get stuck in the scum. The extra cost of the masonry water jacket is relatively modest.

The digester is usually made of brick, concrete or quarry stone masonry with rendering. The gas drum normally consists of 2.5-mm steel sheet for the sides and 2 mm sheet for the cover. It has weldedin braces. These break up surface scum when the drum rotates.

The drum must be protected against corrosion. Suitable coating products are oil paints, synthetic paints and bitumen paints. Correct priming is important. There must be at least two preliminary coats and one topcoat. Coatings of used oil are cheap. They must be renewed monthly.

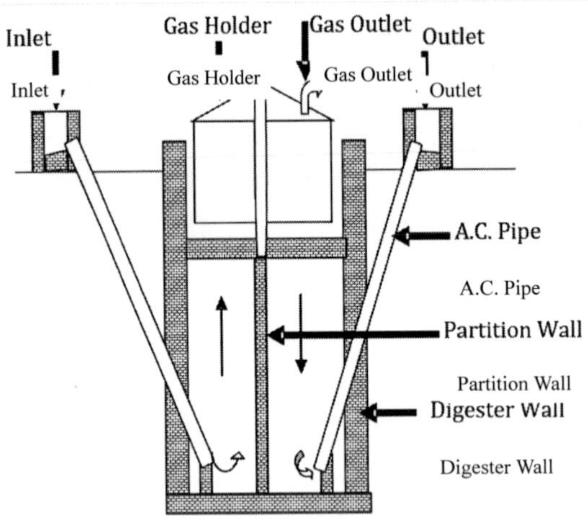

Fig. 11.6: Floating – Drum Plants (KVIC Model)

Plastic sheeting stuck to bitumen sealant has not given good results. In coastal regions, repainting is necessary at least once a year, and in dry uplands at least every other year. Gas production will be higher if the drum is painted black or red than with blue or white, because the digester temperature is increased by solar radiation. Gas drums made of 2 cm wiremeshreinforced concrete or fibre cement must receive a gastight internal coating.

The gas drum should have a slightly sloping roof, otherwise rainwater will be trapped on it, leading to rust damage. An excessively steeppitched roof is unnecessarily expensive. The gas in the tip cannot be used because the drum is already resting on the bottom and the gas is no longer under pressure.

Floating drums made of glass-fibre-reinforced plastic and high-density polyethylene have been used successfully, but the construction cost is higher then with steel.

Floating drums made of wire mesh reinforced concrete are liable to hairline cracking and are intrinsically porous.

The side wall of the gas drum should be just as high as the wall above the support ledge. The floating drum must not scrape on the outer walls. It must not till, otherwise the paintwork will be damaged or it will jam.

For this reason, a floating drum always requires a guide. This guide frame must be designed so that the gas drum can be removed for repair. The drum can only be removed if air can flow into it, either the gas pipe should be uncoupled and the valve opened, or the water jacket emptied.

The floating gas drum can be replaced by a balloon above the digester. This reduces construction costs (channel type digester) but in practice problems always arise with the attachment of the balloon at the edge. Such plants are still being tested under practical conditions.

Types of floating drum plants

There are different types of floating drum plants:

- KVIC model with a cylindrical digester
- PRAGATI model with a hemisphere digester
- GANESH model made of angular steel and plastic foil
- Floating drum plant made of prefabricated reinforced concrete compound units.
- Floating drum plant made of fibre glass reinforced polyester.

(C) Balloon Plants

A balloon plant consists of a heat sealed plastic or rubber digester bag (balloon), in the upper part of which the gas is stored. The inlet and outlet are attached directly on the skin of the balloon. The requisite gas pressure is achieved by weighing down the bag. When the gas space is full, the plant works like a fixeddome plant i.e. the balloon is not inflated, it is not very elastic. Since the material has to be weather resistant, specially stabilized, reinforced plastic or synthetic plastic is given preference. The useful life amounts to 25 years.

The fermentation slurry is agitated slightly by the movement of the balloon skin. This is favourable to the digestion process. Even difficult feed materials, such as water hyacinths, can be used in balloon plant. The balloon material must be UV resistant. Materials which have been used successfully include RMP (Red Mud Plastic), Trevira and Butyl.

Some of advantages of the balloon plants is standardized prefabrication at low cost, shallow installation suitable for use in areas with a high ground water table, high digester temperatures, uncomplicated cleaning, emptying and maintenance.

In this biogas plant, low gas pressure requires extra weight burden and scum cannot be removed. The plastic balloon has a relatively short useful life & is susceptible to damage by mechanical means, and usually not available locally, In addition, local craftsman are rarely in a position to repair a damaged balloon.

Inflatable biogas plants are recommended, if local repair is or can be made possible and the cost advantage is substantial.

(D) Horizontal Plants

Horizontal biogas plants are usually chosen when shallow installation is called for area having high groundwater & rock. They are made of masonry or concrete. In this plant shallow construction are required despite large slurry space. However, problems with gasspace leakage, difficult elimination of scum.

(E) Earthpit Plants

Masonry digesters are not necessary in stable soil (e.g. laterite). It is sufficient to line the pit with a thin layer of cement (netting wire fixed to the pit wall and rendered) in order to prevent seepage. The edge of the pit is reinforced with a ring of masonry that also serves as anchorage for the gasholder. The gas holder can be made of metal or plastic sheeting.

If plastic sheeting is used, it must be attached to a quadratic wooden frame that extends down into the slurry and is anchored in place to counter its buoyancy. The requisite gas pressure is achieved by placing weights on the gas holder. An overflow point in the peripheral wall serves as the slurry outlet.

Low cost of installation (as little as $1/5^{th}$ as much a floating drum plant), including high potential for self-help are few advantages of this type of plant.

However, this type of plant has short useful life and serviceable only in suitable & impermeable types of soil. Earthpit plants can only be recommended for installation in impermeable soil located above the groundwater table. Their construction is particularly inexpensive in connection with plastic sheet gas holders.

(F) Ferro Cement Plants

The ferro cement type of construction can be executed as either a self supporting shell or an earthpit lining. The vessel is usually cylindrical. Very small plants (Volume under 6 m^3) can be prefabricated. As in the case of a fixeddome plant, the ferro cement gas holder requires special sealing measures (provenly reliable and cemented on aluminium foil).

Low cost of construction, especially in comparison with potentially high cost of masonry for alternative plants is also attention for this type of plant.

In this type of plant substantial consumption of necessarily good quality cement, participating craftsmen must meet high standards, uses substantial amounts of steel, construction technique not yet adequately time tested, and special sealing measures for the gasholder are required. Ferro cement biogas plants are only recommended in cases where special ferro cement knowhow is available.

(G) Upflow Anaerobic Sludge Blanket (UASB)

Upflow anaerobic sludge blanket (UASB) reactor systems belongs to the category of high rate anaerobic waste water treatment process.

(i) Anaerobic Fluidized Bed (AFB)

In an anaerobic fluidized bed, the wastewater is mixed with an appropriate amount of chemical for meeting the requirements of anaerobic digestion and introduced at the bottom of a colum with 0.3 - 0.4 m dias and particles at a rate sufficient to fluidise the medium. The fluidised sand provides an extremely large surface area for the attachment of biogas and prevent microorganisms washout even with a large HRT.

(ii) Upflow Anaerobic Sludge Blanket (UASB)

UASB process operates entirely as a suspended growth system and consequently utilizes no packing material. The reactor is initially sealed with any digester sludge & then fed in the upflow mode. The sludge aggregates settle out in the settler compartment. The biogas stored in a gas collector system placed in the upper portion of the system.

(H) Community Biogas Plant

A plant which supplies gas to members of society or village and work on the contribution of dung from many families is known as community biogas plant. These plants are of floating drum or KVIC designs and of sizes 15, 20, 25, 35, 45, 60, & 85 m^3 capacity. The plant is maintained by the management committee set up by the users. The members who contribute dung get back some amount of digested slurry periodically. The users of the gas have to contribute some amount for operation and maintenance of the plant.

The concept has distinct advantage of providing benefits of biogas to those economically backward families, who don't have sufficient cattle head to own a family biogas plant. It also helps the big and marginal farmers maintaining large cattle herd to process the dung to get readyto use manure through large size plant and at the same time getting benefits of required quantity of the biogas. The cost of setting a community biogas plant is less in comparison to combined unit costs of same capacity biogas plants. The successful operation of a community biogas plant develops a sense of fellowship, understanding, cooperative feeling and sacrifice among users. The setting up of such plants are subsidized by the governments. Many such plants are operating well in Gujarat and Maharastra states.

The management of the community biogas plant successfully poses a serious problem. Many plants could not become operational as the contributors of dung were not ready to pay for gas or wanted everyone to share equally the required quantity of dung. The wastage of gas in the absence of metered billing resulted in interrupted supply of gas creating operational and management problems.

(I) Institutional Biogas Plant

Many dairies owned by government, semi government, trusts, organised sectors, goshalas etc. maintain large cattle herds. Such agencies are also encouraged to set up and own a biogas plant of large capacity depending upon the cattle population. Since these plants are owned by institutions, as such the plants known as Institutional Biogas plants (IBP).

The gas produced is used for cooking animal feed, power generation or on farm applications such as pumping water, chaff cutting, feed mixture, feed grinding, maintaining chilling and freezing units etc. besides supplying gas for domestic use to the staff of the institution. These plants are being maintained successfully in all the states under trusts, agricultural universities, dairies etc.

The IBP provides an opportunity of sanitary recycling of the cattle dung in a short period into usable form of manure. The total energy requirement of the farm can be met through the power produced by operation of biogas based internal combustion engine. The digested slurry can be used on the farm or can be sold as polypack biogas manure after drying in urban areas for kitchen gardening. The setting up of these plants is also entitled to subsidy by the governments.

The successful operation of IBP will only depend on the head of the organisation and his interest in the plant.

(J) Landfill Plants

Landfill gas is produced by the complex anaerobiological and chemical processes (anaerobic digestion) of organic wastes, such as food residues, agricultural waste, agroindustrial wastes, paper etc. Landfill gas is largely a mixture of methane and carbondioxide and very small quantities of Nitrogen, Oxygen, Hydrogen and Sulphide. Today landfill gas is used in over 400 landfill sites in Europe, USA, UK and South Africa. In India one project is going on at Delhi. The landfill gas is being used as direct burning in industries and for power production. European community and U.S.A. have legislated that landfill gas be controlled due to adverse effect it can have on humans and the environment.

As a general rule of thumb 60% mass of the material dumped by municipality is organic waste paper, cardboard, wood residues, garden waste and household waste. The remaining 40% include water 20%, which is mostly linked with the organic material, plastic 5%, builders rubble and metal 5%. In general, methane production per unit volume is relatively high by the wastes having high percentage of decomposable organic materials (food, garden and paper waste). Industrial wastes may contain inhibitory concentrations of common salts of sodium, potassium and ammonium.

The organic material of cellulose, carbohydrates and even protein material is readily decomposed by microbes into methane and carbondioxide. These microbes are every where and multiply rapidly if the conditions are right. Most municipal landfill sites end up as huge piles of refuse 1020 meter high.

As a result, anaerobic conditions exist within the dump and such conditions are ideal for methanogenic bacteria to breed. The process can be expressed in terms of chemical equation.

$$C_6H_{12}O_6 \xrightarrow[\text{Bacteria}]{} 3CH_4 + 3CO_2$$

Where the organic material is glucose $C_6H_{12}O_6$. The ratio of these gases and the concentrations of the minor components (Nitrogen, Hydrogen and Hydrogen Sulphide) will depend on the stage of the biological process. The levels of methane and carbon dioxide in the gas, once the methanogenic state has been reached of (six months to two years depending on depth, temperature, moisture content & composition) are 50% and 40% respectively. Methane gas will only form under truly anaerobic conditions and this is best achieved at depths of greater than 10 meters of compacted landfill. Ideally compaction levels should be greater than 0.8 ton/m^3.

The time taken by new fill bioreactor to begin producing biogas varies from place to place and depends on many factors including:

(i) Size, type and composition of refuse
(ii) The effectiveness of clay layer
(iii) Age of the dump
(iv) Moisture content (rainfall in that area)
(v) Size of the landfill
(vi) Depth of the landfills
(vii) Geology of the site
(vii) pH of the liquid (water, industrial effluent etc.)
(ix) Temperature of the landfills.

Landfill scheme consists of six main stages:

- Collection of refuse from city to landfill site
- Screening of waste, compacting and clay covering
- Gas collection from the landfill
- Gas pumping
- Dewatering and gas drying
- Transmission to the end user

Refuse composition effects both gas production rate and the total gas production (Table 11.2). The majority 60% or more of methane gas produced is derived from the biodegradation of paper wastes. Little or no toxic or inhibitory materials should be present in the wastes. Since such substances readily upset the activities of the methane forming bacteria. Industrial waste may contain inhibitory concentrations of common salts of sodium, potassium, calcium, ammonium or sulphide.

Landfill Gas Production

The refuse from the cities is sent to the municipal dumping site. Where it is plied into heaps and regularly covered, usually daily, with a thin layer of clay (Fig. 11.7). The depth of the refuse and clay should be near by 600 mm & 100 mm. The refuse should be compacted with sheep foot roller and clay by the roller. The compactness of refuse and clay increase the efficiency of the landfill.

Total amounts of methane that can be produced form 1 kg. of biodegradable material found to be 0.37 m^3 per kg dry material. Assuming wastes contain about 20% moisture and that only 60% of the dry material is biodegradable, 1 kg. of municipal waste will produce 0.18 m^3 methane. Alternatively, each tonne will produce 180 m^3 methane at about 3045 °C and 1 atmosphere pressure. Soft plant material is converted into methane in the first year itself. The life of hard tree material is about 10 years.

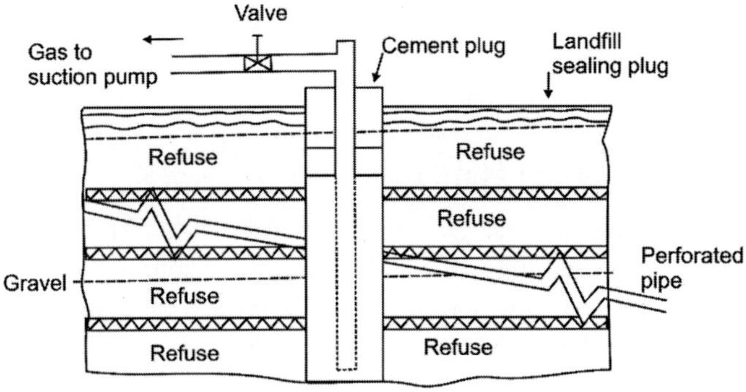

Fig. 11.7: Landfill with a Vertical Gas Collection Well

Table 11.2: Typical Physical Composition of Municipal Solid Wastes and Corresponding Moisture Content Values.

Component	Percentage (Total weight basis)	
Readily Decomposable	0 - 26	50 - 80
Garden waste	0 – 20	30 – 80
Paper waste	28 – 60	4 – 10
Moderately Decomposable	0 - 6	6 – 15
Textiles/leather wood	1 - 4	15 – 40
Refracting Organic		
Plastic / rubber	2 – 10	1 – 4
Non Decomposable		
Glass	4 – 16	1 – 4
Metals	3 – 13	2 – 6
Dirt, ashes, etc.	0 – 10	6 - 12

The gas is drawn from the vertical wells to the power station, cleaned and compressed by a gas delivery unit. The compressed gas is passed through the scrubbing unit in which carbon dioxide is partially absorbed. The scrubbed gas is fed into the compressor which raises its pressure to required level by the engine or directly supply to the house through pipe line fulfilling their domestic need, to industrial unit or to power plant (Fig. 11.8).

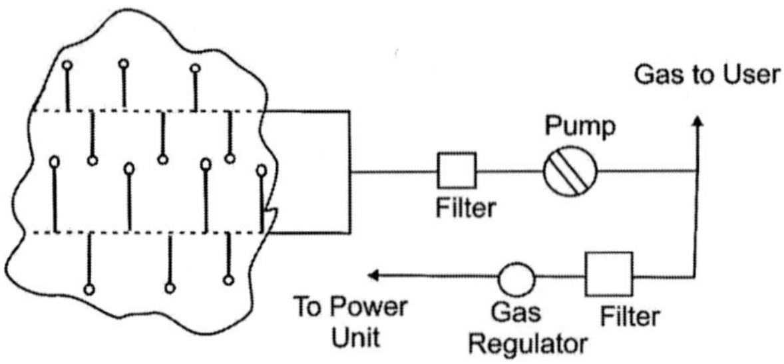

Fig.11.8: Transmission Layout

Development in Landfill Gas Exploitation

There is currently a great deal of interest in India in research into landfill gas enhancement and optimization. Doubling of gas out might be achieved in the medium term and a trebling in the longer term. In addition to optimizing gas production, gas recovery can also be increased. Currently it is believed at the best 70% of gas produced in a site is abstractable. This can be significantly increased using more appropriate gas collection technology. Methane has a specific gravity of 0.55 in relation to air. In other words it is about half the weight of air and so rise when released to atmosphere. Carbon dioxide is more than twice the weight of air and so the resultant mixture of gases when collected moves slowly.

Landfill gas can be used in the following purpose:

- Direct use boilers, furnaces, kilns and household.
- Electricity generation Using engine and turbines

Gas cleaning upgrading to a higher value fuel (by removing carbon dioxide gas and other inert gases making landfill gas 100% methane) cost of gas clean-up has proven prohibitive and exploitation.

If a landfill site is filled up at the rate of 2000,000 tonnes of refuse per month then ideally one should obtain (0.6 x 200 x 400) cum methane per month which is equivalent about 700 M Watts. This is equivalent a reasonable sized power station. Assuming only 10% is trapped and used, that still leaves a lot of energy available, is 70 MW a small power station.

Advantages wise, the landfill plants have:

- The installation costs of the system would be about 10% of normal water borne sewage system.
- The system does not depend on the availability of water land on the house.
- No smell is available near the site.
- Kitchen water can also pass into the site.
- In time of civil unrest, strikes the digester keep on working.

11.9 Design of a Biogas Plant

The biogas plants are designed on the basis of type of waste, material available for digestion, the quantity of gas required, purpose for which gas is required such as cooking, lighting and running of IC engine and pattern of gas consumption .

The type of wastes and its characteristics effect the HRT, thus size of the digester and its type, whereas use and consumption pattern decide the volume of storage chamber. But in all the cases, the availability of waste materials necessary to produce the required quantity of gas will be limiting factor on the size of the digester.

The biogas plant system requires construction of digester, inlet tank, gas holder, slurry pit, therefore, secondary factors which influences the biogas design are selection of site, availability of land, water resources, geography of the area and soil properties.

(A) Design of Digester Tank

The digester tank or digester is main portion of biogas plant which retains biodegradable waste for specific period (HRT) during which gas is produced. The tank should have adequate provision for charging &discharging of slurry. The digester must meet following requirements:

1. It must be absolutely leak proof for liquid or gas.
2. It must have strong structure to withstand slurry pressure.
3. It should have property for resistance to corrosion.
4. It must have property of heat insulation.
5. It must have provision for installation of fittings.

(B) Design of Digester Volume

The digester volume is determined from two parameters namely daily volumetric feed rate and hydraulic retention time. The process of anaerobic digestion is a function of time under controlled conditions. The organic matter needs to be retained for a period till it generates an optimum quantity of gas. If V is digester volume in m^3, v charge of slurry in m^3/day and T is hydraulic retention time in days then:

The volume of daily charge is controlled by various factors such as quantity of gas required daily, number of cattles (availability of total organic waste), number of family members (requirement of gas for cooking) and pattern of gas utilization. In all cases, the limiting factor is the quantity of dung available.

Let us say in a farmer's family having A adult members &cattles equivalent N (which includes total of N_1 Cow, N_2 Buffalos, N_3 Bullock and so on...) and the dung availability from cattle is M (which includes total of M_1 from Cow, M_2 from buffalos, M_3 from Bullock and so on....).

$$\sum_{i=1}^{n} M_i N_i \text{ ; say MN kg}$$

With a conventional digester on an average the gas production rate is 0.04 m^3/kg fresh dung charged. On this basis, daily gas production will be 0.04 MN m^2. Now calculate the total gas requirement on the basis of gas consumption pattern.

a. Daily Gas required for cooking requirement for all family members = Gas required per person per day (0.3 cum)× No.of family member (A)

   $= 0.3\ A\ m^2$.

b. Daily gas required for lighting = No.of lamps required (B lamps of 100 candle power)× Total duration of lighting per day (say 4 hours)× Total gas consumption per lamp (0.15 cum per lamps per hr)

   $= 0.15 \times 4 \times B = 0.6\ B\ m^3$.

c Daily gas required for motive power= Hours Power of the system (HP)say C × Total duration of run per day (say 2 hrs per day)× Gas consumption (0.5 cum per hp per hr)

   $= 0.5 \times 2 \times C = 1.0\ C\ m^3$.

Hence total daily requirement of gas for cooking, lighting and motive power

$= 0.3\ A + 0.6\ B + 1.0\ C$ or say AB m^3.

Now find out which is less, total amount of gas produced from available quantity of waste or the quantity of gas required per day. This is the limiting parameter for determining the volume of daily charge of the digester, that means if:

(i) MN is larger than ABG as consumption rate will decide the quantity of charge.

(ii) MN is less than AB Quantity of dung available will fix the charging rate.

Therefore, the following two general expressions are developed to determine the quantity of dung to be charged daily.

(A) When the number of Cattle is less, then the quantity of dung available will also be less and it will decide the quantity of daily charge as follow:

   W = MN

   where W is quantity available for daily charge to the digester.

(B) When number of Cattle available is large but amount of gas required is less, the amount of gas to be produced daily will decide the quantity of dung to be charged as follows:

If g is gas produced in m^3 per kg of fresh dung charged.

$$W = \frac{AB}{g}$$

It S is the density of slurry (dung + water) in kg/m^3 than

v = daily charge of slurry m^3/day

$$= 2\frac{W}{S}$$

Now one can calculate the digester volume from the product of hydraulic retention time and daily charge of slurry. But the actual volume of the digester for construction should be about 10% more than the theoretical volume. This is to compensate the empty space and volume of partition wall.

Once the actual digester volume is fixed, the next step is to decide the depth to diameter ratio, from which the values of depth & diameter can be obtained from the following relationship.

$$V = 0.785\ D^2\ H$$

Where D is inner diameter of the digestion tank and H is depth or height of the digester, both in meter.

As a general guideline it can be said that for a large capacity plant the H/D ratio may be taken around unity and for a small capacity plant this ratio is more than 1.0 say about 3. The H/D ratio determination also depends upon physical requirement, say whether the digester is horizontal or vertical. In horizontal design diameter dominates than depth where as in vertical design it is other way. It should be noted that diameter and height ratio of the digester in case of Janta model is fixed at 1.75:1.

The shape of the plant should be round, cylindrical or a spherical segment. This type of shape will distribute the load on the surface uniformly. It will also reduce the stresses which develop at sharp corners / turnings.

(C) Design of Gas holder

The gas holder holds the gas till the time of consumption. It should be leak proof. The size of the gasholder depends on the storage capacity of the gas, which in term depends on the gas consumption pattern. As a general guideline, if gas is needed for a longer duration or it is consumed at low rate, the small holding capacity can serve the purpose, such as for cooking and lighting purposes. Whereas for other purpose say for motive power generation, where the system consumes gas at a faster rate, more gas storage capacity will be required. The holding capacity of fixed dome plant i.e. volume of the gas storage and dome is 40-60% of the plant capacity.

In the Indian context, in family plants, generally gas in needed for cooking & lighting purposes. In these applications the holding capacity of gasholder is taken as 40% of the plant. The structural requirement in case of fixed dome plants is same as that of digester. Whereas in case of floating drum plants, the drum should be made of thicker gauge, preferably 10 gauge.

(D) Design of Mixing Tank

The substrate is diluted in the mixing tank with water to yield homogeneous slurry for the desired solid content. A plug can be used to close off the inlet pipe during the mixing process. Also, the impurities liable to clog the plant are removed from mixing tank. The fibrous and foreign matter, if any is raked off the surface. Any stone, and or soil settling at the bottom are cleaned out after the slurry in admitted into the digester.

The normal size of the mixing tank is about 1.5 to 2.0 times the daily input quantity. For reasons of hygiene, toilets should have direct connection to the digester without any need of mixing tank.

(E) Design of Inlet Pipe

The function of inlet pipe is to feed material from inlet tank flows through this into the digester. For this Asbestos cement (AC) concrete or plastic pipe are preferred. For liquid substrate and for fibrous substrate the pipe diameter should be 1015 cm and 2030 cm respectively. The pipe should lead straight into the digester at a steep angle. Also, the pipe should penetrate the digester wall at a point below the slurry level.

(F) Design of Outlet Pipe

It carries the digested slurry to the outlet tank. Its design is same as inlet pipe. However, it is used only in KVIC plants.

(G) Design of Outlet Tank

The outlet tanks hold the digested slurry coming out of the plant. It should allow slurry only upto some level. Volume is designed with a over flowing tube and it should be covered with concrete slabs to avoid accidents.

(H) Design of Gas Pipe and Valves

The gas pipe and valves carry the biogas to the point of consumption. They should be airtight. The bottom of the gas pipe support fitted along with the inner surface of the dome. Otherwise, there is a danger that slurry will enter the pipe. Half-inch diameter pipes are normally adequate for small plants and pipe length up to 30 meters. However, for longer gas pipe a detailed pressure calculation may be done. Also, the condensed water collected in a gas pipe is removed by a water tap at the lowest point of the sloping pipeline.

11.10 Selection of Model

The merits and demerits of fixed dome and floating drum type biogas plant need to be considered while selecting a model. The selection also depends on the following factors:

(A) Technical

Before deciding a type of model, the merit and demerit of each plant based on technical points be considered.

The floating gasholder type can be constructed with moderate skill by village masons, whereas the construction of fixed dome type plant involves greater skill and care.

Provision of a device to break the scum formed on the slurry is also important. In floating drum type biogas plant the gas holder has provision for breaking scum by rotating gas holder half way in both directions for 23 minutes daily. Only stirring is possible by inserting bamboo through outlet in fixed dome biogas plant. In floating gasholder type biogas plant slurry comes out automatically due to gravitational force i.e. due to the different of inlet and outlet levels. While in fixed dome type biogas plant the slurry comes out only when the plant is full of gas or gas is at maximum pressure. Often the slurry does not come out automatically and it has to be removed by buckets. The fixed dome plant has to be charged daily at rated capacity, whereas the gas produced in floating dome plant is proportional to daily feed. Future, good quality of bricks, cement, fine aggregates are required for fixed dome type biogas plant, while a leak proof gasholder is required in case of floating plant.

(B) Climate

Climate is playing an important role in the productions of biogas. Hence, adequate consideration is given towards climate of place where plant is to be constructed.

It has been noted that the gas production goes down drastically in winter. This is a serious problem, especially in the northern part of India. The rate of biogas production is highly depended on the atmospheric temperature, as the plant does not have heating provision and automatic control of temperature. Maximum gas yield has been reported when the slurry temperature is between 30-35°C.

The country can be divided into five climatic zones based on the mean atmospheric temperature during winter months (Table 11.3).

Table 11.3: Different Climatic Zones

Zone	Mean Temperature During Winter Months (°C)	Recommended Retention Period
Zone I	More than 25	30 days
Zone II	Between 20-25	30-40 days
Zone III	Between 15-20	40 days
Zone IV	Between 10-15	55 days
Zone V	less than 10	Not suitable

The above temperature range decide the retention period of biogas plant. The retention period is the period in days for which the slurry remains in the digester for getting 80% the total gas .

(C) Geographical

- Both floating drum and fixed dome type biogas plant can be constructed where digester pit can be excavated to about 3.0 meter without blasting and removing underground water.
- Horizontal and Vertical designs are available.
- The plant should not be constructed within a radius of 15 meter from drinking water sources. This would ensure that no contamination of water with dung slurry due to seepage takes place.

(D) Economics

Economic of a plant is also important in the selection of biogas model. The cost of the biogas plants very according to the prevailing prices of materials from place to place.

- The cost of installation of floating drum type plant is about 50% higher than of fixed dome type plant.
- Plant selected should be cheaper provided other factors are also favourable. Fixed dome type plants are cheaper than any other models as locally material like stone, lime can be used for the construction of digester.
- The maintenance cost should also be as low as possible. Steel gasholders required painting with black paint once in a year. Fixed dome type has little or no maintenance cost.

11.10 Selection of Size

The number of animals largely determines the size of the biogas plant and human those are contributing wastes for its operation. This would decide the maximum capacity of the plant, which can be selected.

- One average cow, bullock and buffalo stall fed produce about 10,12 and 15 kg of dung per day respectively. In case of grazing cattle, the collection efficiency is 60-70%, followed by 50% dung per day (14).
- One kilogram of fresh dung (Containing about 20% solids) generates about 0.036 cum of gas under average climatic conditions.
- Average human excreta per person per day is 0.4 kg and gas yield is 0.07 cum/kg. After deciding the maximum capacity, the actual capacity is based on the number of users and the type of use.
- The quantity of biogas consumed per person per day is 0. 24 cum per day for rural person for cooking and for lighting 0.13 cum/hour for 100 candle power lamp. In dual fuel engine (75-80% replacement of diesel) biogas consumed 0.50 cum per BHP hour.

In order to decide a suitable size of the plant actual availability of cattle dung be assessed. If requirement of gas is less than the gas produced from actual available dung than total requirement of gas would decide size of plant. However, if gas requirement is higher than the gas actually could be produced from available dung than available quality of dung would decide the size of plant. In winter gas requirement increases and production reduces, therefore, it would be better to install a little over sized plant so that gas requirement can be fulfilled in winter season also. This however, depends on availability of dung.

11.11 Selection of Beneficiaries

The successful running of the biogas plant depends upon the selection of beneficiary. The beneficiary must have the sufficient cattle population for the availability of required quantity of cattle dung. An accurate estimate of available dung should be made by measuring for a week. For example, to estimate the amount of dung available, actual weighing of dung collected for seven days should made and an average taken. Estimates based on the number of cattle is often unreliable for many reasons:

- The size of the cattle
- The quality and quantity of the feed given
- The period for which the cattle are stable bound also the grazing pattern and the system of working of the bullocks vary from region to region. In

same way estimating the number of persons using a latrine is calculated, only the number of permanent users should be taken into account.

11.12 Land for Plant

- The beneficiary must have sufficient land for the construction of the plant where sunrays fall on it for maximum time of the day.
- The plant should not be constructed on the Government, Cooperative land etc.
- The beneficiary should also have sufficient land for drying of the slurry as it is not possible to transport wet slurry to the fields every day.
- The plant construction should be avoided in the premises of Government quarter.
- The plant must be near to the kitchen so that pressure drop in supply of gas can be minimized.
- The plant must be near to cattle shed so that labour for collection of dung can be reduced.

11.13 Water Requirement

- It has been found through experiments that 8-10% total solids in slurry give optimum gas production. Thus, cattle dung and water are mixed in 1:1 ratio. This will bring the slurry to 8-10% solid contents. Therefore, sufficient amount of water is required for the operation of the biogas plant i.e. 50, 75 and 100 liters for 2, 3 and 4 cum size plant. Hence beneficiary must have a source of water near the house.
- The beneficiaries who are collecting drinking water from a distance should not be benefited with the biogas plant.
- The beneficiary whose house come under water logged areas should not be chosen.
- The beneficiary who has been benefited with the conventional source of energy should not be pressed for the biogas plant. If the beneficiary is interested only the plant be installed.
- The quality of water should be potable in nature. Too much alkaline or acidic water may change the gas production.

11.14 Selection of Site

While selecting a site following points must be kept in mind.

- The site should be close to the kitchen or the place of use, it will reduce the cost of gas distribution system.
- The site should be near to the cattle shed. It will reduce the transportation of the cattle dung.
- Enough space for storage and drying of slurry should be there.
- Plant should be established at least 15 meters away from a drinking water wall.
- The land for the plant should be levelled and slightly above the ground level to avoid in flow or run off water.
- Plant should and get clear sunshine during most part of the day.
- Soil should not be too loose. It should have a bearing strength exceeding 2 kg per sq.cm.
- There should be easy access to water supply.
- There should not be any big tree near plant site to avoid damage to digester wall by roots of tree and shadow on plant.
- Water table should not be very high, preferably it should be more than 3 meter deep.

11.15 Do's and Don'ts of Biogas Plants

Inspite of careful construction of the plant, in due course of use the user may face with certain problem. The trouble shooting should be done as early as possible to avoid any further damage or loss to the plant. The common problem, that may be faced after the initial gas production started, the reason, diagnosis and how to correct the same, are given below with possible remedies:

S. No.	Do's	Don'ts
1.	Select the size of biogas plant depending on the daily availability of dung.	Don't install a bigger size biogas plant, if you do not have sufficient dung to feed it.
2.	Make sure that the biogas plant is installed in an open place and gets plenty of sunlight during the day round the year	Do not install the biogas plant under or near a tree or water source,, inside the house or at any other shady place.
3.	The outer side of the digester wall must be firmly compacted with soil.	The soil around the outer wall of the digester should not be loosely compacted,, otherwise the digester may get damage.
4.	The digester should be cured for 10 to 12 days after construction.	Cracks may be developed,, if not cured properly.
5.	Feed the plant with homogenous dung water mixture in correct proportion.	Inadequate dung or water may affect the gas production.
6.	Make sure that the mixture of cow dung and water is free from soil and other foreign particles.	Do not allow any soil particles to enter into the digester otherwise they will clog the inlet pipe at bottom.
7.	Rotate the gasholder once or twice a day in order to break the scum.	Do not allow scum to form in the digester otherwise the gas production will stop.
8.	As soon as gas production starts after charging the plant fix the pipeline from the gasholder to the kitchen.	The gas should not be stored for a long period.
9.	Open the gas regulator cock only at the time of its actual use.	Do not leave the gas regulator cock open when the burner is not in use.
10.	Adjust the flame by turning the air regulator till a blue flame is obtained.	Do not use the gas if the flame is yellow. Adjust the flame till it is blue in colour.
11.	Drain out water, which may have accumulated in the gas supply pipe in water trap, at every 1015 days interval.	Do not allow any water to accumulate in the gas pipeline.
12.	Check the outlet pipe periodically during summer to avoid clogging.	Do not allow the slurry to dry or clog at the top of the outlet pipe.
13.	While filling slurry in bigger size plants, make sure that it is filled equally on both sides of the partition wall upto the guide frame.	Do not fill the slurry unequally,, it may cause the central wall to collapse.
14.	When the digester is full with the homogeneous dung water mixture, place the gas holder on the central pipe of the guide frame. Fit a heavy-duty valve on the gasholder.	Do not keep the valve loose, otherwise gas will escape from the plant.

11.16 Biogas Storage

The storage of biogas in balloon or in cylinders can be made by increasing pressure. However, it can not be liquefied as such. More gas is required for any operation thus required larger size of storage tank.

When the gas is stored in balloon, it cannot be used in burner or lamp, unless pressure equivalent to 7-10 cm of water column is applied over balloon. However, it can be used to run an engine, which sucks the gas. The standard gas cylinder can be used to store the biogas upto a pressure of 120 kg/cm^2. However, at the time of use of biogas, its pressure should be reduced through use of a suitable pressure regulator.

11.17 Biogas for Vehicles

The utilization of biogas in vehicles requires a method of compact storage to facilitate the independent movement of the vehicle for a reasonable time. Larger quantities of biogas can only be stored at small volumes under high pressure, e.g. 200-300 bar, or purified as methane in a liquid form at cryogenic conditions, i.e. -161 °C and ambient pressure. The processing, storage and handling of compressed or liquified biogas demand special and costly efforts.

Compression is done in reciprocating gas compressors after filtering of H_2S. At a medium pressure of about 15 bar the CO_2 content can be "washed out" with water to reduce the final storage volume. Intermediate cooling and removal of the humidity in molecular sieve filters are essential as the storage containers should not be subjected to corrosion from inside. The storage cylinders, similar to oxygen cylinders known from gas welding units, can be used on the vehicle as "energy tank" and in larger numbers as refilling store.

One cylinder of 50 l volume can store at a pressure of 200 bar approximately

- 15 m^3 un-purified biogas (CH_4 = 65% Vol) with an energy equivalent of 98 kWh or 101 diesel fuel, or
- 13 m^3 purified biogas (CH_4 = 95% Vol) with an energy equivalent of 125 kWh or 12.51 liter diesel fuel.

The storage volume thus required on the vehicle is still five times more than is required for diesel fuel. Purification of biogas to CH_4 increases the storage efficiency by 25 - 30% but involves an extra gas washing column in the process.

Purified biogas, i.e. methane, has different combustion features than biogas because of the lack of the CO_2 content. It combusts faster and at higher temperatures; this requires different adjustments of ignition timing. Dual fuel methane engines are prone to increased problems with injector nozzle overheating and have to operate on higher portions of diesel fuel (about 40%) to effect sufficient cooling of the jets.

Liquification of biogas requires drying and purification to almost 100% CH_4 in one process and an additional cryogenic process to cool the CH_4 down to -161 °C where it condenses into its liquid form. Storage is optimal at these conditions as the volume reduction is remarkable, i.e. 0.6 m^3 with an energy content of 6 kWh condense to one liter of liquid with an energy equivalent of 0.61 diesel fuel. The required tank volume is only 1.7 times the volume needed for diesel fuel.

This advantage is opposed by a more sophisticated multistage process, the handling of the liquid in specially designed cryo-tanks with vacuum insulation and the fact that for longer storage, it has to be kept at its required low temperature in order to prevent evaporation. This requires additional energy and equipment.

The use of biogas as a fuel for tractors on farms has been elaborately researched. The processing of the gas does not only require about 10% of the energy content of the gas, mainly for compression, but also involves considerable investment. The tractor itself needs to carry four gas cylinders at least for a reasonable movement radius. A 40-kW tractor can then operate for about six to seven hours at mixed/medium load. The modification of the tractor has to include a three-stage pressure reduction system as the fuel gas is fed to the mixer at low pressure, i.e. about 50 mbar.

Modification into an Otto gas engine includes the risk of non-availability of the tractor at biogas shortage. It therefore needs LPG as spare fuel or another diesel tractor standby. Dual fuel tractor engines, on the other hand, are difficult to control, especially because of their frequent speed and load changes during operation in the field.

11.18 Planning a Biogas Engine System

The supply of mechanical or electric power from biogas is only feasible using a biogas engine. The installation of a biogas engine however requires an appropriate planning of the fuel production and also the consumption/ operation procedures. This is a crucial exercise which can usually be-avoided when the power is purchased from an electric grid.

As an engine in general does not supply energy, but rather transforms one form of energy, here biochemical, into another form, mechanical energy, its operation requires a source of energy on one side and a consumer of the energy on the other. The coordination of the energy source (biogas production plant), the transformer (engine) and the consumer (driven machine) is therefore of utmost importance for a technically and economically satisfactory performance of the whole system.

The following parameters have an influence on the system's performance:

a) Technical Parameters

- Biogas production in the biogas plant under consideration of the plant's size, inputs and operation as well as the reliability of the gas supply system.
- Power demand of the driven equipment with regard to its anticipated fluctuation or the anticipated point of continuous operation.
- Demand of low and medium temperature heat from engine's waste heat (cogeneration).
- Daily schedule of operation with regard to biogas consumption, plant size and necessary gas storage capacity.
- Speed or speed range of the driven machine and the engine.
- Mode of control, manual or automatic.
- Local availability of engine service, spare parts, technical expertise and sufficiently competent operating personnel.
- Anticipated development of energy supply and demand in the future.

b) Economic Parameters

- Price of biogas plant cum ancillaries.
- Price of engine cum modification.
- Price of driven machine and energy distribution system (electrical wiring, water system, etc.) unless already existing.
- Operational cost of biogas system, i.e plant, engine and driven machine.
- Cost of the system's service and maintenance.
- Capital costs (interest rates, pay back periods, etc.).
- Expected revenue from provision of selling energy or services, including the use of the engine's waste heat.
- Savings by the omission of cost for other fuels or forms of energy.
- Anticipated development of economic parameters (inflation, laws, regulations, fuel taxes, etc.).

c) Alternative Possibilities of Power Supply

- Electric motors under consideration of availability, reliability and price of electricity from another (e.g. public) supplier.

- Small hydropower in favourable areas for direct drive of machines or generation of electricity.
- Wind power in favourable areas under consideration of the schedule of power demand and the wind regime.
- Diesel, petrol, alcohol or LPG as engine fuels under consideration of availability, price and given infrastructure for a reliable supply.

To summarize, a biogas engine is only one module in a system and can only perform to satisfaction when all other components are well integrated. Furthermore, the economic and boundary conditions, realistically assessed, have to be more favourable than for alternative solutions.

Example 3.1: Design a Biogas plant to supply 100 kWh of electricity per day. The engine and generator efficiency are 25% and 80% respectively. The methane content in biogas is 60 %. Determine the number of cattle, biogas required and greenhouse gas mitigation potential.

Solution

Overall efficiency = 0.25 × 0.80 = 0.20

$$\text{Total Energy Required} = \frac{100}{0.2} = 500 \text{ kWh}$$

$$\text{The Calorific Value of methane is} 35\frac{\text{MJ}}{\text{m}^3}; {}^{35}\!/_{3.6}\,\text{kWh} = 9.97\text{kWh}$$

But, Biogas contains 60% methane,

Therefore, The calorific value of 1 m^3 biogas = 9.97 × 0.6=5.98 kW/m^3

$$\text{The total volume of biogas require to produce} 500\frac{\text{kW}}{\text{day}} = \frac{500}{5.98} = 83.61\ m^3$$

But,biogas production from 1kg of cattle dung = 0.04m^3

$$\text{Therefore, The Cattle dung Required} = \frac{84\ \text{m}^3/\text{day}}{0.04\ \text{m}^3/\text{kg}} = 2100\text{ kg / day}$$

Assume 1 cattle will produce 10kg of dung /day

$$\text{Therefore, total number of cattle} = \frac{2100}{10} = 210 \text{ cattles}$$

Daily charge of slurry is about 2100 kg per day for the biogas production we require to add equal amount of the water to maintain 8-10 % total solid content

Total daily charge slurry = 2100 dung + 2100 water = 4200 kg slurry

Now,the density of slurry = 1090 $^{kg}/_{m3}$

Therefore, Daily charge of slurry = 2100kg dung + 2100kg water = 4200kg slurry

Daily volumetric charge of slurry into the digester(v) = $\frac{4200}{1090}$ =3.85 $^{m3}/_{day}$

Let us,now consider Hydraulic Retention Time (HRT) = 45 Days.

Therefore,volume of the digester = v × HRT = 3.85 × 45 = 173.25 $m^3 \approx 174\ m^3$

As the 10 % extra space for the gas collection,

Actual Volume of digester = 174 + (0.1 × 174) = 191.4 m^3

Volume of digester = C/A × Height

Considering the Diameter to Height ratio is equal of the digester,

$$V = A \times H$$

$$V = \frac{\pi}{4} d^2\ h$$

$$V = \frac{\pi}{4} d^3$$

$$d = \sqrt[3]{\frac{191.4 \times 4}{3.14}} = 6.24\ \text{m}$$

Dimeter = 6.24 m

Height = 6.24 m

Now, Design the Volume of gas holder = 40 % of plant capacity

= 191.4 × 0.4 = 76.56 m^3

1 kWh of electricity produces the following amount of CO_2 in the environment

= 0.2 kg CO_2/kWh of Natural Gas

= 0.23 kg CO_2/kWh of LPG

= 0.25 kg CO_2/kWh of Petrol

= 0.26 kg CO_2 /kWh of Kerosene

= 0.27 kg CO_2 / kWh of Diesel

= 0.34 kg CO_2 / kWh of Coal

So,the this 100kWh/day plant will reduce the CO_2 emission

= 20 kg CO_2 / kWh of Natural Gas

= 23 kg CO_2 / kWh of LPG

= 25 kg CO_2 / kWh of Petrol

= 26 kg CO_2 / kWh of Kerosene

= 27 kg CO_2 / kWh of Diesel

= 34 kg CO_2 / kWh of Coal

11.19 Parameters Affecting Anaerobic Digestion

There are several parameters which affect the anaerobic digestion / gas yields and they can be divided into two parts:

11.19.1 Environmental Factors

There are a few environmental factors which limit the reactions if they differ significantly from their optimum levels. Factors of most interest are s **(a)** temperature, **(b)** pH and **(c)** nutrient contents of the raw materials,

(a) **Temperature:** It is a factor which affects most small & medium size biogas installations in developing countries. There are three zones of temperature in which biogas is produced by anaerobic fermentation of organic matter, viz.: 1) Mesophillic, 2) Thermophillic and 3) Psycrophillic zones. The optimum temperature of digester slurry in Mesophillic zone is 35°C, 55°C in Thermophillic zone and 10°C in Psycrophillic zone. In different temperature zones different sets of microbes, (bacteria) especially the methanogens remain active; whereas the other two groups of microbes either remain dormant and thus more or less inactive as far as the anaerobic digestion is concerned or get killed. However, the rate of fermentation is much faster at high temperature. Most rural household biogas plants (digesters) in developing countries operate at ambient temperatures, thus digester slurry temperature is susceptible to seasonal variation but is more dependent on the ground temperature than the atmospheric temperature. As a result, gas output in winter falls by up to 50 %. Below a slurry temperature of 10°C all the reactions cease to take place but revive gradually with the rise in temperature.

(b) **pH:** The pH range suitable for gas production is rather narrow i.e. 6.6 to 7.5. Below 6.2 it becomes toxic. pH is also controlled by natural buffering effect of $\mathbf{NH_4^+}$ and $\mathbf{HCO_3^-}$ions. pH falls with the production of volatile fatty acids (VFAs) but attains a more or less constant level once the reaction progress.

(c) **Nutrient Concentration:** Biogas producing raw materials can be divided into two parts i.e. 1) Nitrogen rich and 2) Nitrogen poor. Nitrogen concentration is considered with respect to carbon contents of the raw materials and it is often depicted in terms of C to N ratio. Optimum C/N ratio is in the range of 25 to 30:1. In the case of cattle dung the problem of nutrient concentration does not exist as C/N ratio is usually around 25:1.

11.19.2 Operational Factors

Operational factors contributing to the gas production process are: (a) Hydraulic Retention Time (RT) - also referred as detention or residence time, (b) Slurry Concentration and (c) Mixing.

(a) **Hydraulic Retention Time (HRT):** The number of days the feed material is required to remain in the digester to begin gas production, is the most important factor in determining the volume of the digester which in turn determines the cost of the plant. The larger the retention period, higher the construction cost. In India, the different HRTs are recommended for three different temperature zones as shown in Table 11.4.

Table 11.4: Different HRT Zones in Country

Zone	Average ambient temperature	HRT (days)	Approximate regions
I	>20°C	30	Andhra Pradesh, Goa, Karnataka, Ketala, Maharashtra, Tamil Nadu, Pondicherry and Andaman & Nicobar Islands
II	15-20°C	40	Bihar, Gujrat,Haryana, Jammu region of J&K, Madhya Pradesh, Orrisa, Punjab, Rajasthan, Uttar Pradesh and West Bengal
III	<15°C	55	Himanchal Pradesh, North-eastern states, Sikkim, Kashmir region of J&K, and hill districts of UP

(b) **Slurry Concentration:** This is denoted by dry matter concentration of the inputs. The optimum level for cattle dung slurry in the range of 8 to 10% and any variations from this result in lower gas output. Mixing four parts of dung with five parts of water forms a slurry with dry matter concentration of about 9%, whereas 1 part of dung to 1 part to water would give a slurry concentration of 10%. This also affects the loading rate etc.

(c) **Mixing & Stirring**: Proper mixing of manure to form an homogenous slurry before it is fed in the digester, is an essential operation for better efficiency of biogas systems, whereas proper stirring of digester slurry ensures repeated contact of microbes with substrate and results in the utilization of total contents of the digesters. An extremely important function of stirring is the prevention of formation of scum layer on the upper surface of the digester slurry which, if formed, reduces the effective digester volume and restricts the upward flow of gas to the gas storage chamber. Mixing results in premature discharge of some of the input and a perfectly unmixed system is likely to result in better reaction rate but for the problem of scum formation.

12

Biomass Energy

12.1 General

The term "biomass" generally refers to renewable organic matter generated by plants through photosynthesis. Materials having organic combustible matter is also referred under biomass. Biomass is an important fuel source in our overall energy scenario. Biomass is produced through chemical storage of solar energy in plants and other organic matter as a result of photosynthesis. During this process conversion of solar energy in sugar and starch, which are energy rich compounds takes place. The chemical reaction of photosynthesis can be written as:

$$6CO_2 + 6H_2O + \text{sun light} \rightarrow C_6H_{12}O_6 + 6O_2 + 636_{kcal}$$

It indicates that the storage of 636 kcal is associated with the transfer of 72 gm carbon into organic matter. Biomass can be directly utilized as fuel or can be converted through different routes into useful forms of fuel. In fact, biomass is a source of five useful agents, which start with `F' like Food, Fodder, Fuel, Fiber and Fertilizer. Further, biomass has many advantages like.

(a) It is widely available

(b) Its technology for production and conversion is well understood.

(c) It is suitable for small or large applications

(d) Its production and utilisation require only low light intensity and low temperature (5-35°C)

(e) It incorporates advantage of storage and transportation

(f) Comparatively, it is associated with low or negligible pollution.

12.2 Biomass Classification

Biomass includes plantation that produces energy crops, natural vegetable growth and organic wastes and residues. The biomass classification is illustrated in Fig. 12.1. It can be grouped as:

(1) **Agricultural & Forestry Residues:** Silviculture Crops.

(2) **Herbaceous Crops:** Weeds, Napier grass.

(3) **Aquatic and marine biomass:** Algae, Water hyacinth, Aquatic weeds, Plants, Sea grass beds, Kelp and Coral reep etc.

(4) **Wastes:** Municipal solid waste, municipal sewage sludge, animal waste and industrial waste etc.

India produces about 320 million tonnes of agricultural residues every year. Similarly, 273 million cattle population produces on an average about 433 million tonnes of dung annually. Fuel wood is another major source of biomass in India. The fuel wood consumption in India is estimated to be about 227 million tonnes per year. Some of biomass sources are given below:

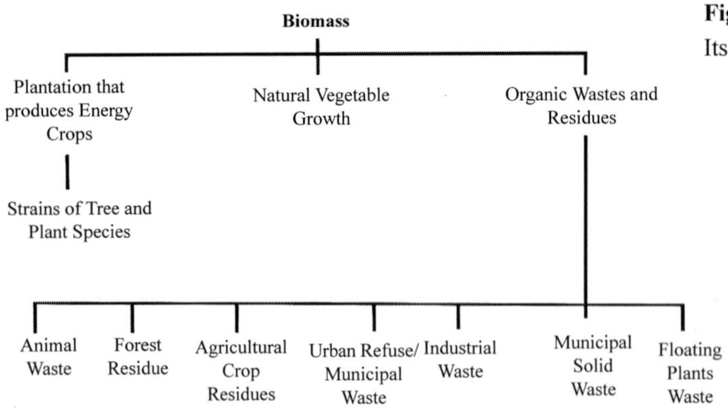

Fig. 12.1 Biomass and Its Classification

12.2.1 Energy Plantation

This term refers to an area that is used to grow biomass for energy purposes. The idea behind energy plantation programme is to grow selected strains of tree and plant species on a short rotation system on waste or arable land. The sources of energy plantation depend on the availability of land and water and careful management of the plants. As far as suitability of land for energy plantation is concerned following criterion is used

(1) It should have minimum of 60-cm annual precipitation

(2) Any arable land having slope equal to or less than 30% is suitable for energy plantation.

The economics of energy plantation depends on the cost of planting and availability of market for fuel. Whereas these two factors are location specific and these varies from place to place. Further productivity of this programme depends on the microclimate of the locality, the choice of the species, the planting spacing, the inputs available and the age of harvest. There are many suitable species for energy plantation, few of them are:

(a) *Acacia nilotica* (Babul),

(b) *Acacia auriculiformis* (Bengali Babul),

(c) *Dalbargia sissoo* (Shisham),

(d) *Eucalyptus comaldulusis* (Eucalyptus),

(e) *Leucaena lencocephala* (Subabul),

(f) *Prosopis chilensis* (Perdesi Babul),

(g) *Prosopis Juliflora* (Vilayati Babul),

(h) *Tamarix articulate* (Jungle jalebi),

(i) *Tamarindusindica* (Imali),

(j) *Albizzalebbek* (siris).

12.3 Biomass Characteristics

Biomass can be characterized for its utility and different energy usage. The various characteristics of biomass falls under following categories

(a) Proximate analysis

(b) Ultimate analysis

(c) Ash deformation and fusion temperature

(d) Calorific value

(e) Rate of devolatilization

(f) Bulk density.

These properties varies from the species to species on their moisture content as well as fuel preparation method employed.

12.3.1 Proximate Analysis

It is the method for measuring various properties of biomass such as moisture content, fixed carbon, volatile matter and ash. The moisture content is one of the important properties of biomass, over which its heating value depends. The moisture content is determined by drying the weighed amount of sample in an open crucible kept at 110°C in an oven for one hour. Always biomass sample is first grind to form fine powder, then this powdered sample is kept for determination of proximate analysis. In fact, for most of the biomass, the consistency in weight is obtained within one hour at 110°C or so. If required, the period of heating may be increased till the consistency of weight is obtained. The difference between the initial and final weights is taken as moisture content of the fuel.

The experimentation of moisture content determination is extended for measurement of ash content of biomass. The sample, so obtained after determination of moisture content, is then gradually heated to 750°C in a muffle furnace and is kept for two hour or were till constant weight is recorded. The weight of the residue represents the ash content of the biomass.

Volatile matter is determined by keeping the dried sample in a closed crucible at 600°C for six minutes and then at 900°C for another six minutes. The difference in the weight due to the loss of volatiles is taken as the total volatile matter present in the biomass. The fixed carbon content is found by applying the mass balance for the biomass sample. The carbon content determined through this method is not the actual carbon content present in biomass but only the nonvolatile part of carbon content, as same of the carbon present in biomass also escapes along with the volatile.

Measurement

(a) Weight of empty silica crucible A

(b) Weight of crucible and sample B

(c) Weight of crucible + sample after drying at 110°C C

(d) Weight of crucible + ash after ignition D

The sample calculation for measurement of proximate analysis is as follows:

(a) Moisture content % = $\frac{B-C}{B-A}\times 100$

(b) Total solid % = $\frac{C-A}{B-A}\times 100$

(c) Total volatile solids % = $\frac{C-D}{B-A}\times 100$

Moisture content on wet basis

$$MC_{wb} = \frac{\text{Wet weight matter} - \text{Dry weight matter}}{\text{Wet weight matter}}\times 100\%$$

Moisture content on dry basis

$$MC_{db} = \frac{\text{Wet weight matter}}{\text{Dry weight matter}}\times 100\%$$

Example 12.1:*Prosopis Juliflora* is cutting around 35% moisture content on wet basis. The ideal moisture content to gasified it is about 12%. Calculate the amount of water is to be removed from per tons of *prosopis juliflora.*

Solution

$$MC_{wb} = \frac{1000 - \text{Dry weight matter}}{1000} \times 100$$

Dry weight of *prosopis juliflora* = 650 kg

Weight after drying at 12% moisture on wet basis is

$$12 = \frac{\text{Wet weight matter} - 650}{\text{Wet weight matter}} \times 100$$

Wet weight is 738.6 kg

Hence amount of water to be removed from prosopis juliflora to make it suitable for gasification is:

= Weight at 35 % moisture content-Weight at 12% moisture content

= 1000-738.6

= 261.4 kg of water

12.3.2 Ultimate Analysis

The ultimate analysis gives carbon, hydrogen, oxygen, nitrogen, and sulphur contents of the fuel. C-H-O analyser determines the carbon and hydrogen contents by standard method. Further, knowing the ash content, oxygen is determined by difference. However, the samples must be dried prior to analysis. Nitrogen and sulphur are normally negligible.

CHO analyser is essentially consisting of (i) an electric furnace (ii) a sample column and (iii) absorbent column. The dry matter is powdered and weighed (w_1) before putting it in the sample column. The absorbent column is filled with a weighed quantity (w_2) of calcium hydroxide. Subsequently the furnace is started and oxygen from a separate oxygen cylinder is supplied to the sample column at a pressure of 4 PSL. A temperature of more than 1400°C is maintained for about 20 minutes. Then the furnace is switched off and the fused sample is taken out and weighed (w_3). The calcium hydroxide from the absorbent column is also taken out and also weighed (w_4). From these observations the carbon content of the sample can be determined using the following relationship. The difference ($w_4 w_2$) will give carbon dioxide formed.

$$\text{Carbon in absorbent } (w_5) = \frac{w_4 - w_2}{w_3} \times 12$$

$$\text{\% Carbon in the sample} = \frac{w_5}{w_1} \times 100$$

12.3.3 Ash Deformation and Fusion Temperature

There is a standard test for measurement of fusibility of coal and coke. It is based on ASTM D1857. In this method, first of all biomass is dried and grounded and finally placed in the muffle furnace at 750°C in the presence of air till constant weight is obtained. The residual ash so obtained is then finely ground. Then this ash is converted in plastic mass by adding a solution containing 10% dextrin, 0.1% salicylic acid and 89.9% H_2O by weight. This plastic mass is moulded to a cone shape by pressing it into a suitable mould. Then, these canes are taken out and allowed to dry. Those dried canes placed on a refractory base are then inserted in a high temperature furnace to about 800°C. After about 15 minutes interval the temperature of the sample is raised at an increment of 50°C. During each interval the shape of the cane is observed. The temperature at which initial rounding off or bending of the apex of the cone is observed, is known as ash deformation temperature. If this temperature is further increased, the same sample would fuse into a hemispherical lump. The temperature during this phenomenon is known as "ash fusion temperature".

12.3.4 Heating value or Calorific value

The heating value of the oven dried biomass samples is determined by Bomb Calorimeter method. The heat evolved when unit mass of fuel is burnt, is known as higher calorific value (HHV). The Bomb Calorimeter method is used for determining higher heating value of biomass. Whereas, the lower heating value (LHV) is calculated by subtracting the heat liberated during condensation of the water vapour formed due to combustion of hydrogen content of the fuel. The hydrogen content is known by the ultimate analysis. The lower heating value is also known as net heating value. Normally net heating value is utilized in combustion and gasification process. The lower heating value can be estimated using following formulae:

$$LHV = HHV - h_g\left(\frac{9H}{100} - \frac{M}{100}\right)$$

where H, M, and hg are hydrogen percentage, moisture percentage, and latent heat of vaporization respectively.

12.3.5 Bulk Density

It depends upon the moisture content, shape and size of the biomass. As this property normally varies depending upon the fuel preparation method employed, therefore it should be determined in site for specific applications.

12.4 Biomass Production Technique

The biomass production should be viewed with number of points starting from preparation of soil to planting seedlings and upto harvesting of the biomass. Infact, these all operations are combined for biomass production to achieve multiple benefits. Thus, a careful planning is required for biomass production, which consists of integration of different techniques and improved methods. The general sequence for biomass production, which consists of integration of different techniques and improved methods. The general sequence for biomass production is illustrated in Fig. 12.2. It includes step starting from site survey, nursery techniques, transplanting techniques and maintenance of the plantation.

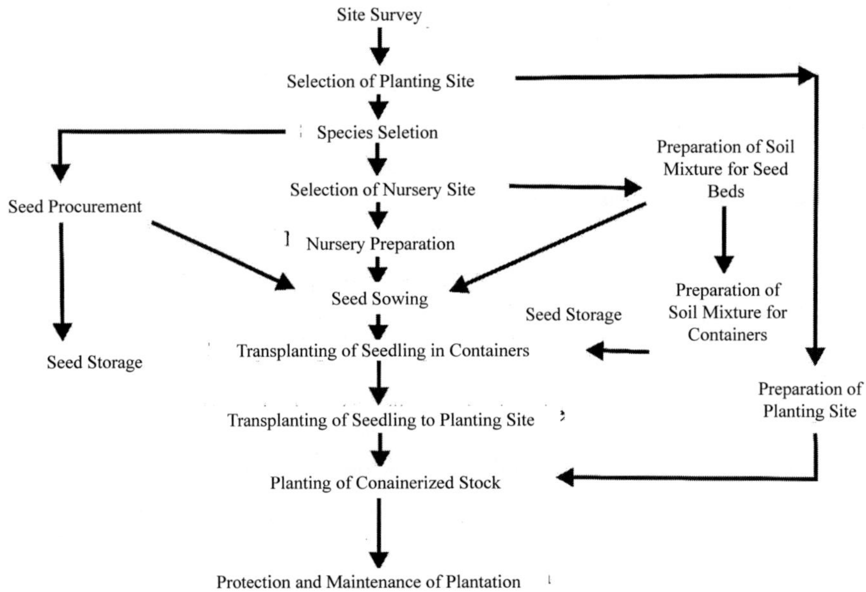

Fig. 12.2: Biomass Production Techniques

12.4.1 Site Survey

As far as site survey for biomass production is concerned, it should watch with the species of tree and shrubs best suited to the area. Many information are needed for proper site survey, such as climate, soil, topography, vegetation, biotic factors, water table levels, availability of supplementary water sources and distance from nursery etc.

12.4.2 Planting Site Selection

It is based on the information of the site survey. The best site is one, which on plantation will lead to the establishment of a successful plantation.

However, the choice of the planting site is limited to lands, which are not suited for agriculture or livestock production. The boundaries of the planning site should be marked with boundary posts after the area has been chosen. When there is a danger of trespassing and damage by grazing animals a boundary fence should be established. Of course, fencing is costly and therefore it should only be built when other means of protection are not effective. Once the forest plantation is well established with sufficiently tall trees, the fences can be removed and reused at another planting site. When roads and other passageways traverse the planting site, they also should be contained with fences.

12.4.3 Species Selection

Once the selection of site plantation is over, then following points should be consider in the selection of the tree and species.

1. A species selected for a particular site, should suit to the site and should remain healthy throughout the anticipated rotation and should exhibit acceptable growth and yield.
2. For a successful planting, performance data can have to be extrapolated from one locality to another. Results from a locality where a tree or shrub species is growing (either naturally or as an exotic) strictly apply only to that locality.
3. When reliable information shows a close similarity between the site to be planted and that on which the species already is successful, this particular site is recognized for largescale planting.
4. The selection of tree and shrub species through the use of analogous climates is important as a first step in selection. This step is amplified by an evaluation of localized factors (for example soil, slope and biotic factors).
5. The ability to match closely a planting site and a natural habitat may not preclude the need for species trials, since climatological or ecological metering may not reveal the adaptability of a species.

12.4.4 Preparation of the Planting Site

The preparation of the planting site is more important in the biomass establishment programme. The site preparation is meant for removing competing vegetation from the site, creating conditions that will enable the soil to catch and absorb as much rainfall as possible and providing good conditions for the planting, including a sufficient volume of routable soil. The site preparation also includes the creation of conditions where the dangers from fire and pests are minimized. The site preparation is directed toward giving the seedlings a good start with rapid early growth. It reduces surface runoff for increasing the moisture in the soil and at the same time it helps in elimination of hard pans of the soil.

12.4.5 Preparation of the Soil Mixture

The soil mixture for the containers as well as for seedbeds should have following characteristics.

(a) It should be relatively light and cohesive

(b) It should have a good water retention capacity

(c) It should possess a high organic and mineral content

(d) It should contain adequate nutrients, which usually are supplied in the form of artificial fertilizers.

(e) The quantity of soil needed in a containerized nursery operation is directly related to the size of the containers used. For example, to fill 1,00,000 small containers, 200 cubic meters of soil are needed, where as 442 cubic meters of soil are needed to fill 1,00,000 of the largest containers (16 times more).

(f) The Soil used in nursery work should be sufficiently acidic in nature. For most tree and shrub species, the pH should not be higher than 6.0.

Damping off is a common and serious disease in many forest nurseries. It can occur either in seedbeds or in containers after transplanting. This disease is caused various fungi. A watery appearing construction of the stem at the ground line generally is visible evidence of the disease. Damping off is favored by high humidity, damp soil surface, heavy soil, cloudy weather, an excess of shade, a devise stand of seedlings and alkaline conditions.

One of the best preventive measures for this disease is to maintain dry soil surface through cultivation, to reduce the sowing density and to thin the seedling to create better aeration at the ground line.

12.4.6 Sowing of Seed

Sowing of seed should match to the ideal sowing time. In order to determine the ideal sowing time, one counts backwards from the beginning of the planting season to identify the number of months required to raise high quality planting stock. The time to raise high quality planting stack depends on number of factors like; type of tree or shrub species, climatic conditions and nursery conditions.

It is essential for each planting project to prepare its own sowing schedules for the locally important tree and shrub species. With large amount of seeds, not all should be sown at the same time. It is better to spread out the sowing dates with one or two week intervals.

A common method of raising seedlings has been to sow the seed in seedbeds or seedling trays, and then to transplant into containers as soon as the plants are sufficiently large to handle. The seedbeds can be either broadcast sown or sown in lines. Small seed should be mixed with some kind of inert fine material to facilitate even distribution.

12.4.7 Method of Sowing

Seedling trays or boxes commonly used for the production of seedlings, which subsequently are transplanted into containers. The trays can be easily moved to the beds for the transplanting operation. A good tray can be made from kerosene or diesel tins by cutting them lengthwise into halves. The bottom of the tray should be perforated to allow drainage or irrigation by absorption. In order to assist drainage, the bottom of the tray can be filled with a 2 cm layer of gravel or charcoal.

Once the seeds are sown in seedlings or seedling trays, they should be covered with sand, which should be pressed down gently to establish a good contact between seeds and soil. Depth of sowing depends on the size of the seed. The normal depth of sowing is from 1 to 3 times the diameter of the seed, but in many instances, it can be desirable to sow slightly deeper to avoid the washing of seeds from the soil by irrigation or heavy rains.

12.4.8 Transplanting of Seedling into Containers

A plant that is grown either in seedbed or seedling trays, in which it was originally sown and than it has to lifted for final replacement, is technically known as a "seedlings". When a seedling is lifted and replanted in the nursery in another bed or container, it is, therefore, termed a "transplant". The transplanting of seedlings, also referred to as "pricking" which is done primarily to induce better development of the root system by increasing the number of fine absorbing root lets.

Transplanting should be made before the seedling has acquired a large, heavy root system, but after it has developed a strong stem. Normally, this stage occurs after the complete unfolding of the cotyledons and during the unfolding of the first leaves. Transplanting is a delicate and time-consuming operation and requires precise organization and sufficient labour. Therefore, there is a tendency to sow directly into the containers when possible, which has the advantage of saving time and labor, and reducing the losses and retardation in plant growth caused by transplanting.

12.4.9 Transport of Seedlings to the Planting Site

Generally, plants are damaged during transport to the planting site. Therefore, adequate care must be taken to avoid mishandling of plants during loading and unloading from vehicles. Even, plants require protection during transportation, as the airflow can cause breakage of stems or drying. Therefore, it is also important that the containers as packed tightly, so that they cannot move. When possible, plants be transported in the planting season on cool, cloudy or even rainy day to prevent desiccation during transport. Following are few suggestions, which should be kept in mind during transportation of seedlings to the planting site.

(a) Always prefer the packing of container raised plants for transportation.

(b) Put packed container raised plants in trays the load into vehicles.

(c) The tins, which have been used for seedling trays, can be used for transporting container plants.

(d) Shipping schedule should be planned to avoid delays and to allow proper disposition of the plants immediately upon arrival.

(e) Normally, plants should arrive one day ahead of planting.

(f) Where shape and watering facilities are available, supplies can be brought in advance of several days.

(g) As soon as the plants arrive at the planting site, they must be watered and if necessary, heeled in a cool, moist, shaded place until they are needed for planting.

12.4.10 Maintenance of the Plantations

Once a plantation has been established, then it is necessary to protect the plantation against damaging. A variety of cultural treatments also can be required to meet the desired purpose of the plantationprotection against weather effect, fire damage, pests and insects, wild animals is essential. The cultural treatments include weeding, thinning, pruning and maintaining desired spacing between trees.

12.5 Harvesting of Biomass

Biomass harvesting is decided when trees and shrubs attain the "optimum size" for the wood production. From a biological point of view, trees and shrubs should not be cut until they have at least grown to the maximum size required for product utilisation. In general, the average annual growth of trees and shrubs increases slowly during the initial year of establishment, reaches a maximum and then falls more gradually in subsequent years. Hence, trees and shrubs usually should not be allowed to grow beyond the point of maximum average annual growth, which is the age of maximum productivity. As far as harvesting of biomass is concerned following factors be considered.

1. Biological Factors: Age of maximum productivity, average growth rate.
2. Pathological factors: Growth in terms of mortality and the amount of defect in living trees.

3. Entomological factors: Forest composition, age, structure and vigour.
4. Silvicultural factors: Seed production, characteristics, methods of obtaining regeneration, competition from less desirable tree species and maintenance of desirable soil conditions.
5. Local harvesting techniques.
6. Available man power.
7. Existing market outlets.

Biomass harvesting should be well planned and organized in order to make the best use of the raw material while keeping labour input and production cost low and minimising damage to the environment. A variety of different harvesting systems can be applied. The factors on which harvesting systems depend are as follows

1. Species of wood
2. Size and assortment (fuel wood, poles or logs)
3. Type of forest (man made or natural)
4. Type of cut (thinning or clear cut)
5. Kind of regeneration (artificial, coppice, natural)
6. Terrain (flat, steep, swampy)
7. Accessibility (roads, waterways) and
8. Means of transport (manual, animal, motorized)

In all these cases, good planning and organization of work depends on following factors

1. The assessment of the volume to be harvested.
2. The determination of the assortments to be produced.
3. The determination of wood storing places, skidding lines and felling direction.
4. Clearly instructed and skilled supervision and workers.
5. Availability of the necessary hand tools and maintenance and other equipment.
6. Clear separation of working areas for individual work teams and different operations (filling and transport).

12.5.1 Harvesting Methods

These are several different harvesting methods that allow the plant to regenerate through sprouting. These are as follows

(A) Coppicing

It is one of the most widely used harvesting methods in which the tree is cut at the base, usually between 1575 cm above the ground level. New shoot developed from the stamp or root. These shoots are some times referred to as sucker or sprouts. Management of sprouts should be carried out according to use. For fuel wood the number of sprouts to container to grow, should depend on desired sizes of fuelwood. If many sprouts will be allowed to grow for a long period, the weights of the sprouts will become heavy and the sprouts may tear away from the main trunk. Several rotations of coppicing are usually possible with many species. The length of the rotation period depends on the required tree products from the plantation. It is suitable method for production of fuelwood. Most eucalyptus species and many species for leguminous family, most of naturally accessing shrubs can be harvested by coppicing (Fig. 12.3a).

(B) Pollarding

It is the harvesting systems, in which the branches including the top of the tree are cut, at a height of about 2 meter above the ground and the main trunk is allowed standing. The new shoots sprout or emerged from the main stem to develop a new crown. This results into continuous increase in the diameter of main stem although not in height. Finally when the tree losses its sprouting vigour, the main stem is also cut for use as large diameter poles. An advantage of this method over coppicing is that the new shoots are high enough off the ground that they are out of reach of most grazing animals. The neem tree (*Azadirachata Indica*) is usually harvested in this manner. The branches may be used for poles and fuel wood (Fig. 12.3 b).

(C) Lopping

In this method most of the branches of the tree are cut. The fresh foliage starts sprouting from the bottom to top of the denuded stem in spite of severe defoliation, surprisingly quickly. The crown also regrows and after a few years, the tree is lopped again. The lopped trunk continues to grow and increase in height, unless this is deliberately prevented by pruning it at the top (Fig. 12.3 c).

(D) Pruning

It is very common harvesting method. It involves the cutting of smaller branches and stems. The clipped materials constitute a major source of biomass for fuel and other purposes, such as fodder mulching between tree sows. It is also often required for the maintenance for fruits and forages trees, alley cropping and lives fences. The process of pruning also increases the business of trees and shrubs for bio fencing. Root pruning at a distance of 23 meter from the hole is effective to reduce border tree competition for water and nutrient with the crops (Fig. 12.3 d).

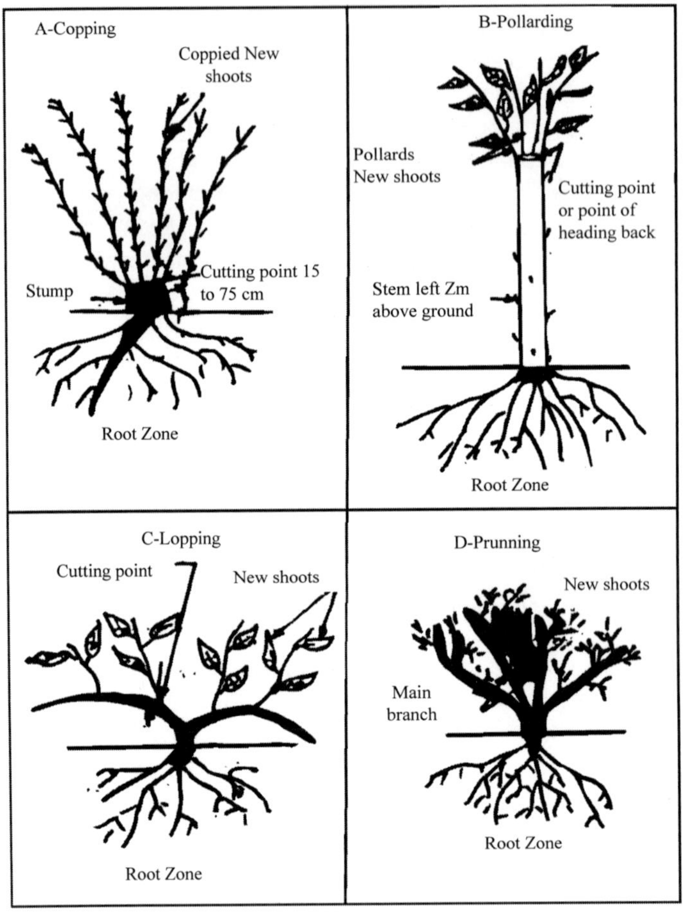

Fig. 12.3: Different Harvesting Methods

(E) Thinning

It is a traditional forestry practice and in fuelwood plantation, it can also be of importance. The primary objectives of the thinning are to enhance diametric growth of some specific trees through early removal of poor and diseased tree to improve the plantation by reducing the competition for light and nutrients. Depending on initial plant density, initial thinning can be used for fuelwood or pole production. If maximum biomass production is the main objectives, of the plantation, regardless of quality, thinning may not be needed.

12.6. Biomass Processing for Fuel Use

Biomass can be processed into a number of ways for fuel use. The biomass can be used as an energy source material for heating, mechanical work or for electricity generation. It can be combusted directly to produce heat or it can be converted into a secondary fuel or into an energy carrier, such as steam, compressed air, electricity etc. Secondary conversion is through physical, chemical or biological processes or a combination of these. General biomass processing routes is shown in Fig.12.4.

12.6.1 Physical Processes

Physical processes involve many unit operations such as drying (change in the moisture constant), size reduction (change in surface area to volume ratio) and densification (change in density).

(a) Drying

It is method of removing moisture content of biomass. The heat generation of biomass is dependent on the moisture content, therefore, drying of biomass is essential for enhancing its calorific value. It has been observed that about 9% of energy value of biomass are lost in reducing the moisture content from 30% to 9%, but if it is not reduced, the decrease in calorific value is about 26 percent.

(b) Size Reduction

The size reduction of biomass is made to convert it to a convenient transportable, storable and usable form. These processes include tree cutting (removing stems and branches from tree), log cutting to domestic size or convenient size for use and log cutting to small billet form for use.

(c) Densification

As the name indicates, this process is to increase bulk density of biomass for efficient and convenient transportation and handling. It also reduces requirement of bulk storage space. The physical dimensions and the combustion characteristics of the fuel become homogeneous and uniforms because of the required particle size, porosity and density and as a result of it, efficient energy conversion is established. Densification may be conversion of loose biomass into pellets, briquettes and cubs. Depending upon type of biomass and its end use, the biomass can be densified into these three forms.

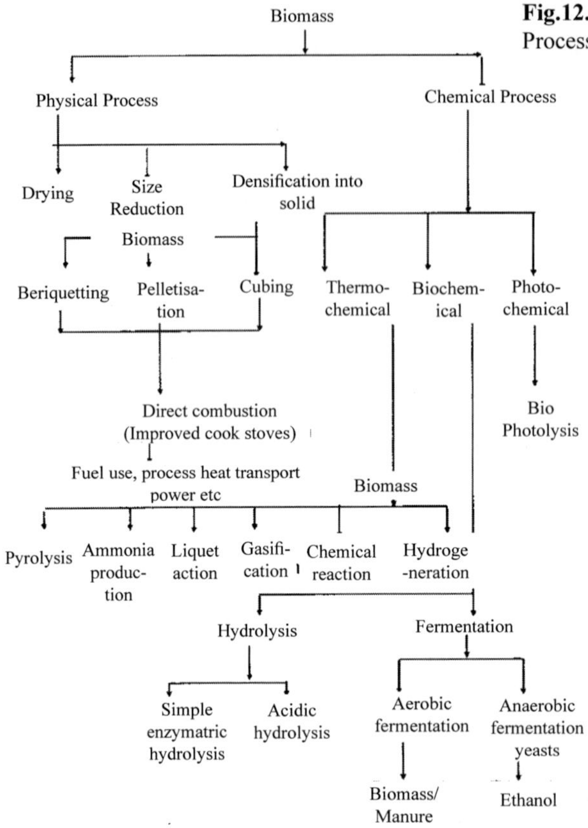

Fig.12.4: Different Biomass Processing Processes

Densification processes are basically of three types

(i) Binderless densification technique Where no external binders are added for compacting the materials. However, it is done at very high temperature (200°C) and at high pressure (1500 kg/cm^2).

(ii) Binder densification technique Where pressure requirement is less through same types of binding materials such as tar, molasses, sodium bentonite, resins or wax.

(iii) Pyrolysis densification technique In this technique, firstly the biomass is carbonized and the charcoal obtained is powdered and then compacted in required shape.

Depending upon final shape of product obtained through densification process, the process is terms as briquetting, pelletization and cubing.

(d) Biomass Briquetting

Bulk of the energy in the rural areas for cooking and heating purposes is derived from firewood, agro and forest residues and to some extent from cow dung. Use of firewood leads to deforestation and inefficient combustion of cow dung, agro and forest residues leads to pollution problems. In addition, materials like pine needles are major causes of forest fires.

To solve these problems and to save wood, the agro and forest residues can be upgraded to convenient and smokeless briquetted fuels. These materials can be converted into small sized briquettes with holes (beehive briquettes) through partial pyrolysis process. These briquettes can burn with a smokeless and clean flame, ideal for domestic and small scale heating purposes. The agro and forest residues can also be directly briquetted by compaction at high pressures. These briquettes are suitable for large scale and industrial applications. The agro and forest residues in briquetted fuel form has added advantage of easy transportation to the place of work.

Example 1.2: Estimate the amount of crop residue generated from 15 ha of rice filed per year. The crop residue ratio for rice straw is to be taken 1:1.5. If the yield is about 5 tons per ha per year.

Solution: The waste generated during rice harvesting

Total production of rice = Field size (ha) × yield (tons per ha per year)

Total production of rice = 15×5

= 75 tons per year

Waste generated

= Total yield (tons per year) × CRR

= 75×1.5

= 112.5 tons per year

12.7 Carbonisation of Biomass

Conversion of biomass into carbon rich product by heating at high temperatures (300 – 600 °C) in a closed reactor containing no to partial levels of air. Under these conditions, biomass undergoes thermo-chemical conversion into biochar. Because of its numerous potentials uses in agriculture, energy, and the environment, much attention has been given to biochar in both political and academic areas. Biochar can be used in a variety of applications such as energy production, agriculture, carbon sequestration, waste water treatment, and bio-refinery; additionally, biochar provides an alternative strategy for managing organic waste. These advantages have renewed the interest of agricultural researchers in producing biochar from bio-residues and using the product as a soil amendment. Fig.12.5 summarizes the thermo-chemical conversion routes of biomass, including direct combustion to provide heat, liquid fuel and other elements for thermal and electrical generation.

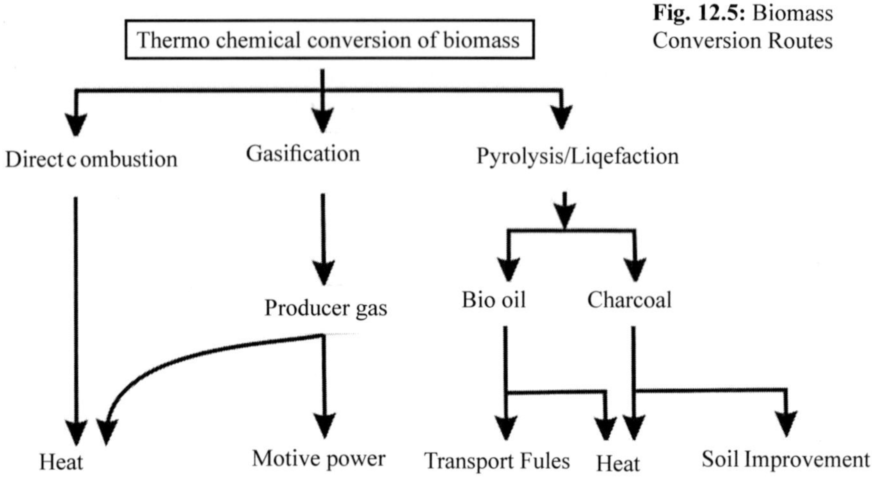

Fig. 12.5: Biomass Conversion Routes

12.7.1 Biochar and Sustainability

Biochar plays a major role in mitigating climate change, promoting environmental sustainability and increasing agricultural productivity, facilitating soil carbon storage, and improving soil fertility to increase plant and overall crop yield. There are four motivational objectives of biochar application, i.e., soil improvement through enhancing soil fertlity, waste management, climate change mitigation, and energy availability. Either individually or in combination, these objectives can have either social or financial benefits or both. Biochar always draws attention as a potential input for agriculture, as it can improve soil fertility, aid sustainable production and reduce contamination of streams and groundwater.

12.7.2 Biochar production technologies

Biochar is derived from a wide variety of biomass including crop residues that have been thermally degraded under different operating conditions. It exhibits a correspondingly immense range of composition. Longer the residence period (up to 4 hours) with moderate temperature (up to 500 °C) biochar yield varied from 15 %– 35 %, while bio-oil yield varied between 30% - 50%. On other hands, with lesser residence time (up to 2 second) higher bio-oil (50% - 70%) yield.

Thermochemical processes like pyrolysis and carbonization convert the biomass into bio fuels and other bio energy products. In the pyrolysis process, thermochemical conversion of biomass is carried out in absence of air and at a temperature above 400°C to form a solid product known as biochar. The biochar mainly consists of carbon (C), hydrogen (H), oxygen (O), nitrogen (N), sulfur (S) and ash. Generally, there are three modes of pyrolysis: slow, intermediate and fast. A higher biochar yield was found with a slow pyrolysis process as compared to others. The Biochar produced from rice husk using a top-lit updraft gasifier and found that such technology is relatively simple for farmers to produce biochar in the field, with an efficiency of 15–33%. Biochar produced from on-farm available crop residues is sufficient to amend 6.3 – 11.8 % of the production area annually.

Carbonization is a slow pyrolysis process that has been in use for thousands of years, and the its main goal is the production of biochar. In slow pyrolysis, a biomass is heated slowly in the absence of air to a relatively low temperature (≈ 400°C) over an extended period of time. Energy can drive the process in the following different ways: (i) directly as a heat of reaction (ii) directly by flue gases from the combustion of feedstock (iii) through indirect heating of the reactor wall using a hot gas; or (iv) through indirect heating of the reactor wall using sand or other non-gas materials. The biochar production process can be classified as illustrated in Fig. 12.6.

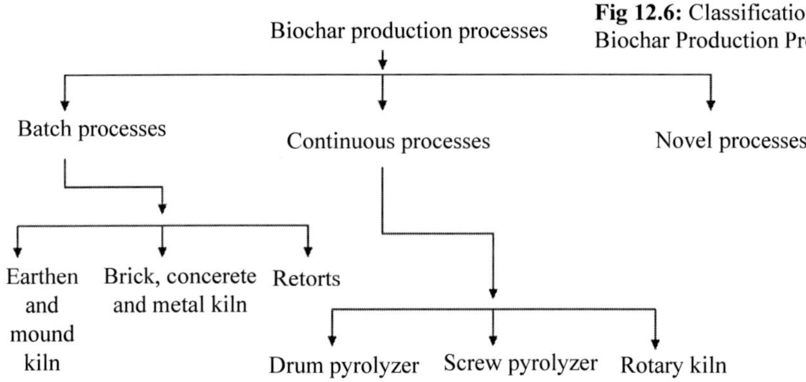

Fig 12.6: Classification of Biochar Production Processes

12.7.2.1 Batch processes

The batch process is an ancient practice and is still practiced in rural areas for biochar production. Though the charcoal yield in such a process varies over the low range of 12.5 to 30 %, it is still preferred in the countryside because of its low operational and construction cost. Batch process for biochar production includes,

(a) Earthen and mound kiln,

(b) Brick, concrete, metal kiln

(c) Retorts

1.1.1.2 Continuous process for production of biochar

At present, the continuous process for production of biochar is widely adopted in the commercial sectors due to maximum yield, energy efficiency, and its quality. The biochar yield found between 25 to 35%. As a major benefit the continuous production of biochar is ideal for medium to large-scale production along with a greater flexibility towards biomass feedstock, which are major benefits. Continuous process for biochar production includes,

(a) Drum type pyrolyzer

(b) Screw type pyrolyzer

(c) Rotary kiln

12.7.3 Factors affecting biochar production

The performance results of biochar production which occur via different production technologies broadly depends on the various types of feedstock used, the moisture content of said feedstock, and the operating temperatures and pressure points at which experiments were conducted. A Biomass has three main groups, for example: cellulose, hemicellulose, and lignin with trace amounts of extractive and minerals. These propositions are varied depending on the feedstock, a variation which highly affects the biochar yield. Moisture content is another factor which affects the biochar properties and char yield. The moisture content affects the char reaction and is extensively used to produce activated carbon. In fast pyrolysis processes, around 10% moisture content is fairly desirable during the charcoal making process.

Production of biochar is a thermochemical process and temperature plays a major role in the properties of biochar and its suitability for soil health. A lab scale study on pyrolysis' ability to produce biochar from pin, mixed larch and spruce chips, and softwood pellets was conducted by Masek (2013) with temperatures between 350 °C and 550 °C and reported that the stability of biochar increases as temperature increases, and the yield of biochar is independent of temperature. Angin and Sensoz (2014) also reported that the chemical and surface properties of biochar are affected by pyrolysis temperature. As the pyrolysis temperature is increased from 400°C to 700 °C, the volatile matter, hydrogen, and the oxygen contents of the biochar were decreased, but the value of fixed carbon was increased. Biomass can't be converted into biochar at low pyrolysis temperature (300 °C) because at this temperature desired carbon frame structure does not developed.

Reactor operating temperature play vital role in deciding of fixed carbon and oxygen content of biochar. It has been found that higher operating temperature have higher fixed carbon content and lower oxygen content.

12.7.4 Classification of Biochar

The International Biochar Initiative (IBI) broadly classifies biochar based on Carbon storage value, Fertilizer value (P, K, S, and Mg only), Liming value, and Particle size distribution. Further, three general classes of biochar on the basis of organic carbon content have been proposed. In class I type biochar the C_{org} mass fraction is about ≥60%, in class II it would be in the range of 30% to < 60% while in class III it would be < 10%.

12.7.5 Stability of biochar in soil

The stability of biochar depends on the conditions of its production and biomass feed stock. Spokas (2010) conducted a study on the stability of biochar in soil and found that lower oxygen-to-carbon (O : C) ratio resulted in a more stable biochar material. Conclusively, when the oxygen - to - carbon molar ratio (O : C) is greater than 0.6, biochar will probably possess a half-life on the order of less than 100 years, if the range is 0.2 -0.6, the accepted range of half-life is between 100 and 1000 years. If the molar oxygen-to-carbon ratio is less than 0.2, the half-life will be greater than 1000 years. In this wise, the process temperature, i.e. pyrolysis temperature, is highly responsible for biochar stability.

12.8 Pyrolysis Process

Pyrolysis is the thermal decomposition of any organic material at a specific temperature in absence of air/oxygen; the process ends with three resulting products, namely liquid (bio-oil), solid (biochar), and syngases. Recently, bio-oil produced through pyrolysis process and further it up grading has been attracted a significant attention due to its major use as biofuel and as a precursor material for making chemicals. The pyrolysis process also classified in three different types as per its operating conditions; slow, fast and flash pyrolysis.

The primary end product of the pyrolysis process is liquid oil known as bio-oil or pyrolytic oil, which has a dark brown colour and potential as an alternative fuel for multiple applications. According to International Energy agency News, the consumption of biofuels around the world will increase from the current 2 % of the total global share of fuel to 27 % in 2050. Bio-oil or pyrolytic oil is composed of organic substances like phenol, amines, ketones, ethers, esters, furans, aromatic hydrocarbons, alcohols, and water. Despite advantages that include being eco-friendly, having a low cost requirement, and having high conversion efficiency, bio-oil is facing some technical challenges and limitations for use in the commercial sector due to its high water content (15-30 %) and high composition of oxygenated compounds (35-60%) such as acids, aldehydes, ketones, and alcohols. Excess level of some components can result in unfavourable effects on bio-oil characteristics: such as lower calorific value, which decreased the combustion efficiency: and instability, which is sometimes due to oxygenated compound corrosion occurs.

12.8.1 Classification of pyrolysis process

Pyrolysis process can efficiently operate under different operating conditions so there are different types of pyrolysis process such as slow, intermediate, fast, flash, vacuum, ultra-flash, catalytic pyrolysis etc. Vacuum pyrolysis is carried out at a very low pressure (up to 4 kPa), while atmospheric pyrolysis of biomass is conducted under atmospheric pressure. Waste material including forest waste, woody biomass and agricultural waste are considered as most suitable primary feedstock for the pyrolysis process. A complete biomass pyrolysis refinery is illustrated in Fig. 12.7

Fig. 12.7: Biomass Pyrolysis Refinery

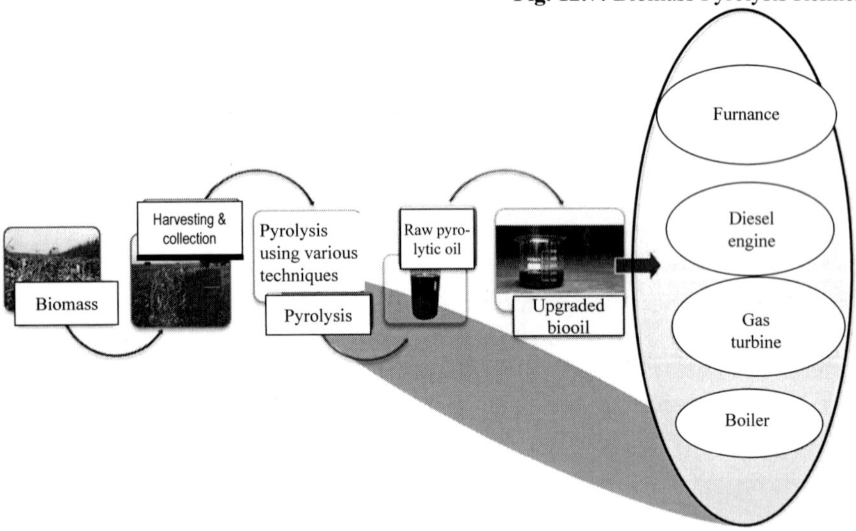

12.8.1.1 Slow pyrolysis

The slow pyrolysis of biomass conveys a number of advantages because it produces a primary end product, biochar, which is an organic carbon-rich product that can act as carbon sequester and improve the soil quality. Slow pyrolysis is performed at a moderate temperature, with a long residence time and a low heating rate; therefore, biochar is the primary end product of this process. The most commonly used reactors for slow pyrolysis are rotary kilns, drums, and screw reactors. Generally, biochar is produced during slow pyrolysis in the absence of air by maintaining the temperature between 300-550 °C; therefore, the complete reaction requires a long residence time (5-30 min) and a very low heating rate (0.1-0.8 °C/s).

12.8.1.2 Intermediate pyrolysis

Intermediate pyrolysis has acquired significant attention in recent years, due to satisfactory raw bio-oil production (up to 50% yield from woody biomass), with less residence time from 10 to 30s, at a temperature nearly 500 °C. Although, this kind of pyrolysis allows making use of larger particle size of feedstock (including chips and pellets) while in case of fast pyrolysis required finely ground raw material. Therefore, this process considered as most robust and reliable because of its suitability in small and medium scale industries.

12.8.1.3 Fast pyrolysis

Fast pyrolysis of biomass for renewable biofuel production is considered as one of the cheapest conversion routes. In fast pyrolysis process thermal degradation of biomass takes place in absence of air/oxygen at a moderate temperature (450-600 °C) with a very short residence time (< 2s). The major resulting end product in fast pyrolysis is an initial dark brown colored viscous liquid known as bio-oil or pyrolytic oil followed by biochar and syngases. The fast pyrolysis process has maximum conversion efficiency and a recorded highest yield of bio-oil about 75 % on a dry mass basis. As compared to other biofuel production technologies, bio-oil produced through a fast pyrolysis process is considered a cost-effective thermochemical conversion route. The profitability of the fast pyrolysis process depends on parameters such as product yield, quality of product, feedstock cost, production scale etc. The fast pyrolysis reactors such as a bubbling fluidized bed, rotating cone, screw/auger, ablative, and vacuum pyrolysis are considered suitable technologies for bio-oil production, although the configuration of fluidized bed and rotating cone reactors are viewed as the most cost effective and commercialized technologies of the group.

In a fast pyrolysis process, the syngas yield has a badly affect on biochar and the bio-oil yield because as syngas yield increases, the yield of biochar and bio-oil drops suddenly. The bio-oil produced from woody biomass shows a higher yield than agricultural by-products, forest residue, and energy crops etc. because clean woody biomass contains a high percentage of cellulose and hemicellulose which is more favourable for bio-oil production. The selection of feedstock particle size in fast pyrolysis process can adversely affect on bio-oil, char and syngas yield, because as used particle size is more, heat transfer rate slightly decreases so it causes raising the biochar yield, while drop in both bio-oil and syngas yield. Therefore, for getting the maximum yield of bio-oil small particle size feedstock should be preferred. Fast pyrolysis experiment using mixture of biomass and waste plastic in a fluidized bed reactor at a temperature range between 525 to 675 °C. during the experiment the maximum yield of bio-oil was recorded about 57.6 % at 625 °C temperature and it was observed that the due to co-pyrolysis resulted bio-oil had a good quality and having higher heating value was about 36.6 MJ/kg.

12.8.1.4 Flash pyrolysis

In flash pyrolysis process, thermal degradation of biomass take place at higher heating rate (from 10^3 to 10^4 °C/s) by keeping a very short residence time (<0.5s), resulted a higher bio-oil yield (75-80 wt%). If maximum bio-oil production is goal then flash/fast pyrolysis is more recognized process as compared to other. Generally, flash pyrolysis reaction take place with a fraction of seconds but it requires a higher heating rate so due to instant heating, biomass particle size should be small. To achieve the higher bio-oil yield (75%wt.) at a higher temperature range from 800 to 1000 °C, then feedstock particle size should be less than 200 micron meter. Owing to this, Jatropha oil cake as a feedstock having a particle size ranged between 0.6-1.18 mm in a fluidized bed reactor for production of bio-oil through flash pyrolysis process and maximum bio-oil yield was obtained when feedstock particle size was 1.0 mm.

12.9 Densification Technology

Densification means compaction of loose material or to increase density of loose biomass so that its volumetric calorific efficiency can be increased. Densification essentially involves two parts; the compaction under pressure of loose material to reduce its volume and to agglomerate the material so that the product remains in the compressed state. The resulting solid may be a briquette, a pellet and a cube. It will be briquettes if roughly, it has a diameter greater than 30 mm. Smaller sizes are normally termed pellets though the distinction is arbitrary. The process of producing pellets is also different from the typical briquetting processes. The densified product can be developed in the cubical shapes as well.

If the material is compacted with low to moderate pressure (0.2-5 MPa), then the space between particles is reduced. Further, increasing the pressure will, at a certain stage particular to each material, collapse the cell walls of the cellulose constituent, thus approaching the physical, or dry mass and more, density of the material. The pressure required to achieve such high densities are typically 100 MPa plus. This process of compaction is entirely depending on to the pressure exerted on the material and its physical characteristics including its moisture content.

The reduction of material density is required for both the savings in transport and handling costs and improvement in combustion efficiency over the original material. Thus, densification saves cost of stage, transportation and increase calorific value per unit volume of material. The ultimate density of a briquette will depend on the nature of the original material and the machine used and its operating conditions. However, the ultimate apparent density of a briquette from nearly all materials is more or less constant; it will normally vary between 1200-1400 kg/m³ for high pressure processes. Lower densities can result from densification in press using hydraulic pistons or during the start-up period of mechanical piston presses, whilst even higher densities are sometimes achieved in pelletisation presses. The ultimate limit for most materials is in between 1450 -1500 kg/m³. The relation between compression pressure, briquetting process and the resulting density of the briquette is illustrated in Fig. 12.8

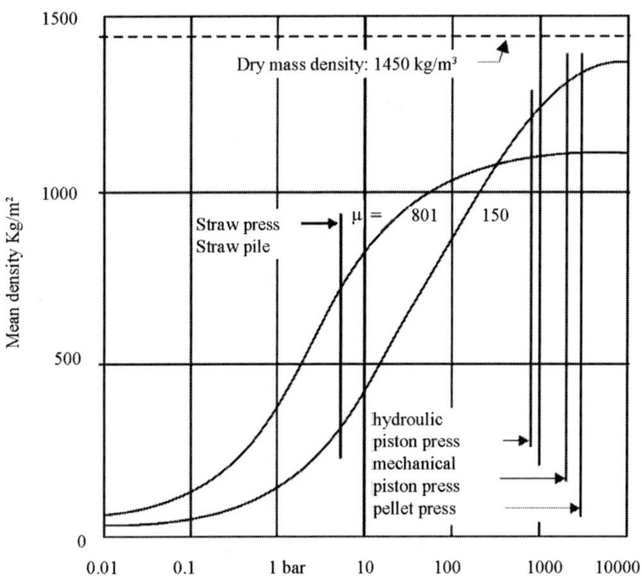

Fig.12.8: The Relation Between Comparison Pressure and Density of Material

The apparent density of briquette are higher than its bulk or packing density as the briquettes are not generally pack perfectly. The usual reduction would be a factor of roughly two depending on the size and shape of the briquette, that is bulk densities of 600-700 kg/m³ are usual, obtained.

The bulk density of the original material like straw which are very easy to compress even manually, is difficult to measure accurately. The lowest bulk densities are around 40 kg/m³ for loose straw and bagasse, up to the highest levels of 250 kg/m³ for some wood residues. Thus, gains in bulk densities of 2-10 times can be expected from densification process. Further, the material have to be dried after compaction in order to facilitate briquetting, thus its increases energy content per unit volume.

A binding agent is also required to prevent the compressed material for returning to its original form. This agent can either be added to the process or, when compressing ligneous material, be part of the material itself in the form of lignin. Lignin, (sulphuric lignin, is a constituent in most agricultural residues) can be defined as a thermo plastic polymer, which begins to soften at temperatures above 100°C and is flowing at higher temperatures. The softening of lignin and its subsequent cooling while the material is under pressure, is the key factor in high pressure briquetting. It is a physico-chemical process related largely to the temperature reached in the briquetting process and the amount of lignin precent in the original material. The temperature in many machines is closely related to the pressure though in some cases, external heat is applied.

There are two ways of classifying briquetting processes:

1) High, intermediate or low-pressure briquetting process: This distinction in principle, dependent on the material used but the following rough classification may be adopted:

   Low pressure up to 5MPa

   Intermediate pressure 5-100 MPa

   High pressure above 100 MPa

2) Whether or not an external binding agent is added to agglomerate the compressed material. Usually high-pressure processes will release sufficient lignin to agglomerate the briquette, however it may not be true for all materials. Intermediate pressure machines may or may not require binders, low-pressure machines require binders for compaction.

12.9.1 Pre-treatment of Biomass

Prior to biomass densification, pre-treatments may be required to optimize the energy content and bulk density of the product.

Pre-treatment can include:

- Size reduction
- Drying to required moisture content
- Application of a binding agent
- Steaming
- Torrefaction

12.9.1.1 Size reduction

Densification process requires specific size of biomass to achieve:

- Lower energy use in the densification process
- Denser products
- A decrease in breakage of the outcome product

12.9.1.2 Drying

Low moisture content is desirable in biomass to improved density and durability of the fuel. The desirable moisture content is in the range of 8%–20% (wet basis). Most compaction techniques require a small amount of moisture to "soften" the biomass for compaction. Above the optimum moisture level, the strength and durability of the densified biomass are decreased.

12.9.1.3 Addition of a Binding Agent

The compaction of biomass during densification process highly depends on inbuilt binding agents of biomass. The binding capacity increases with a higher protein and starch content. Corn stalks have high binding properties, while warm-season grasses, which are low in protein and starch content, have lower binding properties. Binding agents may be added to the material to increase binding properties. Commonly used binders include vegetable oil, clay, starch, cooking oil or wax.

12.9.1.4 Steaming

The addition of steam prior to densification can aid in the release and activation of natural binders present in the biomass.

12.9.1.5 Torrefaction

Torrefaction is a version of pyrolysis processes that comprise the heating of biomass in the absence of oxygen and air. Torrefaction is a pre-treatment process used to improve the properties of pellets. It can also be used as a stand-alone technique to improve the properties of biomass. Torrefaction is a mild version of slow pyrolysis in which the goal is to dry, embrittle and waterproof the biomass. This is accomplished by heating the biomass in an inert environment at temperatures of 280°C–320°C.

12.9.2 Technology or Devices for Densification of Biomass

12.9.2.1 Piston Press

In piston press, as name indicates, pressure is applied discontinuously by the action of a piston on material placed into a cylinder. They may have a mechanical coupling and fly wheel or utilise hydraulic action on the piston (Fig. 12.9). The continuous effect of piston pressure agglomerate the loose material and thus compacted to form briquettes. Briquettes produced through piston press having diameter in the range of 30 mm or greater and are formed when biomass is punched, using a piston press, into a die under high pressure.

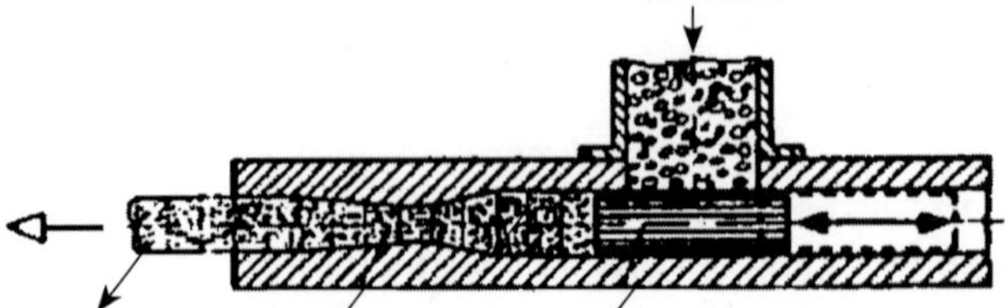

Fig. 12.9: Piston Press

12.9.2.2 Screw extruder

In screw extruder, pressure is applied continuously by passing the material through a screw with diminishing volume. There are cylindrical screws with or without external heating of the die and conical screws. Units with twin screws are also available (Fig. 12.10). The volume of barrel is continuously reduces as the material is extruded outward.

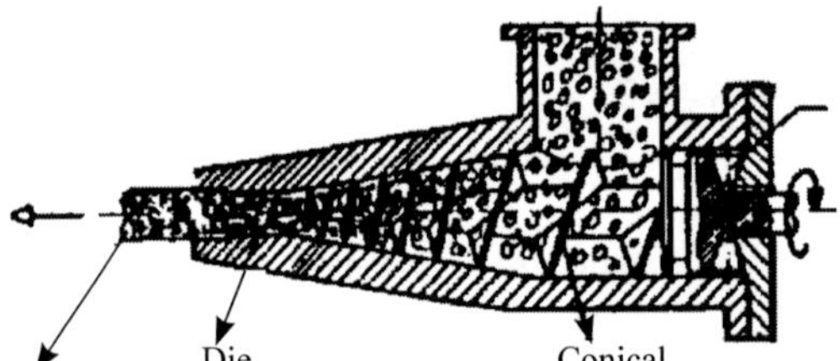

Fig 12.10: Screw Extruders

12.9.2.3 Pellet Press

In pellet press, rollers run over a perforated surface and the material is pushed into a hole each time when roller passes over. This continous rolling of roller over perforated surface forces loose material to form pellets. The dies are either made out of rings or disks though other configurations are possible (Fig.12.11). Pellets are easier to handle and the standard shape of a biomass pellet is a cylinder, having a length about 35 mm and a diameter around 6 to 8 mm. They are uniform in shape but can easily be broken during handling.

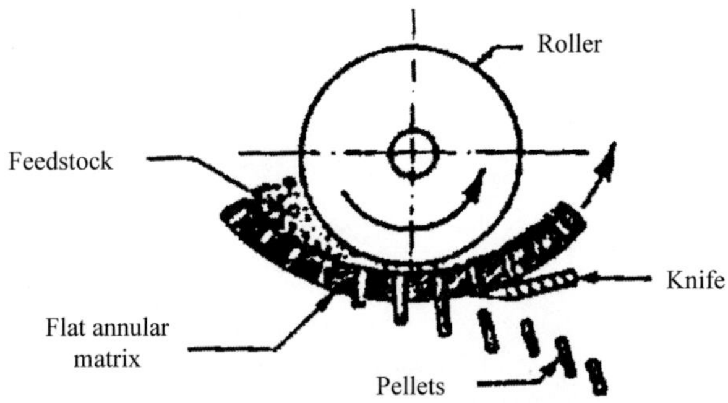

Fig.12.11: Pellet Press

Various other types of roller-presses are also available to form briquettes, especially in making charcoal briquettes from carbonized material. A binding agent is also added in these and the process is more of agglomeration than densification as there is only a limited reduction of volume (Fig. 12.12). These roller presses are generally used for carbonized powder only. There are few other pistons press available for densification process i.e. mechanical screw press and pellet press etc.

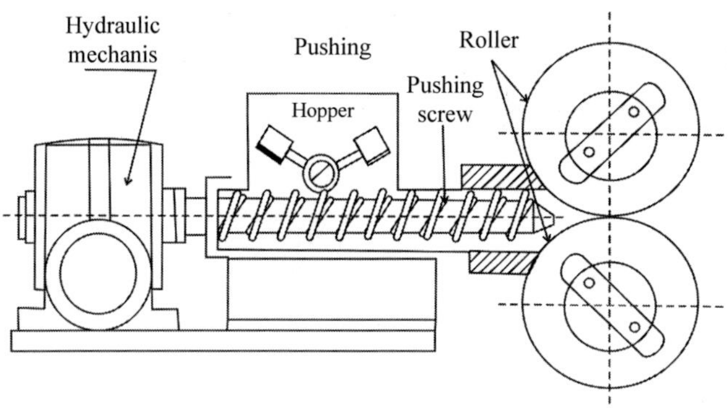

Fig. 12.12: The Roller Press

12.9.2.4 Mechanical Piston Press

A reciprocating piston pushes the material into a tapered die where it is compacted and adheres against the material remaining in the die from the previous stroke. A controlled expansion and cooling of the continuous briquette is allowed in a section following the actual die. The briquette leaving this section is still relatively warm and needs a further length of cooling track before it can be broken into pieces of the desired length.

Fig. 12.13: Mechanical Piston Briquetter

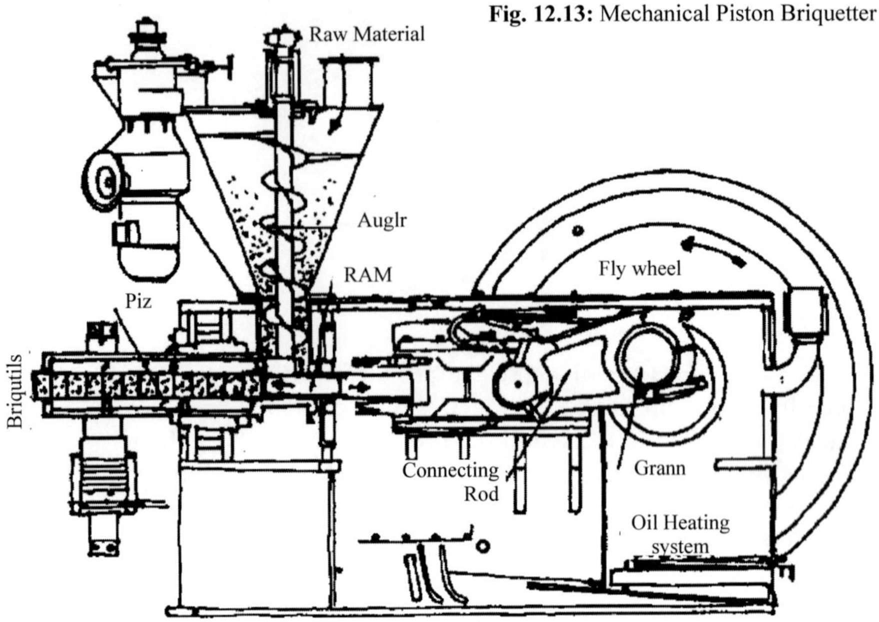

The size of this type of press depends on the quantity of material to be densified and nature of raw material & its heat liberation capacity during agglomeration.

In mechanical systems, the piston gets its reciprocating motion through by mounting eccentrically on a crank-shaft with a flywheel. The shaft, piston rod and the guide for the rod are held in an oil-bath. The moving parts are mounted within a very sturdy frame capable of absorbing high forces acting during the compression stroke.

The most common drive of the flywheel is an electric motor geared down through a belt coupling. A direct-drive system using an internal-combustion or steam engine are also available, which are not changing the basic design of the briquetting machine (Fig. 12.13).

The most common type of briquette press features a cylindric piston and dies with a diameter ranging from 40 - 125 mm. The die tapers towards the middle and then increases again before the end. The exact form of the taper varies between machines and biomass feedstock and is a key factor in determining the functioning of the process and the resulting briquette quality.

The tapering of the dies can be in several designs. However, they may be adjusted during operation by means of narrowing a slot in the cylinder. This is achieved by either screw or hydraulic action. The optimum tapering, and thus pressure, depends on the material to be compressed.

The pressure in the compression section is in the order of 110 to 140 MPa. This pressure together with the frictional heat from the die walls, is in most cases enough to bring the material temperature up to levels where the lignin is becoming fluid and can act as a binder to produce a stable briquette. In fact, heat needs to be extracted from the process to prevent overheating. This is done by water-cooling the die. The process of compaction together with heat liberation and agglomeration coupled with binding and cooling the barrel simultaneously is overall sequence of this press process.

The capacity of a piston press is defined by the volume of material that can be fed to the piston before each stroke and the number of strokes per unit of time. Capacity by weight is dependent on the density of the material before compression. Although the nature of the original material does not alter the physical characteristics of the briquette, it does have a major impact upon the practical output of a machine.

The feed mechanism is crucial feature in the design of piston press. By means of screws or other devices, they try to pre-compress the material in order to get as efficient filling as possible. This is particularly important when using materials whose bulk density is low and which need efficient feeding to achieve more output.

The feed mechanism can, if badly mismatched with the feedstock, cause serious problems in machine operation. If undersized, voids may occur in front of the piston causing damage to the mechanism. The feeder itself may also jam if it is oversized and tries to move too much material into the piston space.

The design parameters of piston machines such as flywheel size and speed, crankshaft size and piston stroke length, are highly constrained by material and operating factors. Table 12.1 shows production capacity variations between materials.

It is likely that consumer acceptance of briquettes is also related to their size. For example, a household user cooking on a open fire would be unlikely to accept a 10 cm diameter briquette and more than a 10 cm piece of wood. Briquettes can be split or broken but this may not be accepted by the consumer and, soft briquettes may lead to crumbling. Industrial customers may, accept large whole briquettes as these conform to their usual wood sizes. This means that in designing plants to receive certain residue volumes, some attention has to be paid to the intended market in deciding, for example, on the number of machines to be used.

Table 12.1: Production Capacity Variation between Materials

Raw material	Bulk density kg/m³	Capacity index	Energy Index
Wood	150	100	100
Shavings	100-110	80	95
Groundnut shells	120-130	90	100

12.9.2.5 Hydraulic Piston Press

The principle of operation of hydraulic piston press is basically the same as with the mechanical piston press. The difference is that the energy to the piston is transmitted from an electric motor via a high-pressure hydraulic oil system. Here the forces are balanced-out in the press-cylinder and not through the frame therefore, compact and the machine can be made very light. The material is fed in front of the press cylinder by a feeding cylinder i.e press-dog which often pre-compacts the material with several strokes before the main cylinder is pressurized. The whole operation is controlled by a programme which can be altered depending on the input material and desired product quality. The speed of the press cylinder is much slower in hydraulic press than with mechanical press thus which results in lower outputs. The briquetting pressures are considerably lower with hydraulic presses than with mechanical systems. The reason is the limitations in pressure in the hydraulic system, which is normally limited to 30 MPa. The piston head can exert a higher pressure when it is of a smaller diameter than the hydraulic cylinder, but the gearing up of pressure in commercial applications is modest. The resulting product densities are normally less than 1000 kg/m³ and durability & shock resistance are naturally suffer compared to the mechanical press.

12.9.2.6 Screw Press

Screw press operate by continuously forcing material into a die with a feeder screw. Pressure is built up along the screw rather than in a single zone as in the piston machines. Three types of screw presses are found in the market. (1) Conical screw press (2) Cylindrical screw press with heated dies and (3) Ditto without externally heated dies.

In this type of press, drying takes place internally in the machine from the frictional heat developed in the process. A system of funnels allows the generated steam to escape from the material and the process can accept raw materials with moisture contents up to 35 %. The energy for the drying will have to be supplied through the mechanical power drive which means that the electric motors are oversized when compared to processes densifying dry material. The higher energy costs for drying with electricity compared to fuel or solar drying, plus the difficulties envisioned in installing the large motor drives in weak electricity grids, make such application unlikely (Fig 12.14).

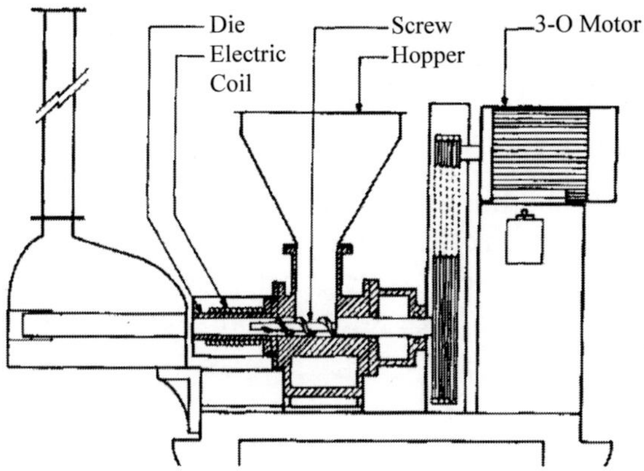

Fig. 12.14: Screw Press with Heated Die

(a) Conical Screw Press

Conical screw presses are available in capacity of 600 to 1000 kg/m^3. It features a screw with a compression die-head. It is reinforced with hard metal inlays to resist the very high wear experienced with this type of extrudes, especially when briquetting abrasive materials. The die is either a single hole matrix with a diameter of 95 mm or a multiple 28 mm matrix. The briquetting pressure is 60 to 100 MPa and the claimed density of the product is 1200-1400 kg/m^3. The machine is equipped with a 74/100 kW 2-speed motor. It is estimated that the actual average energy demand is 0.055-0.075 kW/kg/h. whereas piston press with the same output demand 0.058kW/kg/h.

The mixing and mechanical working of the material in the conical screw press is beneficial to the quality of the product. Continuous operation also aids quality as the briquettes produced do not have the natural cleavage lines as is the case with piston briquettes (Fig. 12.15).

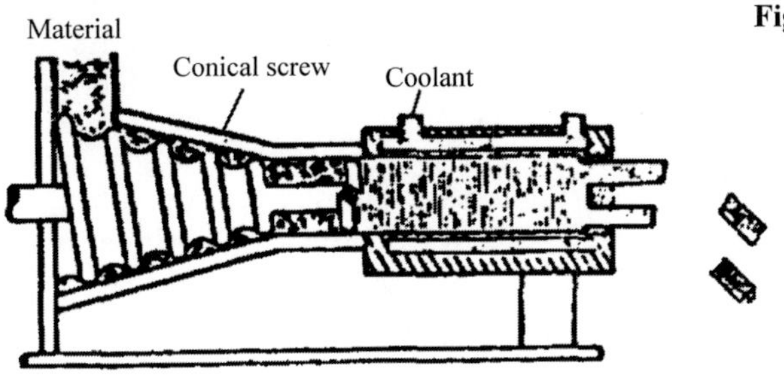

Fig. 12.15: Conical Screw Press

The main disadvantage of this type of press is the severe wear of the die head and die which results in high maintenance costs. The service life of the die head is said to be:

- With groundnut shells 100 h
- With rice husks (estimated) 300 h

(b) Screw Extruders without Die Heating

This type of press is essentially a multiple hole matrix screw extruder for densifying chicken manure. It is a low-pressure process accommodating raw material at moisture content of 30%. The manure is compressed to reduce its volume by half and the product is air dried after compression for use as a boiler fuel.

(c) Screw Extruders with Heated Dies

Essentially this type of press feeds the material from a feeding funnel, compacts it and presses it into a die of a square, hexagonal or octagonal cross-section through a screw. The briquettes have a characteristic hole through the centre from the central screw drive. The die is heated, most commonly by an electric resistance heater wired around the die. The process can be controlled by altering the temperature. The normal operating temperature is in the order of 250 - 300 °C. The central hole of the briquette will act as a chimney for the steam generation on account of generation of the high temperatures in the process. An exhaust is normally mounted above the exit hole from the mould where the briquettes are cut into suitable lengths. A reduction in moisture content is achieved during the formation of the briquettes. The pressure is relatively high which, combines with the high temperatures, limits the moisture content of the raw material to be used. The actual maximum moisture content depends on the raw material but is in the order of 15-20%.

Most models produce a briquette with a diameter of 55 mm and an inner hole diameter of 15 to 25 mm. Variations of the outside diameters between 40 and 75 mm can be found. A common capacity given by the manufacturers of a 55 mm machine is 180 kg/h for wood material and 150 kg/h for rice husk. Variations exist due to differences in screw design and speed.

The energy demand is consistent with these variations in capacity, ranging from 10 kW for a 75 kg/in machine to 15 kW for a 150 kg/in, both based on rice-husk briquettes. For this about 3 to 6 kW of electricity should be added which is used for heating the mould. Assuming the total of 18 kW for a 150 kg/h machine, this results in a specific energy demand of 0.12 kWh/kg.

12.9.2.7 Pellet Press

Pellets are the result of a process which is closely related to the briquetting processes. The main difference is that the dies have smaller diameters (usually up to approx. 30 mm) and each machine has a number of dies arranged as holes bored in a thick steel disk or ring. The material is forced into the dies by means of rollers (normally two or three) moving over the surface on which the raw material is distributed.

The pressure is built up by the compression of this layer of material as the roller moves perpendicular to the centre line of the dies. Thus the main force applied results in shear stresses in the material which often is favourable to the final quality of the pellets. The velocity of compression is also slower when compared to piston presses which means that air locked into the material is given ample time to escape and that the length of the die (i.e. the thickness of the disk or ring) can be made shorter while still allowing for sufficient retention time under pressure. The pellets are still be hot when leaving the dies, where they are cut to lengths normally about one or two times the diameter. Successful operation demands that a rather elaborate cooling or drying system is arranged after the densification process (Fig.12.16).

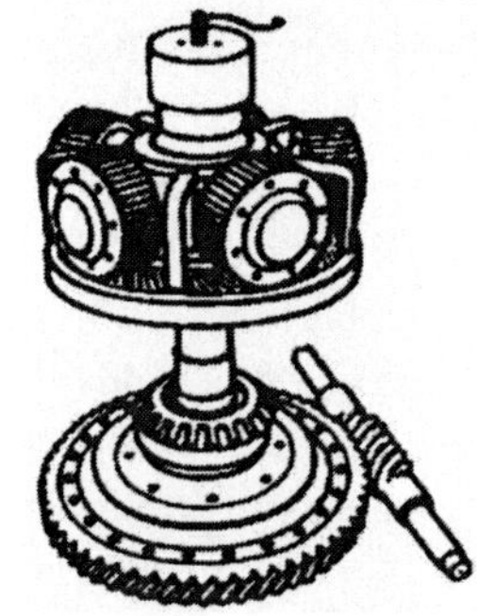

Fig. 12.16 : Flat Die Type Pellet Press

There are two main types of pellet press: Flat and Ring types.

(a) Flate Pettel Press

The flat die type have a circular perforated disk on which two or more rollers rotate with speeds of about 2-3 m/s i.e each individual hole is over run by a roller several times per second. The disks have diameters ranging from about 300 mm up to 1500 mm. The rollers have corresponding widths of 75-200 mm resulting in track surfaces (the active area under the rolls) of about 500 to 7500 cm^2.

(b) Ring Pellet Screw

The ring die press includes a rotating perforated ring on which rollers (normally two or three) press on to the inner perimeter. Inner diameters of the rings vary from about 250 mm up to 1000 mm with track surfaces from 500 to 6000 cm^2.

12.9.3 Advantages of Densification

The main advantages of biomass densification for combustion are:

a) Simplified mechanical handling and feeding

b) Uniform combustion in boilers

c) Reduced dust production

d) Reduced possibility of spontaneous combustion in storage

e) Simplified storage and handling infrastructure, lowering capital requirements at the combustion plant

f) Reduced cost of transportation due to increased energy density

The major disadvantage to biomass densification technologies is the high cost associated with some of the densification processes.

12.10 Torrefaction of Biomass

Torrefaction is a thermophilic process of biomass conversion into a solid, friable, homogenous coal-like material, which has better fuel characteristics than the original biomass. Torrefaction involves the heating of biomass material in the absence of air at 250 to 300°C temperature. Torrefaction produces high-grade hydrophobic nature solid biofuels from various streams of woody biomass or agro residues. Torrefied biomass has high energy density then the original biomass which make torrefied biomass as renewable alternate source for power generation in coal power plant. Torrefied biomass being hydrophobic in nature can be stored easily for longer duration with negligible biological degradation. At 250-300°C temperature volatilization has been taking place and around 30% mass of the biomass is reduced as compared to original untorrefied biomass with net energy loss of approximately around 10%.The end product is a predictable, homogeneous, hydrophobic, high value solid biofuel with far higher energy density and calorific value than the original biomass feedstock. During the torrefaction process a combustible gas is released, which can also utilised to provide heat to the process. The process diagram of torrefaction process is shown in Fig. 12.17.

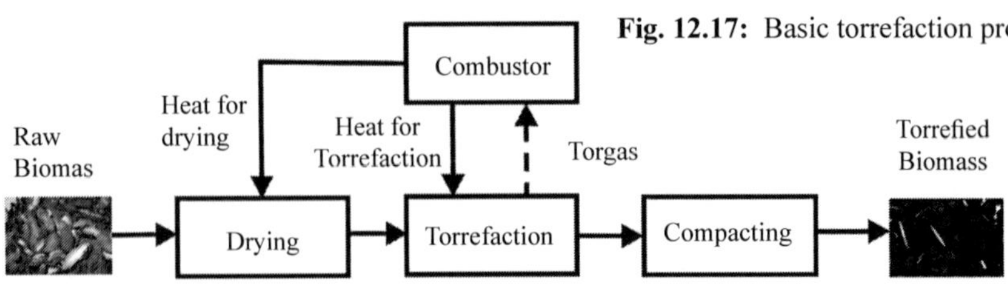

Fig. 12.17: Basic torrefaction process

12.10.1 Benefits of Torrefied Biomass Briquettes over Coal

- Grinds & burns like coal – existing infrastructure can be used
- Lower feedstock costs
- Lower shipping costs
- Minimal de-rating of the power plant
- Provides non-intermittent renewable energy
- Lower sulphur and ash content (compared with coal)

12.10.2 Benefits of Torrefied Biomass Briquettes over Normal Biomass

- Higher calorific value
- More homogeneous product
- Hydrophobic nature/water repellent:
  - Transport and material handling is less expensive & easier
  - Outdoor storage possible
  - Less expensive storage option
  - Significant loss of energy due to re-absorption of moisture in biomass (pellets) is saved
- Negligible biological activity (resistance to decomposition)
  - Longer storage life without fuel degradation
- Low O/C ratio
  - higher yield during gasification
- Higher bulk density
- Excellent grindability
- Higher durability
- Smoke producing compounds removed

12.10.3 Drawbacks of Biomass Torrefaction

- Some of the energy content (around 10 %) in original biomass is loosed due to torrefaction process
- Limited knowledge of torrefcation process temperature, properties of torrefied biomass and composition of volatiles released during the process.

- Torrefcation technology is not yet commercially implemented anywhere in India.

12.10.4 Utility/Application/Market of Torrefied Biomass

- Torrefied biomass pellets can be used as source of fuel in improved cookstove.
- Torrefied biomass briquettes can be used in boiler instead of coal as a source of fuel.
- Torrefied biomass can be used as an alternate source of feedstock in replace of coal for power generation.
- It can be used as source of feedstock to generate transportation fuel by Fischer- Tropsch process.
- Torrefied biomass can also be utilized for heating of houses in cold seasons.

12.11 Biomass Gasification

Biomass gasification is age old technology originated in 1800 century for lighting and cooking purposes. Biomass typically contents biopolymers such as cellulose, hemicellulose and lignin in addition to the presence of average composition of $C_6H_{10}O_5$ which depends upon physical characteristics of biomass. Oxygen is required for combustion of every fuel. For complete combustion of biomass stoichiometric air to fuel ratio required varies from 6:1 to 6.5:1 with CO_2 and H_2O as end product.

Biomass gasification is an efficient and environmentally friendly way to produce energy. Gasification process is nothing but it is a conversion of solid fuel into gaseous fuel for wide applications. This whole process completed at elevated temperature range of 800-1300 °C with series of chemical reaction that is why it come under thermo chemical conversion. Thermo chemical processes are most commonly employed for converting biomass into higher heating value fuels. Major thermal conversion route is including direct combustion to provide heat, liquid fuel and other elements to generate process heat for thermal and electricity generation is summaries in Fig. 12.18.

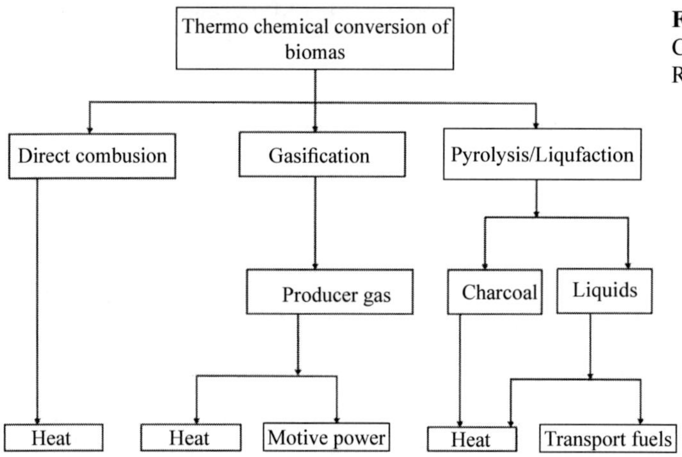

Fig. 12.18: Thermo Chemical Conversion Route of Biomass

Biomass as a feedstock is more promising than coal for gasification due to its low sulfur content and less reactive character. The biomass fuels are suitable for the highly energy efficient power generation cycles based on gasification technology. It is also found suitable for cogeneration. The combustion in gasifier take place in limited supply of oxygen it may be called partial combustion of solid fuel. The resulting gaseous product called producer gas is an energy rich mixture of combustible gas H_2, CO, CH_4 and other impurities such as CO_2, nitrogen, sulfur, alkali compounds and tars[6].

In gasification process incomplete combustion of biomass with less oxygen supply is required with stoichiometric air to fuel ration of 2:1 to 2.3:1. Biomass gasification occurs through a sequence of complex thermochemical reactions in different zones. The design of biomass gasifier depends on fuel availability, shape and size, moisture content, ash content and end user applications. Different types of biomass gasifiers are available depending upon various sizes and design as per requirement which mainly classified in to fixed bed and fluidized bed type of gasifiers. Fixed bed gasifiers are further classified in to updraft, downdraft and cross draft type of gasifiers according to the way of interaction of either air/oxygen or steam with biomass.

12.11.1 Classification of Biomass Gasifier's

Design of gasifier depends upon type of fuel used, air introduction in the fuel column and type of combustion bed as shown in Fig.12.19.

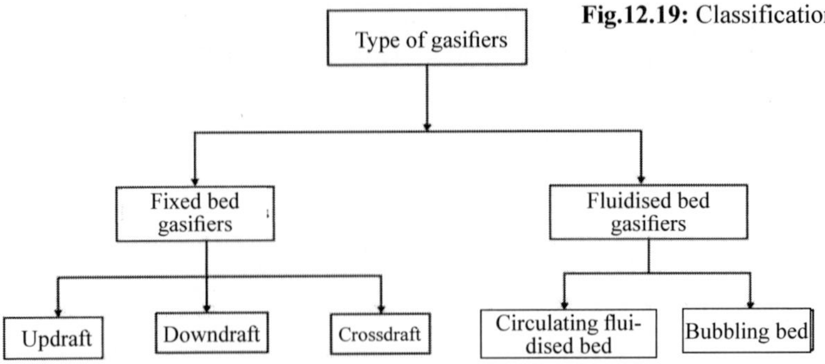

Fig.12.19: Classification of Gasifiers

12.11.1.1 Fixed bed gasifier

Fixed bed gasifiers have a fixed grate to support the biomass fuel fed in the reactor for gasification. Design and operation of these gasifiers are relatively easy as compare to fluidized bed reactor. These are basically classified according to the movement of air or steam in the reactor in to updraft, downdraft and crossdraft type of gasifier. The fixed types of gasifiers are generally preferred for small to medium scale applications and in this case erosion of the reactor body is less.

(a) Updraft gasifier

In updraft gasifier gasifying agent such as air or steam are introduced at the bottom section of gasifier to interact with biomass which is fed from the top of reactor (Fig.12.20). The generated producer gases after gasification exits from the top side of the gasifier hence this gasifier is also called as counter current type of gasifier. After biomass gasification high calorific value syngas as end product exits from the top side of the reactor. This type of gasifier has highest thermal efficiency as the generated hot gases passed through biomass fuel bed which left the gasifier unit at low temperature whereas some part of sensible heat of generated producer gases is used for drying of biomass fuel. Apart from high thermal efficiency updraft gasifiers have some other main advantages such as small pressure drops and slight tendency of slag formation. These gasifiers are suitable for the applications where the high flame temperature is required. However, updraft gasifiers have some drawbacks such as sensitivity to tar and moisture content of the biomass, low gas production, long start up time of the engine and poor reaction capability of the system.

(b) Downdraft gasifier

In downdraft gasifiers air or steam interacts with the solid biomass fuel in the downward direction therefore generated producer gases flow downward in the co-current direction and exits from the bottom side of the reactor. Downdraft type of gasifier as shown in Fig.12.21 is also called as co-current type of biomass gasifier. The end products of the pyrolysis and drying zone are forced to pass through oxidation zone of the reactor for thermal cracking which yield less tar content in the final product and hence better quality of fuel. This gasifier is suitable for small scale decentralize power generation because of low tar and particulate matter content in the finally produced syngas.

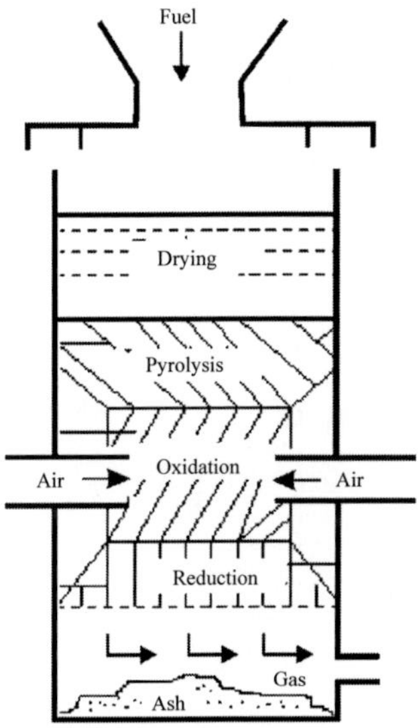

Fig. 12.20: Updraft Gasifier

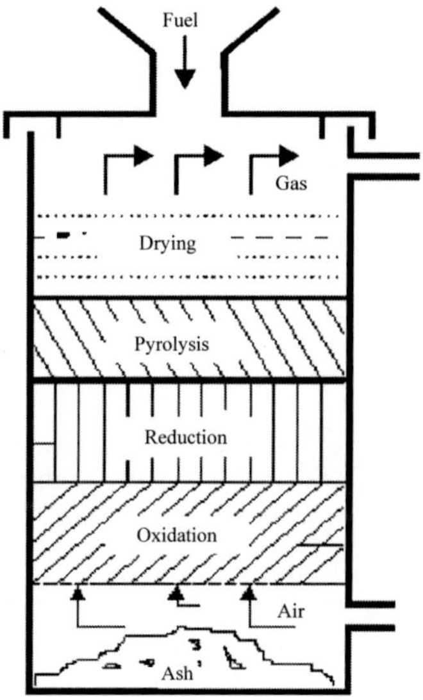

Fig. 12.21: Downdraft Gasifier

(c) Crossdraft gasifier

Crossdraft gasifier as shown in Fig. 12.22, is one of the simplest types of gasifier design in which biomass fuel enters from the top of the reactor whereas air is provided from the side of the reactor instead of top or bottom side of the reactor. Crossdraft gasifier has certain advantages as compare to updraft and downdraft type of gasifiers but it is not of ideal type. The Crossdraft gasifier has separate zones of ash bin, fire and reduction which limit the biomass fuel used for operation with less ash content in the final produced syngas. Crossdraft gasifier has certain advantages such as fast response against biomass fed in the reactor, small start time, compatible with dry air blast,

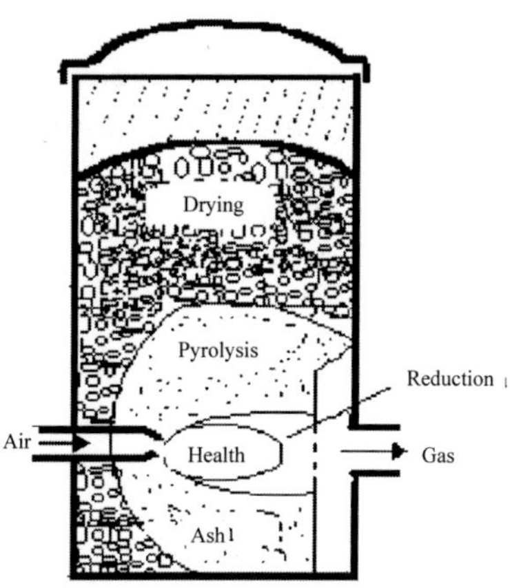

Fig 12.22: Cross Draft Gasifier

flexible syngas production, and comparatively shorter design height. But apart from some major advantages crossdraft gasifier has certain drawbacks such as incapability to handle high tar content and very small biomass particles. This gasifier produces high temperature syngas with poor reduction rate of carbon dioxide. Therefore, Crossdraft gasifier has limited application in the field and not much work has been reported in the literature.

12.11.1.2 Fluidized bed type gasifier

The operation of both up and downdraught gasifiers is influenced by the morphological, physical and chemical properties of the fuel. Lack of bunker flow, slagging and extreme pressure drop over the gasifier are some common problems encountered in these gasifiers.

Fluidized bed gasifier illustrated schematically in Fig. 12.23 is solution of above all problems. In this type of gasifier air is blown through a bed of solid particles at a sufficient velocity to keep these in a state of suspension. The bed is originally externally heated and the feedstock is introduced as soon as a sufficiently high temperature is reached. The fuel particles are introduced at the bottom of the reactor, very quickly mixed with the bed material and almost instantaneously heated up to the bed temperature. As a result of this treatment the fuel is pyrolysed very fast, resulting in a component mix with a relatively large amount of gaseous materials. Further gasification and tar-conversion reactions occur in the gas phase. Most systems are equipped with an internal cyclone in order to minimize char blow-out as much as possible. Ash particles are also carried over the top of the reactor and have to be removed from the gas stream, if the gas is used in engine applications.

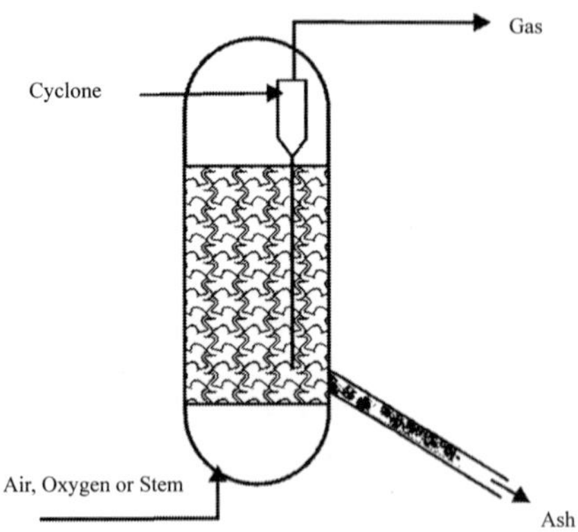

Fig. 12.23: Fluidized Bed Gasifier

(a) Bubbling fluidized bed gasifier

Gasification of different biomass feedstock in bubbling fluidized bed reactor takes place over hot bed of inert material such as sand, dolomite etc. under high pressure with fluidizing medium of air, oxygen or steam. These gasifiers are easy in design and in operation as compared to circulating fluidized bed gasifier. These gasifiers are generally designed to operate at low gas velocity of below 1m/s. Solid particles while moving along the gas flow are separated from the gas in the cyclone and collected at the bottom of reactor. Biomass conversion takes place in the bubbling bed region which gives lesser time for tar conversion in the reactor. These gasifiers are capable to operate at higher temperature of 850°C which results in to higher thermal breakdown of biomass and increase in the quantity of syngas production. However, carbon conversion efficiency is lower as compared to circulating fluidized bed gasifier.

(b) Circulating fluidized bed gasifier

There are numbers of biomass gasifier installed across the country and most of them are fixed bed type which are not suitable for continuous long run. Though downdraft gasifier found suitable for electricity generation but the producer gas leave from it are having very high temperature. It increases excessive cooling load and considerable heat lost reduce the overall efficiency of the gasifier.

Circulating fluidized bed (CFB) gasifer is advanced and emerging technology as compared to fixed bed type gasifier. Producer gas generated using CFB have low tar, low ash and moderate gas temperature, ultimately reduced gas cooling load and increase overall conversion efficiency.

12.11.2 Producer gas and its constituents

Producer gas is nothing but it is the mixture of combustible and non-combustible gases. The quantity and quality of producer gas highly depends on the type of feedstock and gasifier operating condition. The constituents of producer gas are presented in Table 12.2.

Table 12.2: Constituents of Producer Gas

Particulars	Amount (%)
Carbon Monoxide (CO)	15 - 30%
Hydrogen (H_2)	10 - 20 %
Methane (CH_4)	2 - 4%
Nitrogen (N_2)	45 - 60%
Carbon Dioxide (CO_2)	5 - 15%
Water Vapour (H_2O)	6 - 8%
Calorific Value (Higher Heating Value)	4.5 - 6 MJ/m^3

Carbon monoxide gas is toxic in nature and possesses higher octane number of 106. Higher octane numbers retard the ignition speed. The octane number of Hydrogen in producer gas is varied in the range of 60 - 66 and it improve the ignition speed. Amount of available hydrogen and methane in producer gas are responsible for higher heating value. During the gasification process, air is act as gasification media. Atmospheric air having 78.09% Nitrogen and 20.95 % oxygen, during gasification process Nitrogen act as inert gas that is why the Nitrogen percentage in the producer gas is higher than other gases. If the percentage of carbon dioxide gas in producer gas increases, it seems that the gasification process leading to complete combustion. Higher moisture content reflects the higher moisture content in the feedstock.

12.11.3 Gasification process and reaction Chemistry

The gasification process is nothing but it is a thermochemical conversion process through which soil biomass converted into gaseous fuel. There are four steps taking place in entire process i.e., Drying, Pyrolysis, Oxidation, and Reduction.

(a) Drying

Biomass feed prepared for gasification are normally sun dried and having moisture content about 10-15 %, sometime partially dried biomass may be used which have moisture content upto 25%. Moisture content or water present in the biomass feed removed above 100 °C and feedstock completely bone dry. Temperature in this zone ranging during gasification process is about 100-180 °C. There is no thermal decomposition of biomass taking place in this zone.

(b) Pyrolysis

The biomass is heated in absence of air at elevated temperature about 180-700 °C. Biomass thermal decomposed and converted into solid, liquid and gaseous fuel. Mixture of combustible gases are condensable in nature, if condensed properly then liquid fuel i.e., bio-oil or pyrolytic oil can be produced.

(c) Oxidation

Desire air for partial combustion are introducing in this zone. Temperature varied in the rage of 700 - 1500 °C. Heterogeneous thermal reaction takes place between air and carbon produced in pyrolysis zone. Most of thermo chemical reactions are exothermic in nature. Carbon react with air produce carbon dioxide and heat on the other hand hydrogen react with air and produce water vapour. The main reactions are-

$$\mathrm{C+O_2 \rightarrow CO_2+406\ (MJ/kmol)}$$

$$\mathrm{H_2+\tfrac{1}{2}O_2 \rightarrow H_2O+242\ (MJ/\ kmol)}$$

(d) Reduction

Numbers of thermo chemical reactions are taking place in absence of air. These reactions lower the zone temperature and it sustain at 700 - 1100 °C. The principle reactions are as follows-

$$\mathrm{CO_2+C \rightarrow 2CO-172(MJ/kmol)}$$

$$\mathrm{C+H_2O \rightarrow CO+H_2-131(MJ/kmol)}$$

$$CO_2+H_2 \rightarrow CO+H_2O+41(MJ/kmol)$$

$$C+2H_2 \rightarrow CH_4+75(MJ/kmol)$$

12.11.4 Factor affecting the biomass gasification process

The quality of produce gas highly depends of biomass feedstock. The following parameter to be monitor to get desire quality of producer gas:

(a) Gasification media

Gas composition and its calorific value depend on biomass composition, stability of oxidation zone, temperature at different zone, and gasification media. Role of gasification media on its composition and its calorific value are listed in Table 12.3

Table 12.3: Gasification Media, Products and Calorific Value of Producer Gas.

Media	Products	Calorific value of producer gas
Air	CO, CO_2, H_2, CH_4, N_2 and tar	5 – 6 MJ / Nm^3
Oxygen	CO, CO_2, H_2, CH_4, and tar	10 – 12 MJ / Nm^3
Steam	CO, CO_2, H_2, CH_4, and tar	15-20 MJ / Nm^3

(b) Moisture content

Biomass feedstock to be properly dry and kept moisture content below 10-15 percent. A higher moisture content reduced the temperature oxidation zone. Owing to such issue incomplete cracking of hydrocarbons lower the calorific value of producer gas.

(c) Ash content

The amount of ash in different types of feedstock varies widely (0.1% for wood and up to 15% for some agricultural products) and influences the design of the ash removal and handling system.

The oxidation temperature is often above the melting point of the biomass ash, leading to clinkering in the grate and subsequent feed blockages. Clinker is a problem for ash contents above 5%, especially if the ash is high in alkali oxides and salts which produce eutectic mixtures with low melting points.

12.11.5 Equivalence Ratio (ER)

Equivalence ratio is one of the most important parameters, which affect the gasification process. Equivalence ratio is the ratio of the actual air to the stoichiometric air/fuel ratio. Its value is 1 for the complete combustion and 0 (zero) for pyrolysis or carbonisation, and for gasification its values typically ranging from 0.2 to 0.4. Several problem associated including excessive char formation if ER < 0.2. Excessive CO_2 and H_2O generation will take place if ER > 0.4

$$ER = \frac{\text{Actual air}}{\text{Stoichiometric air for complete combustion}}$$

12.11.6 Turndown Ratio

It is the ratio of highest gas generation by a gasifier to the lowest gas generation in actual. The value of turndown ratio of gasifier for vehicle operation is about 3 for Imbert gasifier without insulation while it is about 18 for well insulated gasifier. For running irrigation pumps and electricity generation it is not much important as they running on full load.

12.11.7 Cold and Hot Gas Efficiency

Cold gas efficiency is the ratio of thermal energy of produce gas to the thermal energy of feed stock

$$\text{Cold gas efficiency} = \frac{\text{Heating value of produce gas}}{\text{Heating value of feedstock}} \times 100\ \%$$

Hot gas efficiency: The gas is not cooled before combustion and the sensible heat is also used.

$$\text{Hot gas efficiency} = \frac{\text{Heating value of produce gas} + \text{Sensible heat}}{\text{Heating value of feedstock}} \times 100\ \%$$

12.11.8 Capacity of the gasification system

The gasification system is to be designed on the basis of process heat required kJ per hour

The following assumptions to be made for design

The hot gas efficiency of the gasification system	=	60%
Burner efficiency	=	60%
Specific gasification rate	=	150 kg $h^{-1}m^{-2}$
Calorific value of gas	=	4600 kJ m^{-3}
Calorific value of feedstock	=	12.5 MJ kg^{-1}
Gas output from wood chip (Theoretically)	=	2.2 $m^3\ kg^{-1}$

(a) Feedstock consumption rate

The consumption rate of feedstock was calculated using the following assumptions:

Calculation of Feed Stock Consumption Rate

$$\text{Feedstock consumption rate} = \frac{\text{Gas output x Calorific value of gas}}{\text{Hot gas efficiency x Calorific value of feedstock}}$$

(b) Dimension of the reactor shells

It is calculated by using the following formula:

$$\text{Reactor cross sectional area} = \frac{\text{Feed stock consumption rate}}{\text{Specific gasification rate}}$$

(c) Height of the reactor

The height of the reactor was decided on the basis of required feedstock holding capacity and the duration of operation of the system.

$$\text{Volume occupied by wood chips} = \frac{\text{Holding Capacity}}{\text{Bulk density of wood chips}}$$

$$\text{Height of wood chips holding column} = \frac{\text{Volume occupied by wood chips}}{\text{Reactor cross sectional area}}$$

The height of the reactor to be fixed in order to

Accommodate grate, and Provide space for ash collection at the bottom.

(d) Critical thickness of Insulation

Contrary to common belief that addition of insulating material on a surface always brings about a decreasing in the heat transfer rate, there are instances when the addition of insulation to the outside surfaces of walls (geometries which have non-constant cross sectional area) does not reduce the heat loss. In fact, under certain circumstances, it actually increases the heat flow up to a certain thickness of insulation. The insulation thickness at which resistance to heat flow minimum is called critical thickness of insulation. The critical thickness, designed by R is dependent only on thermal quantities k and h_o

$$R = \frac{k}{h_o}$$

where,

R- Critical thickness of insulation, m

k- Thermal Conductivity of material, W/m K

h_o - Coefficient of convective heat transfer, W/m² K

(e) Design of the burner

The producer gas burner is of simple aspirated type. Various details were worked out based on the gas flow rate which are as follows:

Velocity of the gas at the inlet of burner $\frac{\pi \times D \times N}{60}$ = m s^{-1}

Where,

D – Diameter of the blower, m

N–Revolution per minute of blower

$$\text{Area of the cross section of the gas inlet } (A_b) = \frac{\text{Gas flow rate through burner}}{\text{Velocity of gas}}$$

$$\textit{Diameter of inlet pipe of the burner}\,(d_b^{\,2}) = \frac{\textit{Gas flow rate through burner} \times 4}{(\textit{Velocity of gas} \times 3600 \times \pi)}$$

$$\text{Diameter of Inlet pipe of the burner} = (d_b) = \sqrt{\frac{\textit{Gas flow rate through burner} \times 4}{(\textit{Velocity of gas} \times 3600 \times \pi)}}$$

The diameter of the burner can be designed according to the size of the vessels.

12.11.9 Gasifier and its Applications

(a) Power Applications

Gas engine can be run on100 percent producer gas. Biomass required to produce 1 kWh of power from gas engine is varied from 1.12 to 1.5 kg and it depends on the type of ignition. On the other hand, engine running on dual fuel mode (diesel + producer gas) approximately 70 to 80 percent diesel can be replaced using producer gas. In dual fuel mode the engine likely to de-rated by 20 to 30 percent on rated capacity basis. Biomass and diesel fuel required to produce 1 kWh of power are 0.9-1.2 kg and less than 0.1 liter respectively.

(b) Thermal Applications

Biomass gasifier found most suitable for thermal applications. There is no cleaning and cooling of producer gas is required, it can directly be combust using produce gas burner. It is a viable option to replace the convention fossil fuel such as diesel, light diesel oil, furnace oil and LPG. The biomass required to replace one liter of diesel and light diesel oil are in the range of 3.5 to 4.25 kg respectively, where as one kilogram of LPG and furnace oil are in range of 4 to 4.5 kg and 4 to 5 kg respectively.

Example 12.2: A downdraft biomass gasifier installed at boys hostel to run the canteen. Canteen owner is using groundnut shell briquettes as fuel to run gasifier. The ultimate analysis of groundnut shell is as follows:

C = 41.10 %

H = 4.8 %

N = 1.6 %

O = 39.2 %

S = 0.04%

Determine the chemical formula for groundnut shell and desire air fuel ratio for complete combustion. Further also estimate air required for gasification if only 35% of the stoichiometric air used during gasification. Assuming air density is about 1.2 kg/m^3.

Solution

Chemical formula:

$$C = \frac{41.10}{12} = 3.42$$

$$H = \frac{4.8}{1} = 4.8$$

$$N = \frac{1.4}{14} = 0.10$$

$$O = \frac{39.2}{16} = 2.45$$

$$S = \frac{0.04}{32} = 0.0012$$

Thus chemical formula of groundnut shell is as follows:

$C_{3.42}\,H_{4.8}\,O_{2.45}\,N_{0.10}\,S_{0.0012}$

Air required for complete combustion of groundnut shell briquettes:

$C_{3.42}\,H_{4.8}\,O_{2.45}\,N_{0.10}\,S_{0.0012} + 3.39[(O_2 + 3.76N_2)]$

$\rightarrow 3.42CO_2 + 2.4H_2\,O + 3.39 \times 3.76N_2 + 0.0012S$

$$\frac{\text{Air}}{\text{Fuel}} = \frac{(3.39 \times 32) + (3.39 \times 3.76 \times 28)}{(12 \times 3.42) + (1 \times 4.8) + (16 \times 2.45) + (14 \times 0.1) + (32 \times 0.0012} = 5.38$$

For complete combustion of biomass 5.38 kg of air per kg of fuel is required. For gasification only 35% of desired air is used. Therefore actual air required for gasification is:

$5.38 \times 0.35 = 1.883$ kg

Air fuel ratio for gasification:

$\frac{1.883}{1.2}$ =1.56 m^3 air per kg of groundnhut shell brequitte

Example 12.3: Produce gas combusted in boiler to generate steam with 80 % thermal efficiency. Steam expanded in steam turbine from 500 °C to 100 °C to produce electricity. Calculate the overall efficiency of system.

Solution

Given

Thermal efficiency: 80%

Steam temperature at turbine inlet: 500 °C = 773 K

Steam temperature at turbine outlet: 100 °C = 373 K

Ideal turbine efficiency:

$$\eta = \left(1 - \frac{373}{773}\right) \times 100 = 51.76\ \%$$

Therefore, overall system efficiency:

$$\eta_{overal\,l} = 0.80 \times 0.5176 = 41.39\ \%$$

Example 12.4: Electricity consumption in a food processing industry is about 5 MW. Owner of industry installed a downdraft biomass gasifier to meet electricity demand. Gasifier based power plant operated 300 days in a year and 24 hour per day. Calorific value of biomass used is 16 MJ/kg. The conversion efficiency is about 18%. Calculate the area required for biomass production for smooth running of power plant, if biomass yield is about 30 tons per hectare.

Solution:

Energy require : 5 MW

Conversion efficiency : 18%

Biomass yield : 5 tons per hectare

Operation 300 days, 24 hours per day

$$\text{Energy input} = \frac{\text{Energy Required}}{\text{Conversion efficiency}} = \frac{5}{0.18} = 27.77\ \text{MW}$$

Area requires for cultivating biomass for uninterrupted power supply:

$$27.77\,\frac{\text{MJ}}{\text{s}} \times 300\ \text{days} \times 24\ \text{hours} \times 3600\ \text{second} \times \frac{1\ \text{kg}}{16\ \text{MJ}} \times \frac{1}{30000}$$

$$\frac{\text{hectare}}{\text{kg}} = 1499.58\ \text{hectare}$$

Example 12.5: Calculate the biomass requirement per hour to replace 50 kg of Light Diesel Oil (LDO) operated steam generator. Assume calorific value of LDO and biomass are 40 MJ/kg and 19 MJ/kg respectively, and overall efficiency for both gasification and LDO operated steam generated system is about 20 %.

Solution

Energy supplied by LDO system

$$= \frac{50 \text{ kg} \times 40 \dfrac{\text{MJ}}{\text{kg}}}{0.20} = 10{,}000 \frac{\text{MJ}}{\text{hour}}$$

Energy input desired from biomass:

$$= \frac{10{,}000 \text{ MJ}}{0.20} = 50{,}000 \frac{\text{MJ}}{\text{hour}}$$

Biomass required supplying desired energy:

$$= \frac{50{,}000 \text{ MJ}}{19 \dfrac{\text{MJ}}{\text{kg}} (\text{calorific value of biomass})} = 2631 \frac{\text{kg}}{\text{hour}}$$

Example 12.6: Calculate the diameter and height of a downdraft biomass gasifier to supply 500 kW thermal energy to generate process heat for 4 hours. Biomass gasifying at 60% efficiency with specification gasification rate of 130 kg h^{-1} m^{-2}. Assume heating value of biomass and producer gas are 16.5 MJ kg^{-1} and 5.2 MJ Nm^{-1} respectively.

Solution

Themal output = Producer gas flow rate × Heating value of producer gas

$$\text{Producer gas flow rate} = \frac{\text{Thermal output}}{\text{Heating value of producer gas}}$$

$$\text{Producer gas flow rate} = \frac{500}{1000} \frac{\text{MJ}}{\text{s}} \times \frac{1}{5.2} \frac{\text{Nm}^{-3}}{\text{MJ}}$$

$$\text{Producer gas flow rate} = 0.096 \frac{\text{Nm}^{-3}}{\text{s}}$$

Biomass required to generate desired producer gas flow rate

$$\text{Fuel consumption rate} = \frac{\text{Thermal power output}}{\text{Heating value of biomass} \times \text{Heating value of producer gas}}$$

$$\text{Fuel consumption rate} = \frac{500}{1000} \frac{\text{MJ}}{\text{s}} \times \frac{1}{16.5} \frac{\text{kg}}{\text{MJ}} \times \frac{1}{0.6} = 0.050 \frac{\text{kg}}{\text{s}}$$

$$\text{Fuel consumption rate} = 181.81 \frac{\text{kg}}{\text{h}}$$

Internal diameter of gasifier reactor

$$\text{Cross sectional area of reactor} = \frac{\text{Fuel Consumption rate}}{\text{Speific gasification rate}}$$

$$\text{Cross sectional are of reactor} = 181.81\frac{\text{kg}}{\text{h}}\times\frac{1}{130}\frac{\text{h m}^2}{\text{kg}} = 1.39\ \text{m}^2$$

$$\text{Diamenter of reactor} = \sqrt{\left(\frac{1.39\times 4}{\pi}\right)} = 1.315\ \text{m}$$

If gasifier operated in single feeding, 727.24 kg biomass to be loaded. Assuming bulk density of biomass is 350 kg m^{-3}

$$\text{Height of wood column} = \frac{\text{Fuel consumption rate}\times\text{Duty hour}}{\text{Bulk Density of Biomass}}$$

$$\text{Height of wood column} = \frac{181.81\times 4}{350} = 2.07\ \text{m}$$

Example 12.7: Calculate the Stoichiometric air required for the solid Fuel having the chemical composition: $C_{3.33}\ H_6\ N_{0.071}\ O_{2.38}$

Solution: The molecular weight of the C =12, H=1, N=14 and O=16

So, Actual Proportion of the C, H, N, O in fuel are:

C = 3.33 × 12 = 39.96 % = 0.3996

H = 6 × 1 = 6% = 0.06

N = 0.071 × 14 = 0.994% = 0.00994

O = 2.38 × 16 = 38.08% = 0.3808

Out of all of the above components the oxidation reaction will takes place with Carbon and Hydrogen Only

But, In the atmosphere N_2 = 79 % and O_2 = 21%

Therefore one Oxygen molecule $= O_2 + \frac{79}{21}N_2 = O_2 + 3.76N_2$

Now For the Carbon,

$C + (O_2 + 3.76N_2) \rightarrow CO_2 + 3.76N_2$

adding the moecular weight

$12 + (2 \times 16 + 3.76 \times (2\times 14)) \rightarrow 44 + 3.76 \times 2 \times 14$

$12C + 32O_2 + 105.28\ N_2 \rightarrow 44CO_2 + 105.28N_2)$

Now adding the proportionate in to the equation

$$0.3996C + 0.3996 \times \frac{32}{12} O_2 + 0.3996 \times \frac{105.28}{12} N_2$$

$$\rightarrow 0.3996 \times \frac{44}{12} CO_2 + 0.3996 \times \frac{10.5.28}{12} N_2$$

$0.3996\ C + 1.065\ O_2 + 3.50N_2 \rightarrow 1.465\ CO_2 + 3.50N_2$

For the Hydrogen,

$$H_2 + \tfrac{1}{2}(O_2 + 3.76N_2) \rightarrow H_2O + \frac{3.76}{2} N_2$$

$$2 + \left(\frac{1}{2} \times 32\right) + \left(\frac{1}{2} \times 3.76 \times 28\right) \rightarrow 18 + \left(\frac{1}{2} \times 3.76 \times 28\right)$$

$2\ H_2 + 16\ O_2 + 52.64\ N_2 \rightarrow 18\ H_2\ O + 52.64\ N_2$

$$0.06 + \left(0.06 \times \frac{16}{2}\right) + \left(0.06 \times \frac{52.64}{2}\right) \rightarrow \left(0.06 \times \frac{18}{2}\right) + \left(0.06 \times \frac{52.64}{2}\right)$$

$0.06\ H_2 + 0.48\ O_2 + 1.58\ N_2 \rightarrow 0.54H_2\ O + 1.58\ N_2$

Now, adding all the composition on Left Hand Side and,

$0.3996\ C + 0.06\ H_2 + (0.00994\ N_2 + 0.3808\ O_2) + (1.065\ O_2 + 3.50N_2 + 0.48\ O_2 + 1.58\ N_2)$

$\rightarrow 1.465\ CO_2 + 0.54H_2\ O + 3.50N_2 + 1.58\ N_2$

$+ (0.00994\ N_2 + 0.3808\ O_2)$

$0.3996\ C + 0.06\ H_2 + (0.00994\ N_2 + 0.3808\ O_2) + 1.545\ O_2 + 5.08N_2$

$\rightarrow 1.465\ CO_2 + 0.54H_2\ O + 3.50N_2 + 1.58\ N_2$

$+ (0.00994\ N_2 + 0.3808\ O_2)$

Total amount of air required $(O_2 + 3.76N_2) = 1.545 + 5.08 = 6.625$

For 1 kg of Fuel we required 6.625 kg of total air

12.12 Biodiesel Production

Biodiesel is a mixture of methyl esters of long chain fatty acids like lauric, palmitic, stearic, oleic, and so on. It is produced by the transesterification of animal fats and vegetable oils – all of which belong to a group of organic esters called triglycerides. Typical examples are rape/canola oil, soyabean oil, sunflower oil, palm oil and its derivatives, etc. from vegetable sources, beef and sheep tallow and poultry oil from animal sources and also from used cooking oil. The chemistry is basically the same irrespective of the feedstock. The chemistry of biodiesel is given in Fig. 12.24.

Fig. 12.24: The Chemistry of Biodiesel

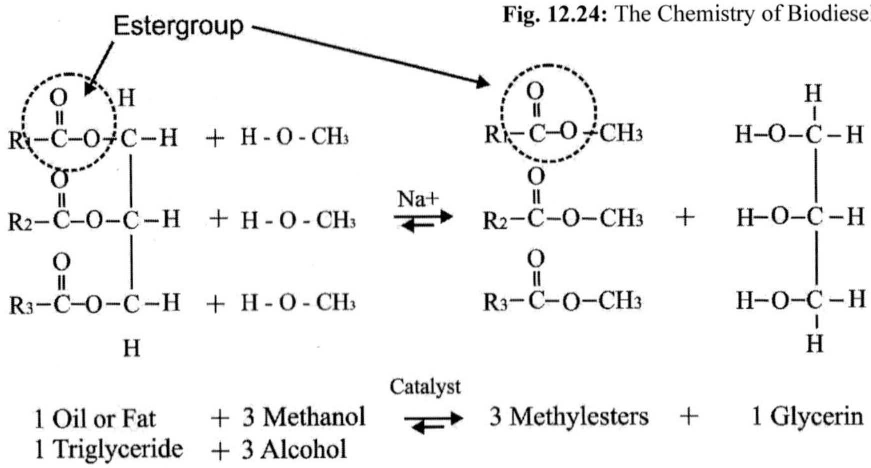

The abbreviation R_1, R_2 and R_3 are symbolic representations of the fatty acid chains, which can vary in molecular chain length from typically C8 to C22, and also their degree of unsaturation. An example is **Oleic acid**, which has 18 carbon atoms and one double bond:

CH_3-CH_2-CH_2-CH_2-CH_2-CH_2-CH_2-CH_2-CH=CH-CH_2-CH_2-CH_2-CH_2-CH_2-CH_2-CH_2-COOH

or more simply: CH_3-$(CH_2)_7$CH=CH-$(CH_2)_7$COOH

The abbreviation C(18:1) is also often used.

12.12.1 Biodiesel Process

The production of biodiesel involves intensively mixing of methanol with oil or fat in the presence of a suitable catalyst, and then allowing the lighter methyl ester phase to separate by gravity from the heavier glycerol phase. However, as with most organic reactions the degree of conversion depends on the equilibrium reached as well as the influence of other reactions. Achieving product quality is also very important. The process flow chart of biodiesel production is illustrated in Fig.12.25

Fig. 12.25: Process Flow for Biodiesel Process

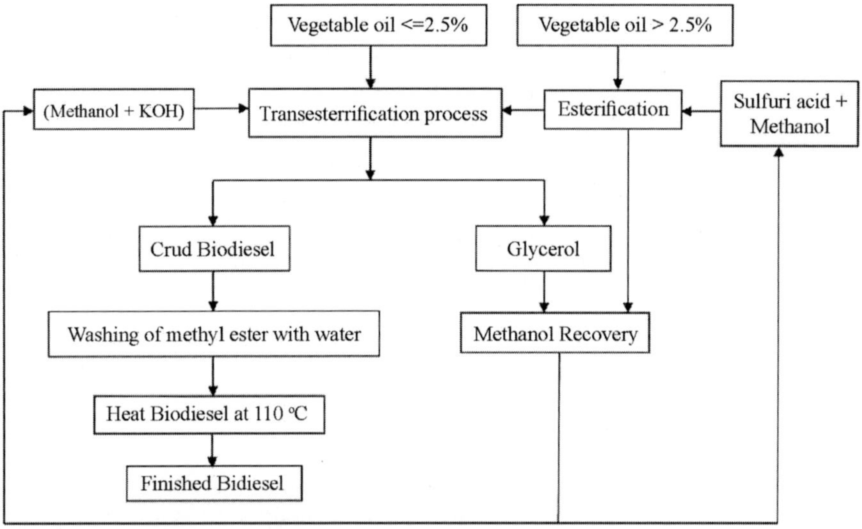

The key features of Biodiesel Process are:

- Technology applicable to multiple feedstocks
- Continuous process at atmospheric pressure and nominally 60°C
- Dual Reactor System operating with a patented Glycerin Cross Flow configuration for maximized conversion
- Reaction using excess methanol, but with full methanol recycle to avoid any losses
- Closed loop water wash recycle to minimize waste water generation
- Clear phase separation by special gravity process (no centrifuges necessary)
- Based on suitable feedstock, this produces biodiesel to current world standards
- Raw Glycerin to BS 2621.

The crude glycerin can be further upgraded to pharmaceutical glycerin standard EU Pharmacopoeia 99.5 by distillation, bleaching if required, and vacuum drying.

12.12.2 Prospective feedstock in India

Oil can be extracted from a variety of plants and oilseeds. Under Indian condition only such plant sources can be considered for biodiesel production which are not edible oil in appreciable quantity and which can be grown on large-scale on wastelands. Moreover, some plants and seeds in India have tremendous medicinal value, considering these plants for biodiesel production may not be a viable and wise option. Considering all the above options, probable biodiesel yielding trees in India are:

(a) *Jatropha curcas* or Ratanjot

(b) *Pongamia pinnata* or Karanj

(c) *Calophyllum inophyllum* or Nagchampa

(d) *Hevea brasiliensis* or Rubber seeds

(e) *Calotropis gigantia* or Ark

(f) *Euphorbia tirucalli* or Sher; and

(g) *Boswellia ovalifololata.*

Among all the above prospective plant for as biodiesel production, Jatropha curcas stands at the top. One hectare Jatropha plantation with 4400 plants under rain fed conditions can yield about 1500 literes of oil. It is estimated that about 3 million hectares plantation is required to produce oil for 10% replacement of petrodiesel. The residue oil cake after extraction of oil from Jatropha can be used as organic fertilizers. It is also estimated that one acre of Jatropha plantation could produce oil sufficient to meet the energy requirement of a family of 5 members and the oil cake left out when used as fertilizer could cater to one acre.

12.12.3 Economic Feasibility

The expected cost of extracted oil in Indian condition would be Rs. 35-45 per liter. However, the price of commercially available bio-diesel is presently Rs. 45 – 55 per liter. With recent increase in oil prices, it is essential to look for substitutes of fossil fuels for both economic and environmental benefits to the country. The bio-fuel extraction will disseminate technology and provide better energy services at the village level.

12.12.4 Environmental Benefits

There are number of environmental benefit through use of biodiesel, which are given as below:

- The use of bio-fuel avoids fossil fuel use and hence avoids CO_2/CO emission in atmosphere
- It is a promising alternative fuels source for future

- Substantial reduction of unburned hydrocarbons, carbon monoxide and particulate matter
- Decrease the solid carbon fraction of particulate matter
- Increase in the green cover as result of plantations would check soil erosion and retain moisture and soil nutrients.
- Positive ecological benefits in terms of lending support to biodiversity, especially since degraded lands are involved

Biofuels also help the environment because the plants grown to make these fuels take greenhouse gases such as carbon dioxide out of the air and fix it in their roots, stems and leaves. Much of this carbon dioxide gets sequestered in the soil, reducing the overall level of carbon dioxide in the atmosphere.

12.13 Bioethanol Production

The rapid growth of industrialization and population significantly affect on the demand of ethanol as an alternate fuel. Generally, conventional crops like sugarcane and corn become unable to meet the demand of bioethanol, due to their primary role in food and feed. Therefore, lignocellulosic biomass from agriculture waste is attractive precursor for bioethanol production. Biomass is renewable organic material, abundant and cost effective. Ethanol production from agriculture waste has been projected an economically attractive option for industrial sector. However, sugar and starch derived ethanol have a more theoretical yield as compared to lignocelluloses, these sources are insufficient for bioethanol production. The major agricultural wastes including rice straw, corn straw, wheat straw and sugarcane bagasse etc. can subsequently use for ethanol production. Rice straw has a highest worldwide potential for bioethanol production (205 giga liter) followed by wheat straw (104 giga liter), corn straw (58.6 giga liter), and finally for sugarcane bagasse (51.3 giga liter).

A bio fuels is environmentally friendly options for power generation. Most of the fossil fuels are biological in nature. These are plant forms that, typically, remove carbon dioxide from the atmosphere, and give up the same amount when burnt. The bio fuels are therefore considered to be "CO_2 neutral", not adding to the carbon dioxide level in the atmosphere. The type of bio fuel used will depend on a number of factors, chief amongst them being the available feedstock and the energy that can be used locally.

India import 70% of the oil it uses, and the country has been hard by the increasing price of oil, uncertainty and environmental hazards that are concerned with the consumption of fossil fuels. In this context, bio fuels constitute a suitable alternative source of energy for India. There are two examples of bio fuels are Ethanol and Bio diesel (Renewable Diesel)

Ethanol can be made from biomass material containing sugar, starches, or cellulose (starch and cellulose are complex from of sugar)

12.13.1 Composition and properties

The alcohol molecular structure includes an (OH), or hydroxyl radical which is responsible for high solubility in water and high latent heat of vaporization. This water like characteristics are most apparent in the alcohol of low molecular weight methanol and ethanol, because the (OH) radical predominates over their short hydrocarbon chains. They are least apparent in the alcohols of high molecular weight, tertiary butyl or heavier alcohols, because their longer hydrocarbon chains predominate over the (OH) radical. These characteristics can be advantageous or disadvantageous depending on what function the alcohol it to serve.

Example 12.8: Estimate the pure sugar required in kilogram to produce a liter of pure ethanol. The density of pure ethanol is about 789 kg/m^3

Solution

The theorical equation to produce ethanol from pure sugar is as follow:

$$C_6H_{12}O_6 + \text{Yeast} \rightarrow 2C_2H_5OH + 2CO_2 + \text{Heat}$$

Molecular weight of pure sugar

$$C_6H_{12}O_6 = (12 \times 6) + (12 \times 1) + (16 \times 6) = 180$$

Molecular weight of pure ethanol

$$C_2H_5OH = (12 \times 2) + (5 \times 1) + (16 \times 1) + (1) = 46$$

One mole of sugar produce two moles of ethanol.

Thus, molecular weight of ethanol in equation $= 94$

Pure glucose required to generate a liter of ethanol:

$$= \frac{\text{Molecular weight of pure sugar}}{\text{Molecular weight of ethanol}} \times \text{Density of ethanol}$$

$$= \frac{180\text{ kg}}{92\text{ kg}} \times \frac{0789\text{ kg}}{\text{liter}} = 1.54 \frac{\text{kg}}{\text{liter}}$$

Hence 1.54 kg of pure sugar is needed to produce 1.0 liter of pure ethanol.

12.13.2 Feedstocks for bioethanol production

Bioethanol is usually produced from the agricultural produces which containing sugar. The agricultural produce extensively used for bioethanol production can be classified in two categories, in first category, agricultural produce containing sugar and starch, and lignicellulosic sugar containing fall in second category. In the present context ethanol produce using yeast to ferment the starch and sugars in sugar beets, sugar cane, and corn. The starch contained in corn kernels is fermented into sugar, which is further fermented into alcohol.

12.13.3 Ethanol from Starch Materials

There is considerable amount of farm crops are capable of being extensively used as raw material for fermentation process leading to ethanol production. A variety of raw material drawn from the vegetable world, and widely grown crops are capable of fermentation into ethanol. Starch based feedstock include a variety of cereals, grains, and tuber crops.

Starch content of some of the major sources are resented in Table 12.4. The starchy feed stocks can be hydrolysed to get fermentable sugar syrup and give an average yield of upto 42 liters of ethanol per kg of feedstock.

Table 12.4: Starch Based Feedstocks for Ethanol Production

S. No.	Feedstock	Starch Percentage
1.	Corn	60 - 68
2.	Sorghum	75 - 80
3.	Rye	60 – 63
4.	Cassava	25 – 30
5.	Rice	70 – 72
6.	Barley	55 – 65
7.	Potato	10 - 25

12.13.4 Ethanol Production Processes

The process of producing ethanol can be schematized as shown in Fig 12.26:

Fig. 12.26: Schematic View of Ethanol Production

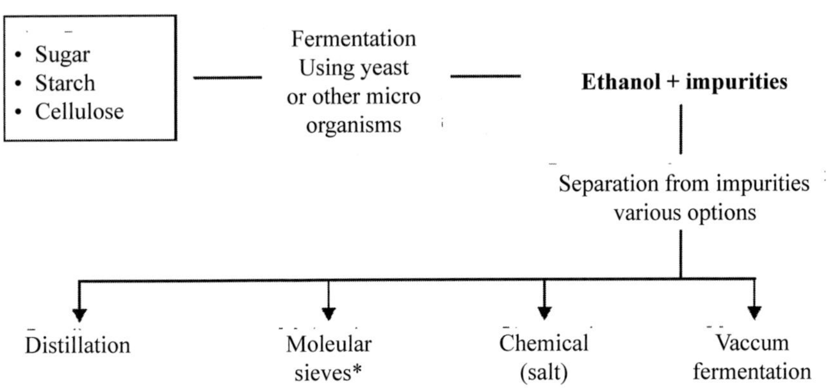

Today, almost all ethanol plants use molecular sieves for dehydration. The technology alone reduces energy use by 10 per cent per litre of ethanol

produced. Two methods are currently used to produce ethanol from grain: wet milling and dry milling.

Dry mills produce ethanol, distillers' grain and carbon dioxide (Fig. 12.27). The carbon dioxide is a co-product of the fermentation, and the distillers' dried grain with solubles (DDGS) is a non-animal based, high protein livestock feed supplement, produced from the distillation and dehydration process. If distillers' grains are not dried, they are referred to as distillers' wet grain (DWG).

Fig. 12.27: Conventional Dry Mill Ethanol Production Process

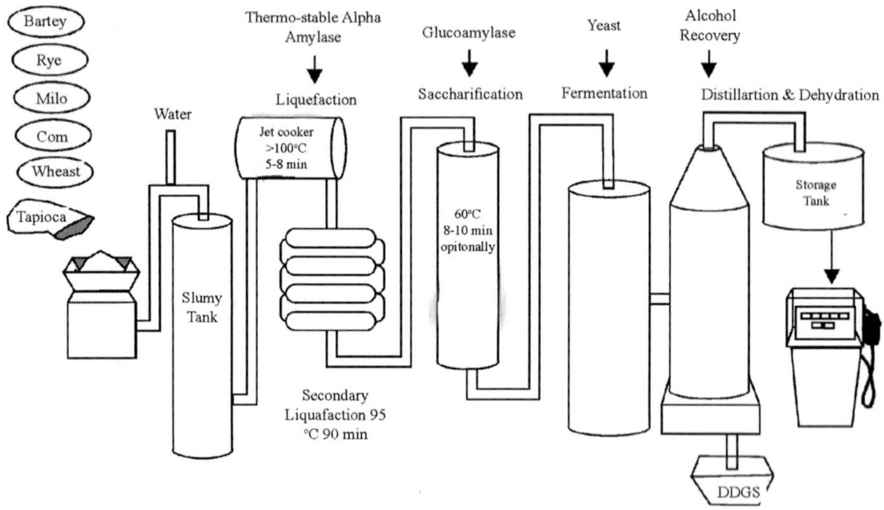

Wet mill facilities are 'bio-refineries' producing a host of high-values products (Fig. 12.28). Wet mill processing plants produce more valuable by-products than the dry mill process. For example, in wet mill plants, using corn as feedstock, they produce:

- Ethanol;
- Corn gluten meat (which can be used as a natural herbicide or as a high protein supplement in animal feeds);
- Corn gluten feed (also used as animal feed);
- Corn germ meal;
- Corn starch;
- Corn oil; and
- Corn syrup and high fructose corn syrups.

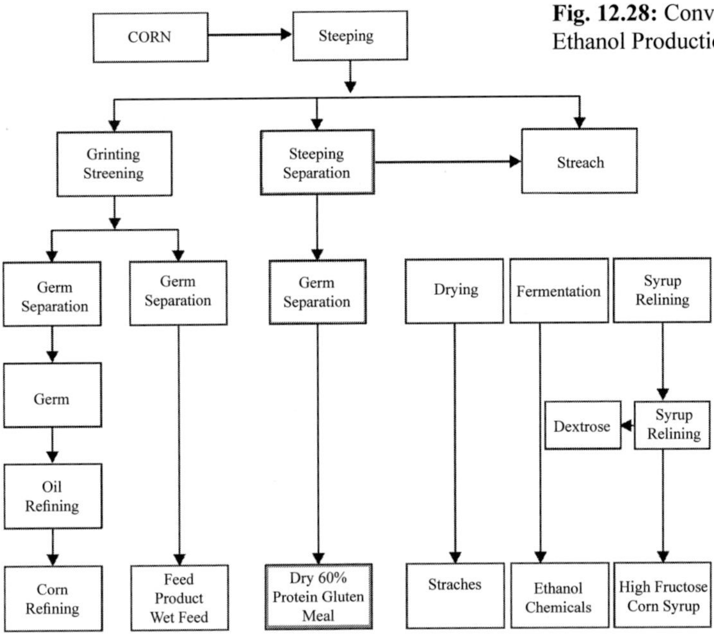

Fig. 12.28: Conventional Wet Mill Ethanol Production Process

12.13.5 Pretreatment of lignocellulosic biomass

The main challenge in biofuel production is the pretreatment of lignocellulosic biomass. Lignocellulosic biomass is basically composed of three main constituents as cellulose, hemicelluloses and lignin. After pretreatment process the solid biomass becomes more accessible for further biological and chemical treatment. In case of raw biomass, constituents of cellulose and hemicelluloses are tightly packed by lignin layer, which restrict them for further enzymatic hydrolysis. Therefore, it is essential to have a pretreatment process to break the lignin layer to expose the constituents mainly cellulose and hemicellulose for further enzymatic action. Following are the main goals of an effectual pretreatment process;

- Minimizes the crystallinity of cellulose, removes the hemicelluloses, improves the biomass surface area, break the lignin layer etc.
- Cellulose became more accessible to enzymes, therefore possible conversion of carbohydrate polymers into fermentable sugars may be achieved
- Formations of sugars by hydrolysis
- To minimize the losses and degradation of sugars
- To reduce the production cost etc.

Pretreatment process mainly classified into three categories as physical, chemical, and biological shown in Fig. 12.29

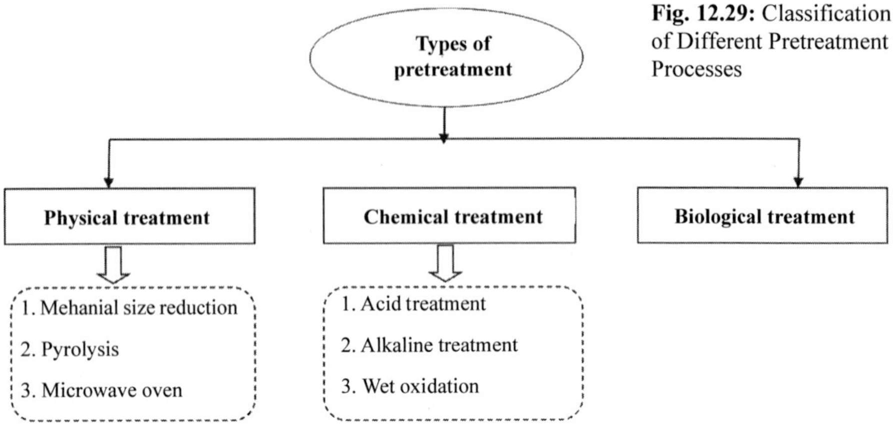

Fig. 12.29: Classification of Different Pretreatment Processes

12.13.6 Ethanol production from biomass

Agro waste mainly composed of cellulose, hemicelluloses and lignin. There are main two routes for conversion of feedstock to ethanol, which can be referred as sugar route and the syngas route. The basic concept of ethanol production is shown in Fig. 12.30. In sugar conversion route, the cellulose and hemicelluloses content initially converted into fermentable sugar, which is again fermented to produce bioethanol. The fermentable sugar mainly composed of glucose, arabinose, xylose, mannose, and galactose. These sugar components were generated after hydrolysis of hemicelluloses and cellulose in the presence of either enzymes or acids. On another side, in case of syngas route, pretreated biomass is subjected to gasification process for the production of syngas or producer gas. In gasification process, biomass is heated at high temperature in no oxygen or limited supply of oxygen required for complete combustion. It subsequently produce a syngases which contains mostly hydrogen and carbon monoxide, normally gas called as synthesis gas or producer gas or syngas. These syngases are further can be fermented using some specific microorganisms or catalytically converted into ethanol. However, in case of sugar conversion route only carbohydrates are used for ethanol production, while in case of syngas platform all the lignocelluloses of the biomass are converted into ethanol.

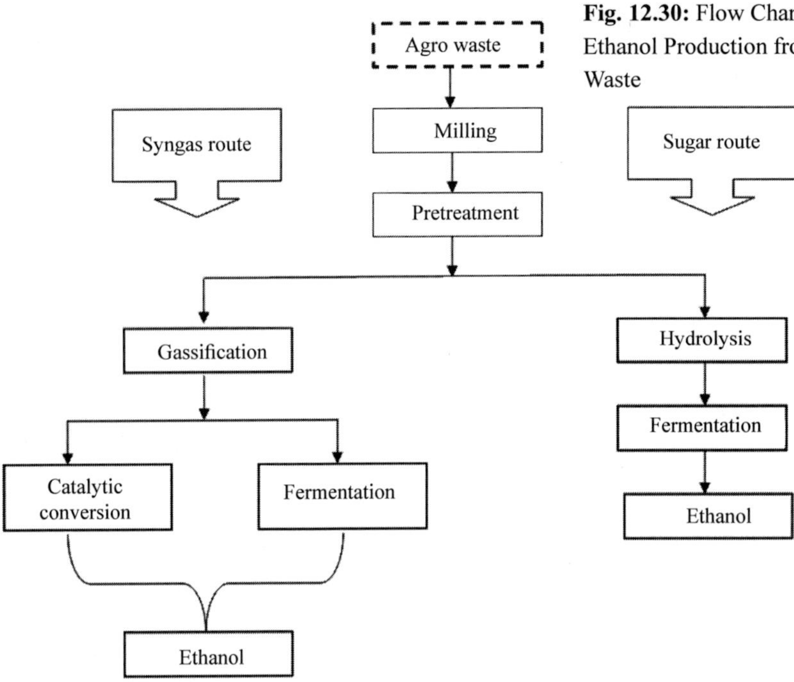

Fig. 12.30: Flow Chart of Ethanol Production from Agro Waste

12.13.7 Applications of Ethanol

Most probably ethanol is considered as cost effective blending fuel with a small portion of volatile fuel like gasoline. Thus, blending of bioethanol with diesel or gasoline has been used including;

- E85G (85% ethanol, 15% gasoline)
- E15D (15% ethanol, 85% diesel)
- E5G to E26G (5-26% ethanol, 74-95% gasoline)

In addition, bioethanol can be used as a;

- As an alternative to gasoline
- As a transport fuel
- For power generation after thermal combustion
- In a fuel cell
- For application in cogeneration system
- Application in chemical industry
- Reduce the greenhouse gas emission

13

Geothermal Energy

13.1 Introduction

Geothermal energy is the nature-based heat generated from within the earth. Underneath the earth's relatively thin crust, temperature range from 1000 to 4000°C & in some areas, pressures exceed 138 MPa is available. Geothermal energy is most likely generated from radioactive thorium, potassium and uranium dispersed evenly throughout the earth's interior which produce heat as part of the decaying process. This process generates enough heating, keep the core of the earth at temperature approaching 400°C. Composed primarily of molten nickel & iron, the core is thought to be surrounded by a layer of molten rock, the mantle at approximately 100°C. Nine major crustal plates float on the mantle & currents in the mantle cause the plates to drift, colliding in some areas and diverting in others. When two continental plates converge a complex series of chemicals in the form of molten magna rises in the earth's crust and depending upon its magnitude, it may break the crust and reach at the earth surface. Volcano, hot spring, geysers & fumaroles are natural geothermal resources near the surface and perhaps where economic drilling operations can tap their heat and pressure.

13.2 History of Geothermal Energy

Humans have used geothermal energy for a variety of uses in a variety of time periods. The Romans used geothermally heated water in their bath houses for centuries. The Romans also used the water to treat illnesses and heat homes. In Iceland and New Zealand, many people cooked their food using geothermal heat. Some North American native tribes also used geothermal vents for both comfort heat and cooking temperatures. Most of these early uses of the Earth's heat were through the exploitation of geothermal vents.

13.3 Classification of Geothermal Energy Sources

There are three main classes of geothermal energy sources. These are direct use, wet steam & dry steam based.

The first is called direct usage. The water that is heated by the magma beneath the Earth's surface can be pumped to buildings and used in heat exchanging systems.

The second way that geothermal energy is harnessed is through using the steam that comes from superheated water. If the steam vents are under sufficient pressure, then they can be used to turn turbines.

The third class of geothermal energy is called dry steam. An outside water source (naturally or otherwise) is applied to fractured rock that has been heated to high temperatures, and then the steam that arises can be used to turn turbines.

13.4 Resources Definition

In the broad sense, geothermal energy is the heat in the earth and released by conduction at an average heat flux of 60 MW/m^2. The four prerequisites necessary to exploit geothermal energy are

(a) A heat source which could be a magma body, or a simple hot rock at depth,

(b) Heat carrier fluid

(c) Permeable or fractured rock acting as a reservoir and

(d) Cap rocks providing an impermeable and insulating cover

The most obviously usable geothermal resources require convective heat transfer i.e. presence of fluid. This occurs at a limited number of locations. Whenever conduction alone prevails (anywhere) heat recovery requires that a fluid be forced through a large fractured heat exchange area to sweep the energy stored in the rocks at depth. This is basically the concept of hot dry rock technology, which is very promising. Hydrothermal resources are classified according to the specific enthalpy of the fluid. Waters with temperatures between 30 - 120°C are called low enthalpy resources (0.03 to 0.4 MJ/kg. Waters with temperatures above 120°C are termed as high enthalpy fluids (0.5 to 3 MJ/kg)

Location of geothermal provinces is dictated by the geodynamic model of the earth's crust, known as the global plate tectonics. This theory accounts for the most of the geodynamic processes affecting the earth's crust. These geodynamic processes include subduction, subsidence, uplift, fracturing etc. These occurrences result in associated geothermal features such as the distribution of heat flow, active tectonics, volcanism and hydrothermal convection.

13.5 World Geothermal Energy Status

The global geothermal power potential is estimated in the range of 35 GW to 2 TW and currently over 14GW of electricity is being generating through geothermal resources world-wide. To put geothermal generation into perspective, this generating capacity is about 0.4% of the World total installed generating capacity. The USA, Philippines, Italy, Mexico, Iceland Indonesia, Japan and New Zealand are the largest users of geothermal energy resources (both direct and indirect). Table 13.1 shows the location of present electric power generation from geothermal energy in order of size per country. The 2018 capacity of 14,369MW electricity was a 47% increase from the capacity installed in 2007.

Other countries with less than 100 MW generation are: Nicaragua, Papua New Guinea, Guatemala, Portugal, Russia, China, Germany. France, Ethiopia, Austria, Australia, and Thailand.

The majority of the earlier geothermal plants were funded and operated by National Power agencies around the world with the exception of California where the development of the Geysers geothermal field was carried out by privately funded utility companies. With the recent international trend towards de-regulation in the power industry, private developers have become more directly involved in both resource assessment and development. This has been particularly so in Indonesia and the Philippines.

Flash steam plants totally dominate the marketplace, but over the past ten years many smaller scale binary cycle plants have been installed while several combined (flash steam/binary plants) have been installed. The majority of the World's geothermal power stations are base load stations meaning that they operate 24 hours a day for 365 days. Allowing for a load factor of about 80% and an average steam cost of Rs. 2 per kWh geothermal power.

Table 13.1: World Geothermal Electricity Generation Installed Capacity (MW)

Country	2007	2010	2013	2015	2018	Share of national eneration (%)
USA	2687	3086	3389	3450	3591	0.3
Philippines	1969.7	1904	1894	1870	1868	27.0
Indonesia	992	1197	1333	1340	1948	3.7
Mexico	953	958	980	1017	951	3.0
New Zealand	471.6	628	895	1005	1005	14.5
Italy	810.5	843	901	916	944	1.5
Iceland	421.2	575	664	665	755	30.0
Kenya	128.8	167	215	594	676	51.0
Japan	535.2	536	537	519	542	0.1
Turkey	38	82	163	397	1200	0.3
Costa Rica	162.5	166	208	207		14.0
El Salvador	204.4	204	204	204		25.0
Nicaragua	79	82	97	82		9.9
Papua New Guinea	56	56	56	50		
Guatemala	53	52	42	52		
Portugal	23	29	28	29		
Russia	79	79	82	82		
China	27.8	24	27	27		
Germany	8.4	6.6	13	27		
France	14.7	16	15	16		
Ethiopia	7.3	7.3	8	7.3		
Austria	1.1	1.4	1	1.2		
Australia	0.2	1.1	1	1.1		
Thailand	0.3	0.3	0.3	0.3		
Total	9,731.9	10,709.7	11,765	12,635.9	14,369	–

13.6 Technology and Resource Type

Geothermal resources vary in temperature from 30-350 °C, and can either be dry, mainly steam, a mixture of steam and water or just liquid water. In order to extract geothermal heat from the earth, water is used as the transfer medium. Naturally occurring groundwater is available for this task in most places but more recently technologies are being developed to even extract the energy from hot dry rock resources. The temperature of the resource is a major determinant of the type of technologies required to extract the heat and the uses to which it can be put. The Table 13.2 lists the basic technologies normally utilized according to resource temperature.

Table 13.2: Basic Technology Commonly Used

Reservoir Temperature	Reservoir Fluid	Common Use	Technology Commonly Chosen
High Temperature >220°C	Water or Steam	Power Generation Direct Use	Flash Steam; Combined (Flash and Binary) Cycle Direct Fluid Use Heat Exchangers Heat Pumps
Intermediate Temperature 100-220°C	Water	Power Generation Direct Use	Binary Cycle Direct Fluid Use; Heat Exchangers ; Heat Pumps
Low Temperature 50-150°C	Water	Direct Use	Direct Fluid Use; Heat Exchangers; Heat Pumps

13.7 Power Generation Technology

13.7.1 High Temperature Resources

High temperature geothermal reservoirs containing water and/or steam can provide steam to directly drive steam turbines and electrical generation plant. More recently developed binary power plant technologies enables more of the heat from the resource to be utilised for power generation. A combination of conventional flash and binary cycle technology is becoming increasingly popular.

High temperature resources commonly produce either steam, or a mixture of steam and water from the production wells. The steam and water is separated in a pressure vessel (Separator), with the steam piped to the power station where it drives one or more steam turbines to produce electric power. The separated geothermal water (brine) is either utilised in a binary cycle type plant to produce more power, or is disposed of back into the reservoir down deep (injection) wells. The following is a brief description of each of the technologies most commonly used to utilise high temperature resources for power generation.

13.7.1.1 Flash Steam Power Plant

This is the most common type of geothermal power plant. The illustration (Fig. 13.1) below shows the principle elements of this type of plant. The steam, once it has been separated from the water, is piped to the powerhouse where it is used to drive the steam turbine. The steam is condensed after leaving the turbine, creating a partial vacuum and thereby maximizing the power generated by the turbine-generator. The steam is usually condensed either in a direct contact condenser, or a heat exchanger type condenser. In a direct contact condenser, the cooling water from the cooling tower is sprayed onto and mixes with the steam. The condensed steam then forms part of the cooling water circuit, and a substantial portion is subsequently evaporated and is dispersed into the atmosphere through the cooling tower. Excess cooling water called blow down is often disposed of in shallow injection wells. As an alternative to direct contact condensers shell and tube type condensers are sometimes used, as is shown in the schematic below. In this type of plant, the condensed steam does not come into contact with the cooling water, and is disposed of in injection wells.

Fig.13.1: Single Flash Power Plant

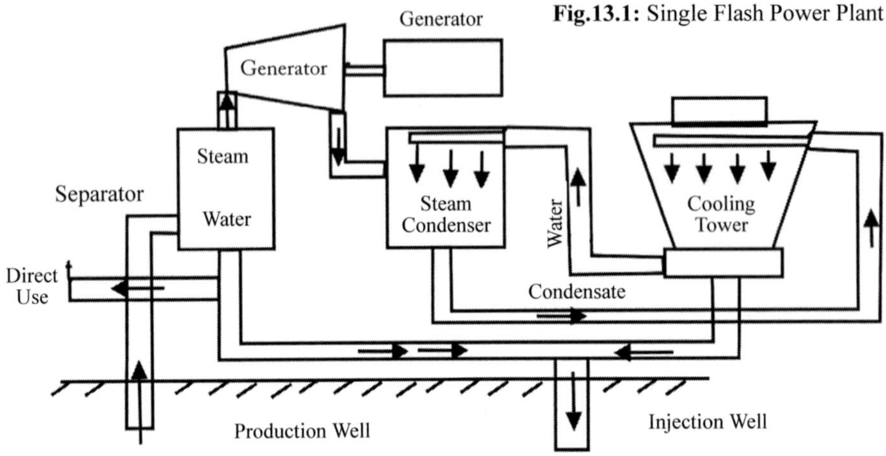

Typically, flash condensing geothermal power plants vary in size from 5 MW to over 100 MW. Depending on the steam characteristics, gas content, pressures, and power plant design, between 6000 kg and 9000 kg of steam each hour is required to produce each MW of electrical power. Small power plants (less than 10 MW) are often called well head units as they only require the steam of one well and are located adjacent to the well on the drilling pad in order to reduce pipeline costs. Often such well head units do not have a condenser, and are called back pressure units. They are very cheap and simple to install, but are inefficient (typically 10-20 tonne per hour of steam for every MW of electricity) and can have higher environmental impacts.

13.7.1.2 Binary Cycle Power Plants

In reservoirs where temperatures are typically less than 220° C. but greater than 100° C binary cycle plants are often utilised. The illustration (Fig. 13.2) shows the principle elements of this type of plant. The reservoir fluid (either steam or water or both) is passed through a heat exchanger which heats a secondary working fluid (organic) which has a boiling point lower than 100° C. This is typically an organic fluid such as Isopentane, which is vaporised and is used to drive the turbine. The organic fluid is then condensed in a similar manner to the steam in the flash power plant described above, except that a shell and tube type condenser rather than direct contact is used. The fluid in a binary plant is recycled back to the heat exchanger and forms a closed loop. The cooled reservoir fluid is again re-injected back into the reservoir.

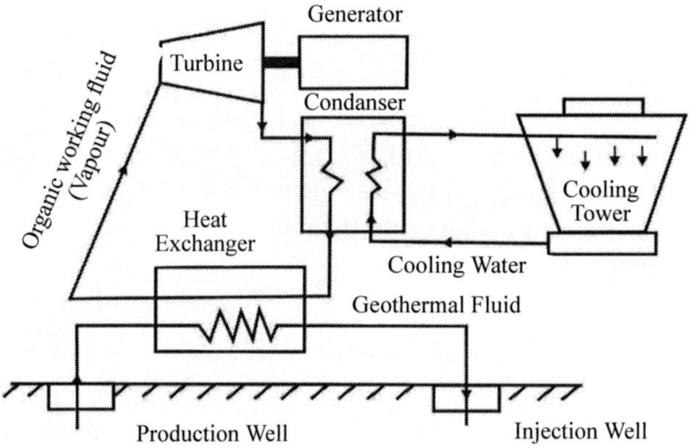

Fig.13.2: Binary Cycle Power Plant

Binary cycle type plants are usually between 7 and 12 % efficient, depending on the temperature of the primary (geothermal) fluid. Binary Cycle plant typically vary in size from 500 kW to 10 MW.

13.7.1.3 Combined Cycle (Flash and Binary)

Combined Cycle power plants are a combination of conventional steam turbine technology and binary cycle technology. By combining both technologies, higher overall utilization efficiencies can be gained, as the conventional steam turbine is more efficient at generation of power from high temperature steam, and the binary cycle from the lower temperature separated water. In addition, by replacing the condenser-cooling tower cooling system in a conventional plant by a binary plant, the heat available from condensing the spent steam after it has left the steam turbine can be utilized to produce more power.

13.7.2 Medium Temperature Resources

Medium temperature resources are normally hot water with temperatures ranging from 100° C to 220° C. The most common technology for utilising such resources for power generation is the binary cycle technology. This technology is described above under high temperature resources.

13.8 Direct Use Technology

Direct use technologies are where geothermal heat is used directly rather than for power generation and are built around the extraction of heat from relatively low temperature geothermal resources, generally of less than 150° C. Because geothermal heat is non-transportable, (except short distances by fluid pipeline) any applications must generally be sited within10 km or less of the resource. For many resources, the relatively low temperatures and/or pressures in the reservoirs means that they have insufficient energy and/or pressure differences to naturally carry the fluids to the surface and pumps are frequently used (either down-hole or at the surface).

The type of technology selected for utilising geothermal heat for direct use applications is dependent on the nature of the geothermal fluid and the type of direct use planned. In many direct use applications, the geothermal fluid cannot be used directly, such as in drying processes or where clean steam or hot water is necessary, as geothermal fluid often contains chemical contaminants. In such cases heat exchangers are utilised to extract the heat from the hot geothermal fluid and transfer it to either clean water, or in the case of drying processes, to air.

There are two main types of heat exchangers commonly used. They are plate heat exchangers and shell and tube type heat exchangers. The heat exchanger technology employed in the geothermal industry is the same as is commonly used over a wide range of industries where heat exchangers are utilised.

Commonly used heat pump technology can also be employed in order to utilise geothermal heat for air conditioning and refrigeration applications.

13.9 Technological Issues with Geothermal Developments

Whether geothermal energy is utilized for power production or for direct use applications, there are issues in geothermal utilization that often have technical implications. Geothermal fluids often contain significant quantities of gases such as hydrogen sulphide as well as dissolved chemicals and can sometimes be acidic. Because of this, corrosion, erosion and chemical deposition may be issues, which require attention at the design stage and during operation of the geothermal project. Well casings and pipelines can suffer corrosion and /or scale deposition, and turbines, especially blades can suffer damage leading to higher maintenance costs and reduced power output.

However, provided careful consideration of such potential problems is made at the design stage, there are number of technological solutions available. Such potential problems can be normally overcome by a combination of utilising corrosion resistant materials, careful control of brine temperatures, the use of steam scrubbers and occasionally using corrosion inhibitors.

Provided such readily available solutions are employed, geothermal projects generally have a very good history of operational reliability. Geothermal power plants for example, can boast of high capacity factors (typically 85-95%)

13.10 Modern Uses of Geothermal Energy

The most common utilization of geothermal energy is through hydrothermal means. Hydrothermal energy is energy derived from heated water. Water that is heated by the Earth can be used directly to supply heat to human homes. The steam generated by heated water can be used to move turbines in under sufficient pressure. A third way of harnessing geothermal energy is through dry steam, which can also turn turbines.

13.11 Applications of Geothermal Energy Sources

Heating and cooling buildings using geothermal energy is the primary use of the earth's heat energy. Much energy is placed into the moderation of temperature inside buildings, especially during times of extreme cold or heat. Using geothermal energy as a way of maintaining temperatures in buildings is one way to continue to provide that comfort but reduce the use of energy sources that are more polluting to the Earth's atmosphere. Geothermal energy can also be used to create electricity and supplement the conventional sources available. By using geothermal energy to heat homes could save between 20%-50% in total emissions, and reduce the load of utilities and appliances (many of which are refrigeration or heating units) on the electrical power grid by 75%. Not only would the use of geothermal energy reduce emissions, it could also reduce energy costs over time. Geothermal energy provides a source of energy that has little subsidiary costs (such as processing, pollution control, and transportation of raw fuels) compared to other fuel sources. The dependency of geothermal energy as an alternate fuel source is also very high. The research indicate that geothermal plants can be on-line an average of 97% of the time. In contrast to nuclear (65% average on-line time) and coal (75% average on-line time), geothermal sources score quite high.

13.11.1 Heating and Cooling Systems

Heating and refrigeration units work by moving heat from one source to another via a gradient system. For example, an air conditioner takes the hot air inside a building and exchanges its heat to the outside air. Generally, such units need large surface areas to dispense an adequate amount of heat to the environment. The same thing can be done with the gradient directed towards the Earth. Using a system of valves (often called a *ground loop*) that are connected to the cooler rock in the Earth, the heat from the air (or a water medium, like in water heaters) can be exchanged into the crust. This cooling process can be done nearly anywhere in the world, and in temperate climates can sufficiently heat a building the majority of the year. For places that are near sites which are geothermally heated to reasonably high temperatures, then this set-up can also be used to heat a building in cold climates. The nice aspect of a geothermal type system is that the system is reversible and it can be used to heat or cool.

Geothermal systems are often hidden in the ground, so unsightly and noisy heating and cooling systems are no longer encountered on a building with a geothermal system. They also take up less physical space when designing a building, and generally provide very low maintenance costs.

13.12 Indian Geothermal Provinces

In India nearly 400 thermal springs occur, distributed in seven geothermal provinces. These provinces include The Himalayas: Sohana: West coast; Cambay: Son-Narmada-Tapi (SONATA): Godavari and Mahanadi. These springs are perennial and their surface temperature range from 37 to 90° C with a cumulative surface discharge of over 1000 l/m. Figure 13.3 shows the location of these geothermal provinces. These provinces are associated with major rifts or subduction tectonics and registered high heat flow and high geothermal gradient. For example the heat flow values and thermal gradients of these provinces are 468 mW/m^2; 234° C/km (Himalayas); 93 mW/m^2; 70° C/km (Cambay); 120 - 260 mW/m^2; 60-90° C/km (SONATA); 129 mW/m^2; 59° C/km (west coast); 104 mW/m^2; 60° C/km (Godavari) and 200 mW/m^2; 90° C/km (Bakreswar, Bihar). The reservoir temperature estimated using the above described geothermometers are 120° C (west coast), 150° C (Tattapani) and 200° C (Cambay). The depth of the reservoir in these provinces is at a depth of about 1 to 2 km. These geothermal systems are liquid dominated and steam dominated systems prevail only in Himalayan and Tattapani geothermal provinces. The issuing temperature of water at Tattapani is 90° C; at Puga (Himalaya) is 98° C and at Tuwa (Gujarat) is 98° C. The power generating capacity of these thermal springs is about 10,000 MW. These resources can be utilized effectively to generate power using binary cycle method. Since majority of these springs are located in rural India, these springs can support small scale industries in such areas. Dehydrated vegetables and fruits have a potential export market and India being an agricultural country, this industry is best suited for India conditions.

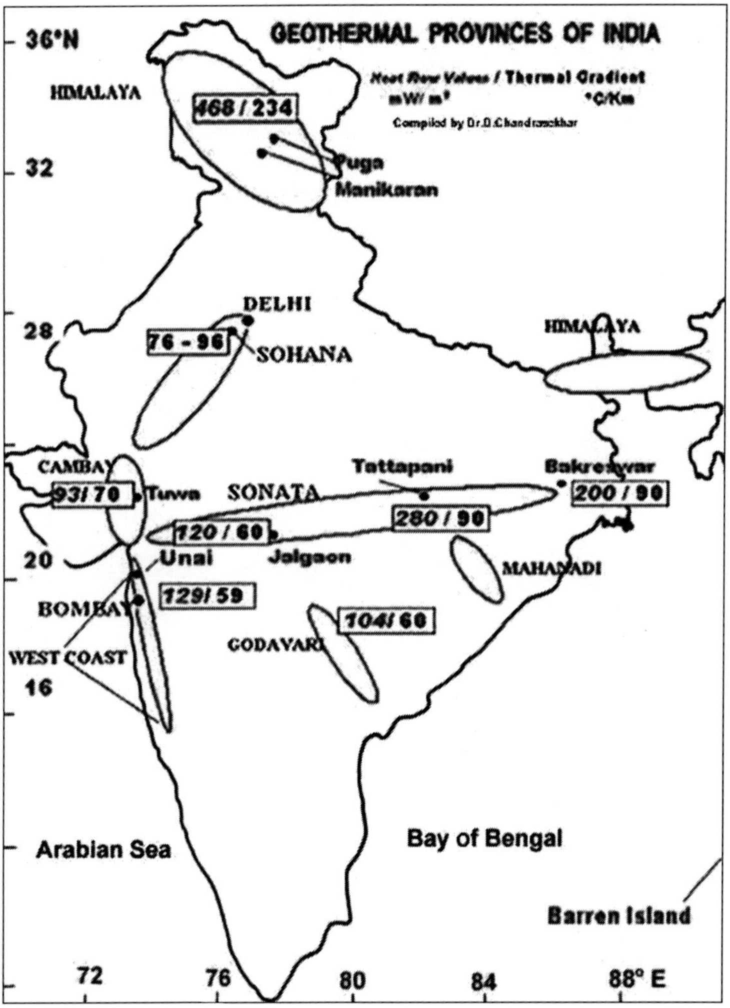

Fig. 13.3: Indian Geothermal Province with Heat Flow and Geothermal Gradient

13.13 Disadvantages of Geothermal Energy

Geothermal energy is generally a highly localized resource, and the processes used to extract energy move at a much higher rate than the processes that restore energy into the geothermal environment. In the same way that wind and hydro power rely upon certain wind speeds and certain levels of water, geothermal energy relies upon an area having a certain level of activity. Areas with very stable geothermal properties (or very unstable, such as near a volcano) may not be able to support a geothermal project. In areas where geothermal activity is present but not at high rates, exhausting the supply of energy (at least temporarily) is possible. Surprisingly (or may be not so surprisingly), a large number of critical view points on geothermal energy are not available on the World Wide Web as of yet. Most of the thorough geothermal sites are so concerned with showing the advantages of the technology that they do not list many of the disadvantages of the systems. This state is bound to change as more and more view points reach the Web in the coming days.

14

Hydro Electric Power

14.1 Introduction

Hydro-electric power is electricity produced by the movement of fresh water from rivers and lakes. Gravity causes water to flow downwards and this downward motion of water contains kinetic energy, that can be converted into mechanical energy, and then from mechanical energy into electrical energy in hydro-electric power stations. ("*Hydro*" comes from the Greek word hydra, meaning water). At a good site hydro-electricity can generate very cost-effective electricity.

14.2 History and Development

The conversion of kinetic energy into mechanical energy is not a new idea. As far back as 2000 years ago wooden waterwheels were used to convert kinetic energy into mechanical energy. The exact origin of water wheels is not known, but the earliest reference to their use comes from ancient Greece.

However, it was much later, in 1882 in the United States, that the first hydro-electric plant was built. This plant made use of a fast-flowing river as its source. Some years later, dams were constructed to create artificial water storage areas at the most convenient locations. These dams also controlled the water flow rate to the power station turbines.

Originally, hydro-electric power stations were of a small size and were set up at waterfalls in the vicinity of towns because it was not possible at that time, to transmit electrical energy over great distances. The main reason why there has been large-scale use of hydro-electric power is because it can now be transmitted inexpensively over hundreds of kilometres to where it is required, making hydro-power economically viable. Transmission over long distances is carried out by means of high voltage, overhead power lines called transmission lines. The electricity can be transmitted as either AC or DC, however AC is used for longer transmission distances.

Unlike conventional coal-fired power stations, which take hours to start up, hydro-electric power stations can begin generating electricity very quickly.

This makes them particularly useful for responding to sudden increases in demand for electricity by customers, known as "peak demand". Hydro-stations need only a small staff to operate and maintain them, and as no fuel is needed, fuel prices are not a problem. Also, a hydro-electric power scheme uses a renewable source of energy that does not pollute the environment. However, the construction of dams to enable hydro-electric generation may cause significant environmental damage.

14.3 How Hydro-Electric Power Stations Operate

The amount of electrical energy that can be generated from a water source depends primarily on two things: the distance the water has to fall and how much water is flowing. Hydro-electric power stations are therefore situated where they can take advantage of the greatest fall of a large quantity of water-at the bottom of a deep and steep-sided valley or gorge, or near the base of a dam (see figure 14.1& 14.2).

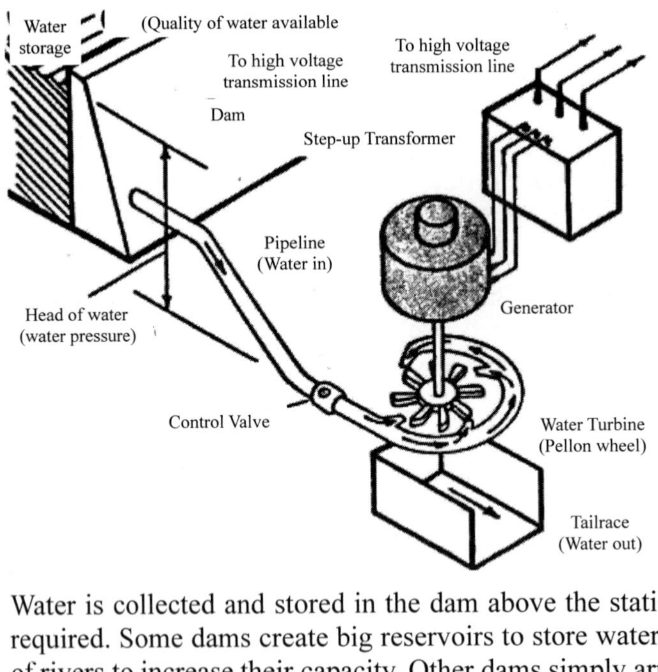

Fig 14.1: Hydro-Electric Generation Scheme

Water is collected and stored in the dam above the station for use when it is required. Some dams create big reservoirs to store water by raising the levels of rivers to increase their capacity. Other dams simply arrest the flow of rivers and divert the water down to the power station through pipelines.

While a water turbine is much more sophisticated than the old water wheels, it is similar in operation. In both cases blades are attached to a shaft and when flowing water presses against the blades, the shaft rotates the effect is the same as wind pressing against the blades of a windmill. After the water has given up its energy to the turbine, it is discharged through drainage pipes or channels called the "tailrace" of the power station for irrigation or water supply purposes or, in some parts of the world, even into the ocean.

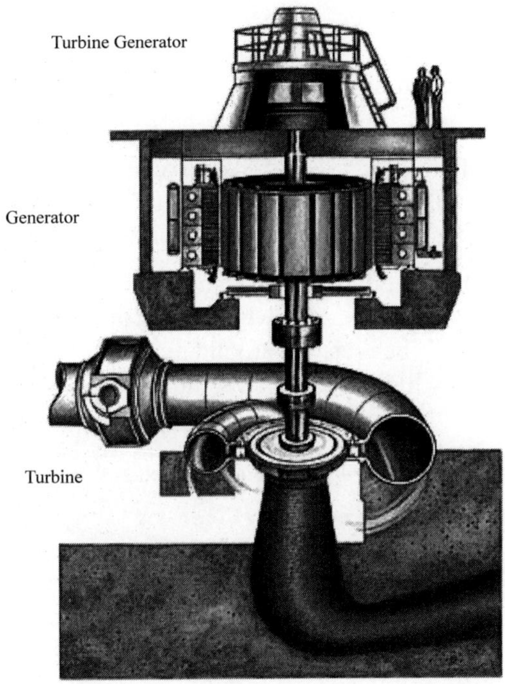

Fig14.2: Cut-away Drawing of a Water-Turbine Generator

In a conventional coal-fired (thermal) power station each "generating unit" consists of a boiler, a steam turbine, and the generator itself. A hydro-electric generating unit is simpler and consists of a water turbine to convert the energy of flowing water into mechanical energy, and an electric generator to convert mechanical energy into electrical energy. The amount of energy available from water depends on both the quantity of water available and its pressure at the turbine. The pressure is referred to the head, and is measured as the height that the surface of the water in the dam/river is above the turbine down near the base at the outlet.

The greater the height (or head) of the water above the turbine, the more energy each cubic metre of water can impart to spin a turbine (which in turn drives a generator). The greater the quantity of water, the greater the number and size of turbines that may be spun, and the greater the power output of the generators.

14.4 Types of Water Turbines

Water for a hydro-electric power station's turbine can come from a specially constructed dam set high up in a mountain range, or simply from a river close to ground level. As water sources vary, water turbines have been designed to suit the different locations. The design used is determined largely by head and quantity of water available at a particular site.

The three main types are Pelton wheels, Francis turbines, and Kaplan or Propeller type turbines (named after their inventors). All can be mounted vertically or horizontally. The Kaplan or propeller type turbines can be mounted at almost any angle, but this is usually vertical or horizontal.

The Pelton wheel (see Fig. 14.3) is used where a small flow of water is available with a 'large head'. It resembles the waterwheels used at water mills in the past. The Pelton wheel has small 'buckets' all around its rim. Water from the dam is fed through nozzles at very high speed hitting the buckets, pushing the wheel around.

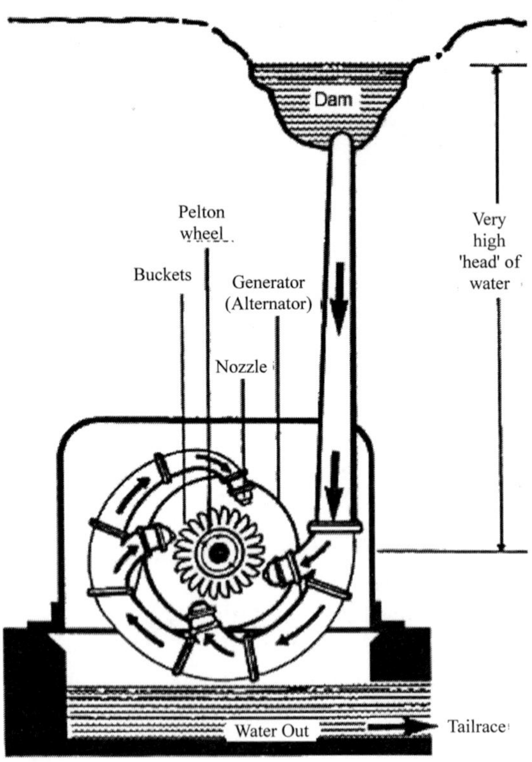

Fig 14.3: Pelton Wheel

The Francis turbine (see Fig. 14.4) is used where a large flow and a high or medium head of water is involved.

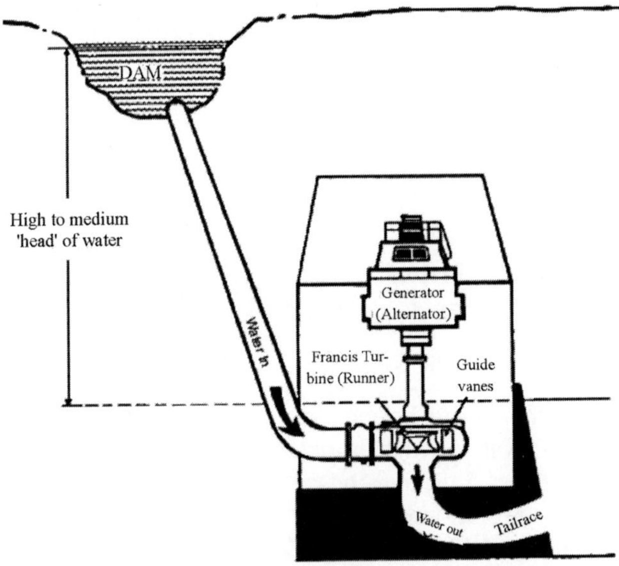

Fig.14. 4: Francis Water Turbine

The Francis turbine is also similar to a waterwheel in that it looks like a spinning wheel with fixed blades in between two rims. This wheel is called a 'runner'. A circle of guide vanes surrounds the runner and control the amount of water driving it. Water is fed to the runner from all sides by these vanes causing it to spin. Propeller type turbines are designed to operate where a small head of water is involved. These turbines resemble ship's propellers. However, with some of these (Kaplan turbines, see Fig 14.5) the angle (pitch) of the blades can be altered to suit the water flow.

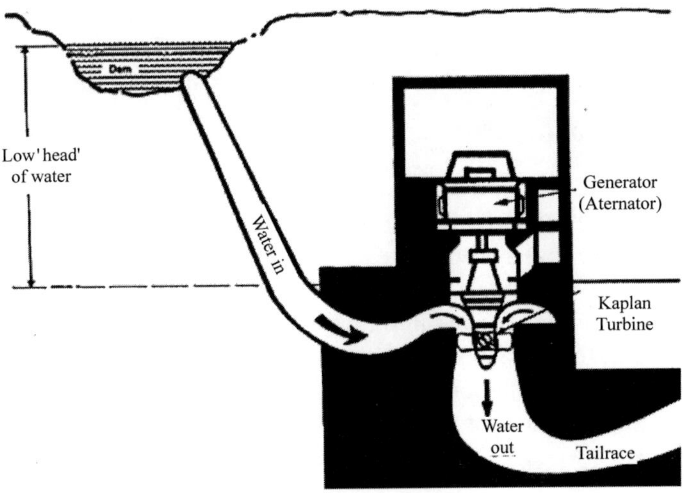

Fig. 14.5: Kaplan and Propeller Type Turbine

The variable pitch feature permits the machine to operate efficiently over a range of heads, to allow for the seasonal variation of water levels in a dam.

14.5 Large Scale Hydro

Large scale hydro-electric power systems have been installed all over the world, with the largest capacity of over 10,000 megawatts (MW) (10 gigawatts (GW)). Each of these large-scale systems requires a very large dam, or series of dams, to store the enormous quantities of water required by the system. The Kariba dam in Zambia holds 160 billion m^3 of water.

14.6 Pumped Storage Hydro-Electric Schemes

A large number of new hydro-electric projects are of the pumped storage type. Each station re-uses the water which is passes through it, by storing it in catchment areas below the station and then pumping it back up to the higher catchment dams above the station in a closed-circuit arrangement. This pumping is carried out in 'off-peak' times when there is a surplus of power available from coal, oil, or gas-fuelled stations to accomplish the task. In many countries nuclear power is used for off-peak pumping.

When pumping is required, a reversal of roles occurs. The generator becomes an electric motor, receiving electricity from a nearby power station, and operates the turbine as a pump. The turbine receives energy instead of delivering it. However, in some pumped-storage schemes there are two sets of equipment. One set is for generating and the other is for pumping. The use of pumped storage increased the total amount of power generated by the hydro power station, but this increase is not renewable. The pumps are run by non-renewable sources allowing excess electrical energy to be stored as the potential of energy of water raised to the height of the dam. The amount of renewable energy produced by the hydro power station remains the same.

14.7 Small Scale Hydro Power

Hydro power is available in a range of sizes from a few watts to over 10GW. At the low end of the spectrum, small hydro power can be divided into four categories. A classification scheme widely used breaks hydro into a range of sizes from a few hundred watts to over hundreds of megawatts. At the low end of the scale, small hydro can be divided into three categories: micro (less than 100 kW), mini (100 kW to <1 MW) and small (1 MW to <10 MW) (AGO, 2001). This section focuses on micro-hydro systems, which are generally stand-alone systems, i.e. they are not connected to the electricity grid.

Micro-hydro systems operate by diverting part of the river flow through a penstock (or pipe) and a turbine, which drives a generator to produce electricity

(Fig. 14.6). The water then flows back into the river. Micro-hydro systems are mostly "run of the river" systems, which allow the river flow to continue. This is preferable from an environmental point of view as seasonal river flow patterns downstream are not affected and there is no flooding of valleys upstream of the system (Harvey, 1993). A further implication is that the power output of the system is not determined by controlling the flow of the river, but instead the turbine operates when there is water flow and at an output governed by the flow. This means that a complex mechanical governor system is not required, which reduces costs and maintenance requirements. The systems can be built locally at low cost, and the simplicity gives rise to better long-term reliability. However, the disadvantage is that water is not carried over from rainy to dry season. In addition, the excess power generated is wasted unless an electrical storage system is installed, or a suitable 'off-peak' use is found.

Micro-hydro systems are particularly suitable as remote area power supplies for rural and isolated communities, as an economic alternative to extending the electricity grid. The systems provide a source of cheap, independent and continuous power, without degrading the environment. It is estimated that in 1990 there was an installed capacity worldwide of small hydro power (less than 10MW) of 19.5GW (World Energy Council, 1994).

Fig. 14.6: A Low-Head Micro-Hydro Installation

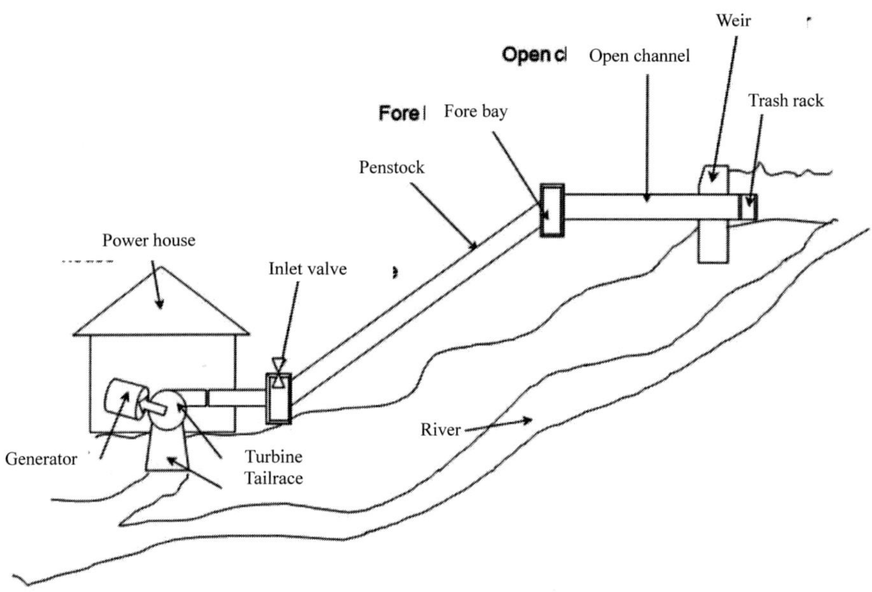

There are two main types of turbines used in micro-hydro systems, depending on the flow and the head, namely impulse turbines and reaction turbines.

Typical impulse turbines are the Pelton wheel and the Turgo wheel, and these are generally used for medium to high-head applications. Reaction turbines are generally used at low (propeller turbine) or medium head (Francis turbine)

Electrical energy can be obtained from a micro-hydro system either instantaneously or through a storage system. In an instantaneous power demand system, the system provides 240V AC power to the load via a turbine that must be sufficiently large to meet the peak power demand. These systems require a large head and/or flow. In a storage system, the micro-hydro generator provides a constant DC charge to a battery system, which then supplies power to the load via an inverter. The battery system must be sized to the daily electrical demand. However, the turbine is significantly smaller than for an instantaneous demand system, and it operates at a constant power output.

Specific Speed

Specific speed is basically defined as the speed of a turbine which is identical in shape, geometrical dimensions, blade angles, gate opening etc., with the actual turbine but of such a size that it will produce unit power when working under unit head. Different types of turbines are compared by using the value of specific speed as every type of turbine will have different specific speed.

The performance or operating conditions for a turbine handling a particular fluid are usually expressed by the values of N (RPM) , P(Power, kW) and H(Head, m) , and for a pump by N (RPM) , Q (Discharge) and Hm (Manometric Head) . It is important to know the range of these operating parameters covered by a machine of a particular shape (homologous series) at high efficiency. Such information enables us to select the type of machine best suited to a particular application.

$$\text{Specific Speed of Turbine} = \frac{N\sqrt{P}}{H^{5/4}}$$

$$\text{Specific Speed of Pump} = \frac{N\sqrt{Q}}{Hm^{\frac{3}{4}}}$$

High specific speed turbine

The specific speed of such turbines will be in the range of 250 to 850. Kaplan and Propeller turbine are the examples of high specific speed turbine.

Medium specific speed turbine

Specific speed of such turbines will be in the range of 50 to 250 and hence such turbines will be termed as medium specific speed turbines. Francis turbine is the best example of medium specific speed turbine.

Low specific speed turbine

Specific speed of such turbines will be in the range of 8 to 30 with single nozzle and up to 50 with multiple nozzles. Hence such turbines will be termed as low specific speed turbines. Pelton turbine is the best example of low specific speed turbine.

Example 14.1: Select a suitable turbine to generate 20 MW power at 300 rpm with 50 meter water head.

Solution

$$\text{Specific Speed of Turbine} = \frac{N\sqrt{P}}{H^{\frac{5}{4}}}$$

$$= \frac{300\sqrt{20\times 10^3}}{150^{\frac{5}{4}}}$$

Specific Speed of Turbine = 319

High specific speed turbine i.e., Kaplan or propeller turbine should be used to generate desire power.

14.8 Hydro Power in India

Hydro power is recognized as a renewable source of energy, which is economical, non-polluting and environmentally benign. Hydro power projects are classified as large and small hydro projects based on their sizes. Different countries have different size criteria to classify small hydro power project capacity ranging from 10 MW to 50 MW. In India, hydro power plants of 25MW or below capacity are classified as small hydro, which have further been classified into micro (100 kW or below), mini (101 kW-2 MW) and small hydro (2-25 MW) segments. There has been a continuous increase in the installed capacity of hydro power stations in India, which presently is 45,699 MW, the share of hydro power in Indian Power generation is about 12.4%.

Ministry of Power in the Government of India is responsible for the development of large hydro power projects in India. Ministry of Non-conventional Energy Sources (MNES) now Ministry of New and Renewable Energy (MNRE) has been responsible for small and mini hydro projects up to 3 MW station capacity since 1989. The subject of small hydro between 3-25 MW has been assigned to MNES w.e.f. 29th November, 1999. In order to maintain the balance between hydro power and thermal power, Ministry of Power has announced a Policy for accelerated development of hydro power in the country. Development of small hydro power at an accelerated pace is one of the tasks in the Policy.

14.8.1 Small Hydro in India

Small and mini hydel projects have the potential to provide energy in remote and hilly areas where extension of grid system is un-economical. Realizing this fact, Government of India is encouraging development of small and mini hydro power projects in the country.

14.8.2 Potential

An estimated potential of about 20,000 MW of small hydro power projects exists in India. Ministry of New and Renewable Energy has created a database of potential sites of small hydro and 6474 potential sites with an aggregate capacity of 4324 MW has been harnessed at 1077 sites for projects up to 25 MW capacity.

14.9 Benefits and Constraints to Large-Scale Hydro Power Use

While hydro power has benefits in terms of carbon dioxide emissions and air pollution, it also has significant negative environmental impacts. Hydroelectric power installations have a detrimental effect on river flows and water supplies. Large-scale hydro schemes result in the flooding of large areas of land, often leading to the displacement of people living in the area, and to negative impacts on local fauna and flora.

14.10 Future of Micro-Hydro Power

As a cheap, renewable source of energy with negligible environmental impacts, micro-hydro power has an important role to play in future energy supply scenarios, particularly in developing countries. It is an attractive alternative to diesel systems in rural and remote areas of developing countries as a means of achieving rural electrification.

15

Ocean Energy

15.1 Introduction

The oceans cover 71% of the earth's surface and it acts a natural collector and store of solar energy. On an average day, 60 million sq. km. of tropical seas absorb an amount of solar radiation equivalent in heat content to about 245 billion barrels of oil, if this energy could be trapped, large scale renewable source becomes available especially for tropical countries. The energy available in the ocean is clean, continuous throughout the year and renewable.

Renewable energy contribution to the available energy resources is very important to the country as it is presently facing an energy resources crunch. Out of the total installed capacity of approximately 370.106 GW as on March 2020, Renewable Power Plants including hydro-electric plant contribute 35.86 % of total installed capacity. However, with the present increase in demand, more attention is required to extract the renewable sources so that at least 50 to 60 % share for the total energy requirement by the renewable energy sources.

There are various forms of ocean energy such as Ocean Thermal Energy Conversion (OTEC), Wave Energy, Tidal Energy, Salinity Gradient Energy, Offshore Wind Energy, Marine Currents, Marine Biomass Conversion. Among these, the first three forms are likely to be technically viable for the future.

More recently Ministry of New and Renewable Energy (MNRE), Government of India hasdeclared Ocean Energy as renewable energy, and clarified to all the stakeholders that energy produced using various forms of ocean energy such as tidal, wave, ocean thermal energy conversion among others shall be considered as renewable energy.

MNRE reported that, the total identified potential of tidal energy is about 8000 - 9000 MW, with about 7000 MW in the Gulf of Cambay in Gujarat, about 1200 MW in Gulf of Kutch and less than 100 MW in Sundarbans. Primary estimates of wave energy potential along Indian coast are about 5 – 15 MW/m, so that theoretical estimates are around 40 – 60 GW. Ocean Thermal Energy Conversion (OTEC) has a theoretical potential of 180,000 MW in India subject to suitable technological evolution. As of date, there is not any installed Ocean Energy capacity in India.

15.2 Ocean Thermal Energy Conversion (OTEC)

OTEC makes use of the temperature difference between warm waters at the surface of the ocean and the cold deep sea water at depths of about 1000 m to operate a heat engine, which produces electric power.Calculations have shown that OTEC plants can be operated continuously without significant environmental effects. There is enough analysis to show that at rating around 4050 MW, floating OTEC plants are likely to be competitive with fossilfueled ones. But for remote islands like Andaman and Nicobar Islands, even small plants of 1 to 5 MW ratings will be cost competitive even today, In tropical regions such as India, the temperature difference is about 20°C between the surface of the ocean and at depths of about 1000 m due to various physical processes.

The temperature differences between warm and cold water is enough to support a thermodynamic cycle and to run a heat engine producing mechanical energy, which can be finally converted into electrical energy. Though the practical efficiency of this conversion is as low as about 2 to 3% because of small OTEC plant provides a continuous generation of power at no fuel cost and can work as a base load power supply system day and night.

The major disadvantage of OTEC system is that they suffer from very high initial cost, temperature difference involved, even though the running cost is less.

15.3 Applications

Ocean thermal energy conversion (OTEC) systems have many applications or uses. OTEC can be used to generate electricity, desalinate water, support deep-water mariculture, and provide refrigeration and air conditioning as well as aid in crop growth and mineral extraction. These complementary products make OTEC systems attractive to industry and island communities even if the price of oil remains low.

OTEC can also be used to produce methanol, ammonia, hydrogen, aluminum, chlorine, and other chemicals. Floating OTEC processing plants that produce these products would not require a power cable, and station-keeping costs would be reduced.

15.3.1 Desalinated Water

Desalinated water can be produced in open- or hybrid-cycle plants using surface condensers. In a surface condenser, the spent steam is condensed by indirect contact with the cold seawater. This condensate is relatively free of impurities and can be collected and sold to local communities where natural freshwater supplies for agriculture or drinking are limited. System analysis indicates that a 2-megawatt (electric) (net) plant could produce about 4300 cubic meters of desalinated water each day.

The large surface condensers required to condense the entire steam flow increase the size and cost of an open-cycle plant. A surface condenser can be used to recover part of the steam in the cycle and to reduce the overall size of the heat exchangers; the rest of the steam can be passed through the less costly and more efficient direct-contact condenser stages. A second-stage direct-contact condenser concentrates the non-condensable gases and makes it possible to use a smaller vacuum exhaust system, thereby increasing the plant's net power.

One way to produce large quantities of desalinated water without incurring the cost of an open-cycle turbine is to use a hybrid system. In a hybrid system, desalinated water is produced by vacuum flash distillation and power is produced by a closed-cycle loop. Other schemes that use discharge waters from OTEC systems to produce desalinated water have also been considered

15.3.2 Electricity Production

Two basic OTEC system designs have been demonstrated to generate electricity: closed cycle and open cycle.

(a) Closed-Cycle OTEC System

In the closed-cycle OTEC system, warm seawater vaporizes a working fluid, such as ammonia, flowing through a heat exchanger (evaporator). The vapor expands at moderate pressures and turns a turbine coupled to a generator that produces electricity. The vapor is then condensed in another heat exchanger (condenser) using cold seawater pumped from the ocean's depths through a cold-water pipe. The condensed working fluid is pumped back to the evaporator to repeat the cycle. The working fluid remains in a closed system and circulates continuously (Fig. 15.1).

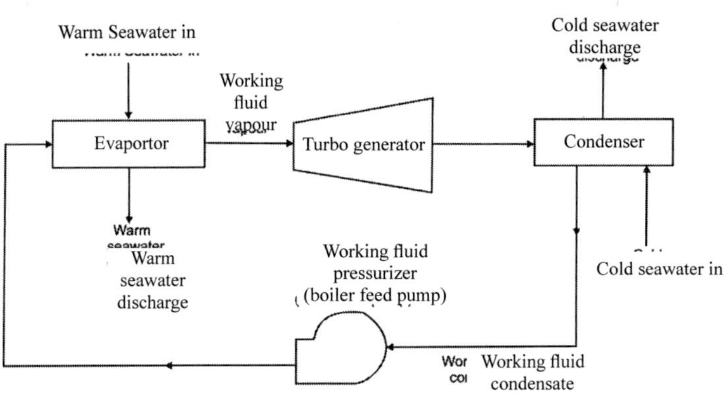

Fig. 15.1: Closed-Cycle OTEC System

(b) Open-Cycle OTEC System

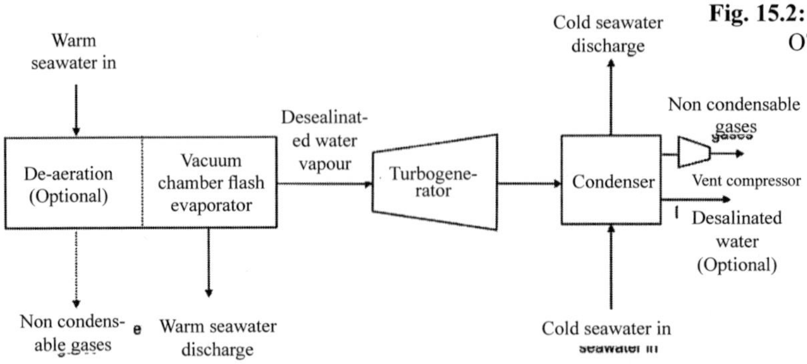

Fig. 15.2: Open-Cycle OTEC System

In an open-cycle OTEC system, warm seawater is the working fluid. The warm seawater is "flash"-evaporated in a vacuum chamber to produce steam at an absolute pressure of about 2.4 kilopascals (kPa). The steam expands through a low-pressure turbine that is coupled to a generator to produce electricity. The steam exiting the turbine is condensed by cold seawater pumped from the ocean's depths through a cold-water pipe. If a surface condenser is used in the system, the condensed steam remains separated from the cold seawater and provides a supply of desalinated water (Fig.15.2)

(c) Hybrid OTEC System

A hybrid cycle combines the features of both the closed-cycle and open-cycle systems. In a hybrid OTEC system, warm seawater enters a vacuum chamber where it is flash-evaporated into steam, which is similar to the open-cycle evaporation process (Fig. 15.3) The steam vaporizes the working fluid of a closed-cycle loop on the other side of an ammonia vaporizer. The vaporized fluid then drives a turbine that produces electricity. The steam condenses within the heat exchanger and provides desalinated water.

The electricity produced by the system can be delivered to a utility grid or used to manufacture methanol, hydrogen, refined metals, ammonia, and similar products.

(d) Heat Exchangers for Closed-Cycle Systems

Heat exchangers are a big part of the major performance and cost issues relating to closed-cycle systems. Surface heat exchangers must be large enough in area to transfer sufficient heat at OTEC's small temperature difference. Several have been designed. One form of conventional heat exchanger uses a shell-and-tube configuration in which seawater flows through the tubes, and the working fluid evaporates or condenses in a shell around them. This design can be enhanced by using fluted tubes: the working fluid flows into the grooves and over the crests, producing a thin film that evaporates more effectively. In an advanced plate-and-fin design, working fluid and seawater flow through alternating parallel plates; fins between the plates enhance the heat transfer.

Titanium was the original material chosen for closed-cycle heat exchangers because it resists corrosion. However, it is an expensive option for plants that use large heat exchangers. Corrosion-resistant copper-nickel alloys, which can be used to protect platforms and cold-water pipes, are not compatible with ammonia, the most common working fluid. A suitable alternative to these materials may be aluminum. In tests conducted by Argonne National Laboratory (ANL), brazed-aluminum tubes and channels similar to those used in full-size heat exchangers performed well under marine conditions. Results indicate that selected aluminum alloys may last 20 years in seawater.

15.3.3 Refrigeration and Air-Conditioning

The cold (5°C) seawater made available by an OTEC system creates an opportunity to provide large amounts of cooling to operations that are related to or close to the plant. Salmon, lobster, abalone, trout, oysters, and clams are not indigenous to tropical waters, but they can be raised in pools created by OTEC-pumped water; this will extend the variety of seafood products for nearby markets. Likewise, the low-cost refrigeration provided by the cold seawater can be used to upgrade or maintain the quality of indigenous fish, which tend to deteriorate quickly in warm tropical regions.

The cold seawater delivered to an OTEC plant can be used in chilled-water coils to provide air-conditioning for buildings. It is estimated that a pipe 0.3-meters in diameter can deliver 0.08 cubic meters of water per second. If 6°C water is received through such a pipe, it could provide more than enough air-conditioning for a large building.

15.4 Benefits of OTEC

We can measure the value of an ocean thermal energy conversion (OTEC) plant and continued OTEC development by both its economic and noneconomic benefits. OTEC's economic benefits include:

- Helps produce fuels such as hydrogen, ammonia, and methanol
- Produces baseload electrical energy
- Produces desalinated water for industrial, agricultural, and residential uses
- It is a resource for on-shore and near-shore mariculture operations

- Provides air-conditioning for buildings
- Provides moderate-temperature refrigeration
- Has significant potential to provide clean, cost-effective electricity for the future.

OTEC's noneconomic benefits, which help us achieve global environmental goals, include:

- Promotes competitiveness and international trade
- Enhances energy independence and energy security
- Promotes international socio-political stability
- Has potential to mitigate greenhouse gas emissions resulting from burning fossil fuels.

In small island nations, the benefits of OTEC include self-sufficiency, minimal environmental impacts, and improved sanitation and nutrition, which result from the greater availability of desalinated water and maricultural products.

15.5 Tidal Power

If there is one thing we can safely predict and be sure of on this planet, it is the coming and going of the tide. This gives this form of renewable energy a distinct advantage over other sources that are not as predictable and reliable, such as wind or solar. The Department of Trade and Industry has stated that almost 10% of the United Kingdom's electricity needs could be met by tidal power.

Why do the tides come and go? It is all to do with the gravitational force of the Moon and Sun, and also the rotation of the Earth. This is displayed in the Fig. 15.4

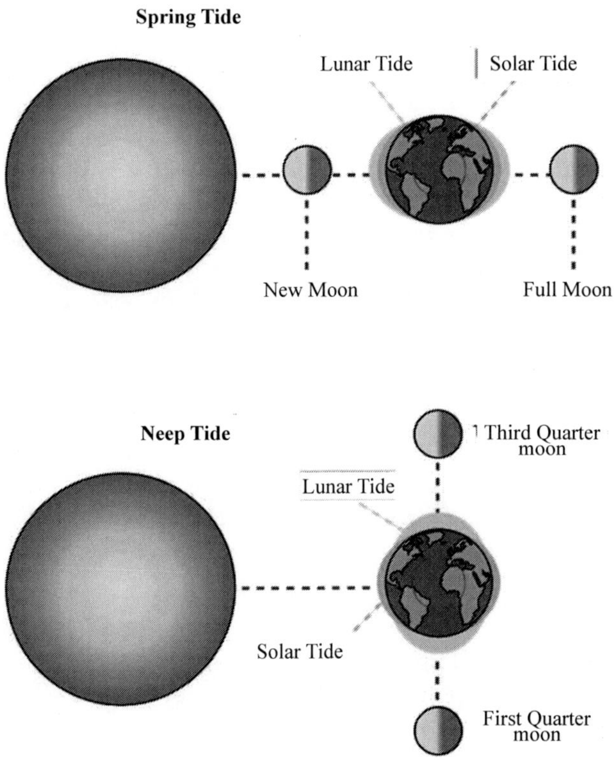

Fig. 15.4: Gravitational Effect of the Sun and the Moon on Tidal Range

The diagram shows how the gravitational attraction of the moon and sun affect the tides on Earth. The magnitude of this attraction depends on the mass of the object and its distance away. The moon has the greater effect on earth despite having less mass than the sun because it is so much closer. The gravitational force of the moon causes the oceans to bulge along an axis pointing directly at the moon. The rotation of the earth causes the rise and fall of the tides. When the sun and moon are in line their gravitational attraction on the earth combine and cause a "spring" tide. When they are as positioned in the first diagram above, 90° from each other, their gravitational attraction each pulls water in different directions, causing a "neap" tide.

The rotational period of the moon is around 4 weeks, while one rotation of the earth takes 24 hours, this results in a tidal cycle of around 12.5 hours. This tidal behaviour is easily predictable and this means that if harnessed, tidal energy could generate power for defined periods of time. These periods of generation could be used to offset generation from other forms such as fossil or nuclear which have environmental consequences. Although this means that supply will never match demand, offsetting harmful forms of generation is an important starting point for renewable energy.

15.5.1 The Tidal Barrage

This is where a dam or barrage is built across an estuary or bay that experiences an adequate tidal range. This tidal range has to be in excess of 5 metres for the barrage to be feasible. The purpose of this dam or barrage is to let water flow through it into the basin as the tide comes in. The barrage has gates in it that allow the water to pass through. The gates are closed when the tide has stopped coming in, trapping the water within the basin or estuary and creating a hydrostatic head. As the tide recedes out with the barrage, gates in the barrage that contain turbines are opened, the hydrostatic head causes the water to come through these gates, driving the turbines and generating power. Power can be generated in both directions through the barrage but this can affect efficiency and the economics of the project.

This technology is similar to Hydropower, something that we have a lot of experience with in Scotland. There is potential for a project of this kind in Scotland, one place in particular which has been looked at is the Solway Firth in south west Scotland, where there is a tidal range of 5.5 metres

The construction of a barrage requires a very long civil engineering project. The barrage will have environmental and ecological impacts not only during construction but will change the area affected forever. Just what these impacts will be is very hard to measure as they are site specific, and each barrage is different.

15.5.2 Current Technology

The Fig.15.5 is a simplified version of a tidal barrage.

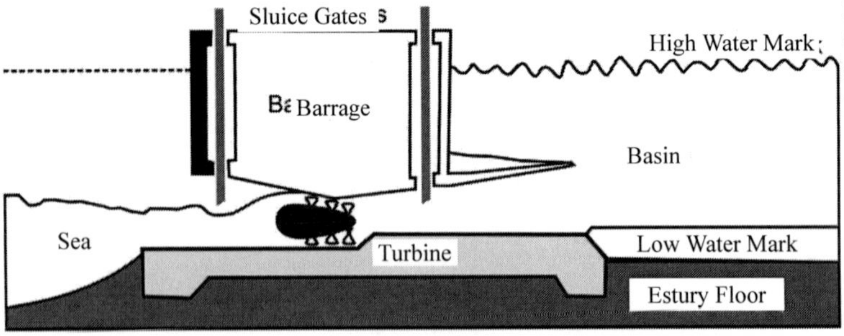

Fig. 15.5: Ebb Generating System with a Bulb Turbine

There are different types of turbines that are available for use in a tidal barrage. A bulb turbine is one in which water flows around the turbine. If maintenance is required then the water must be stopped which causes a problem and is time consuming with possible loss of generation. When rim turbines are used, the generator is mounted at right angles to the turbine blades, making access easier. But this type of turbine is not suitable for pumping and it is difficult to regulate its performance. Tubular turbines have been proposed for the UK's most promising site, The Severn Estuary, the blades of this turbine are connected to a long shaft and are orientated at an angle so that the generator is sitting on top of the barrage. The environmental and ecological effects of tidal barrages have halted any progress with this technology and there are only a few commercially operating plants in the world, one of these is the La Rance barrage in France. The power available from the turbine at any particular instant is given by as Fig.15.6.

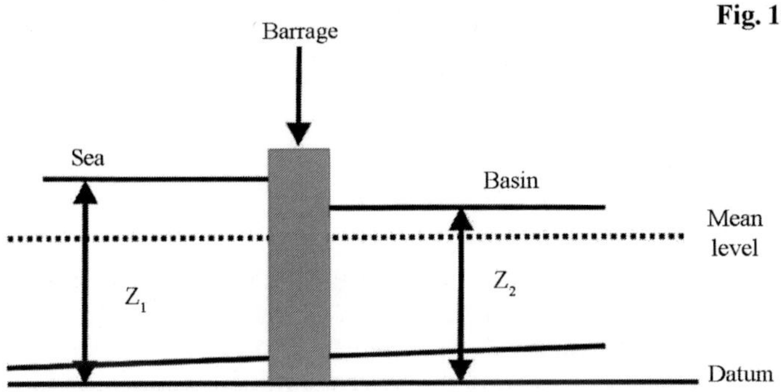

Fig. 15.6: Power from Wave

$$P = \rho \times g \times C_d \times A \times \sqrt{2g(Z_1 - Z_2)}$$

where,

C_d = Discharge Coefficient

A = Cross sectional area (m^2)

g = gravity = 9.81 (m/s^2)

ρ = density (kg/m^3)

The discharge coefficient accounts for the restrictive effect of the flow passage within the barrage on the passing water.

The equation above illustrates how important the difference between the water levels of the sea and the basin, (Z_1-Z_2), is when calculating the power produced.

15.5.3 Pumping

The turbines in the barrage can be used to pump extra water into the basin at periods of low demand. This usually coincides with cheap electricity prices, generally at night when demand is low. The company therefore buys the electricity to pump the extra water in, and then generates power at times of high demand when prices are high so as to make a profit. This has been used in Hydro Power, and in that context is known as pumped storage.

15.5.4 Economics

The capital required to start construction of a barrage has been the main stumbling block to its deployment. It is not an attractive proposition to an investor due to long payback periods. This problem could be solved by government funding or large organisations getting involved with tidal power. In terms of long term costs, once the construction of the barrage is complete, there are very small maintenance and running costs and the turbines only need replacing once around every 30 years. The life of the plant is indefinite and for its entire life it will receive free fuel from the tide.

The economics of a tidal barrage are very complicated. The optimum design would be the one that produced the most power but also had the smallest barrage possible.

15.5.5 Environmental Aspects

Perhaps the largest disadvantages of tidal barrages are the environmental and ecological effects on the local area. This is very difficult to predict; each site is different and there are not many projects that are available for comparison. The change in water level and possible flooding would affect the vegetation around the coast, having an impact on the aquatic and shoreline ecosystems. The quality of the water in the basin or estuary would also be affected, the sediment levels would change, affecting the turbidity of the water and therefore affecting the animals that live in it and depend upon it such as fish and birds. Fish would undoubtedly be affected unless provision was made for them to pass through the barrage without being killed by turbines. All these changes would affect the types of birds that are in the area, as they will migrate to other areas with more favourable conditions for them.

These effects are not all bad, and may allow different species of plant and creature to flourish in an area where they are not normally found. But these issues are very delicate, and need to be independently assessed for the area in question.

15.6 Wave Energy

Waves are movements of ocean water generated by wind blowing across and transferring energy into the surface layer. The waves are of two basic types: Swells, which are like the regular ripples in still pond in to which a stone is dropped, and which result from in major disturbance (such as storm) some distance away, and seas which are localised disturbances in the immediate area being swept by the wind, which are characterized by the formation of more confused pattern of irregular peak and troughs in the surface.

The time elapsed between the passage of two successive arrests passes a fixed point is called the wave period. The number of wave crest which pass a fixed point in a given time is called the waves frequency. The time during which the wind blows over the ocean surface creating waves is called the duration, and the area over which the wind contacts the ocean is called the fetch.

As waves move out of the storm area, sea waves with their random pattern tend to cancel each other out and dissipates relatively quickly, baring a regular train of long wavelength swell waves to transmit the storm energy further a field. Swell waves are especially good at transferring energy from one place to another. Since the energy is concentrated and transferred by the move shape moving across the ocean surface. The energy is transmitted by the vertical rather then horizontal movement of water, except where waves are usually breaking on the shoreline. Thus a swell wave dissipates small amount of energy as it moves.

There is a need for wave data in order to design wave power plant, namely, statistical distribution of wave amplitude, wave period, and direction. A waverider buoy may be installed to monitor wave activity. A typical waverider buoy is 0.9 meter in diameter weighs 166 Kilograms and has an external shell made of stainless steel. The buoy has sensors which monitor wave height and period, energy flux, vertical acceleration, time and buoy location. wave data is required for a number of years in order to characterise typical wave patterns at a given site.

Energy contained in an ideal wave is proportional to the square of wave height, i.e. Power in wave, that is the rate at which energy can be extracted from the wave, is proportional to the product of the wave period and the square of the wave height. The total energy of a series of trochoidal deep sea waves may be expressed as follows: Horsepower per foot of breadth of wave

$$= 0.0329 \times H^2 \sqrt{L\left[1 - 4.938\left(\frac{H^2}{L^2}\right)\right]}$$

where,

H is height of wave (foot)

L is length of wave between successive crests (foot).

Not much more than a quarter of the total energy of such waves would probably be available after reaching shallow water, and apparatus rugged enough for the purpose.

A motor employs a hydraulic ram for raising a portion of the wave to high level. The waves enter a scoop which is connected to the ram by a long drive pipe. The apparatus may be automatically adjusted for vertical level as the wave changes.

Gravity waves may be only a few feet high. Yet develop as much as 50 KW/ foot of wave front.

The waves, if sufficiently regular and powerful, may provide an attractive source of energy. The wave energy potential varies from place to place, depending upon the location, the wave energy availability changes from one season to another, daily, monthly and yearly.

On the basis of operating principles, the devices have been categorised as ramps, float flaps, air bells and wave pumps. According to another classification the devices are rectifiers, turned oscillators and unturned oscillators.

15.7 Ocean Currents

Ocean currents are large rivers of water circulating within and around the world's oceans. The ocean currents are caused by the interaction of a number of geophysical factors such as the rotation of an earth, wind, and tides, distribution and salinity, interaction of cold polar water and warm tropical water and shape of ocean basins.

The power contained within ocean currents varies as the cube of the velocity. The most promising sites for harnessing ocean currents are those relatively few areas where the current is regular and have a consistently high velocity. Most of the currents which fulfil this requirement are the socalled western boundary currents, which flow around the edge of the Atlantic and Pacific oceans and include the Gulf Stream off Eastern North America, the Kuroshio or Black current off Japan, etc.

Technology for extracting energy from fluid motion is well understood and is based on the use of turbines similar to those used in lowhead hydro, tidal power and wave power projects. But the open ocean environment presents particular engineering problems. The turbines must be designed in order to minimise the restriction on the flow otherwise the water will simply flow around them. Ocean current flow characteristics also require very low speed, high volume operation. Thus, the ocean current turbines will be very large and have a very low overall efficiency. For example, a 46 meter diameter surrounded turbine rotating at 2 revolution per minute could generate approximately 1.1 KW in a 1.65 metre per second current. An array of such turbines 1 kilometre wide would generate just 22 KW. The turbines would require equally large anchoring structures and power transmission lines. Moreover, there will be a very high cost of long distance underwater transmission line.

15.8 Offshore Winds

The flow of wind over the ocean is smoother more consistent and generally faster than over land because of the lack of features which cause turbulence and wind shading. Even small difference in wind speed can be significant in terms of the potential for electricity generation as the power of the wind varies as the cube of the velocity. For example, an increase in mean wind speed of only 20 percent from 56 metre per second increase the power availability by 73 per cent.

16

Fuel Cell and MHD Technology

16.1 Fuel Cell

Fuel cell is an electrochemical device that convert chemical energy of a reaction directly into electrical energy. The basic physical structure of a fuel cell consists of an electrolyte layer in contact with a porous anode and cathode on either side. A schematic representation of a fuel cell with the reactant or product gases and the ion conduction flow directions through the cell is shown in Fig. 16.1

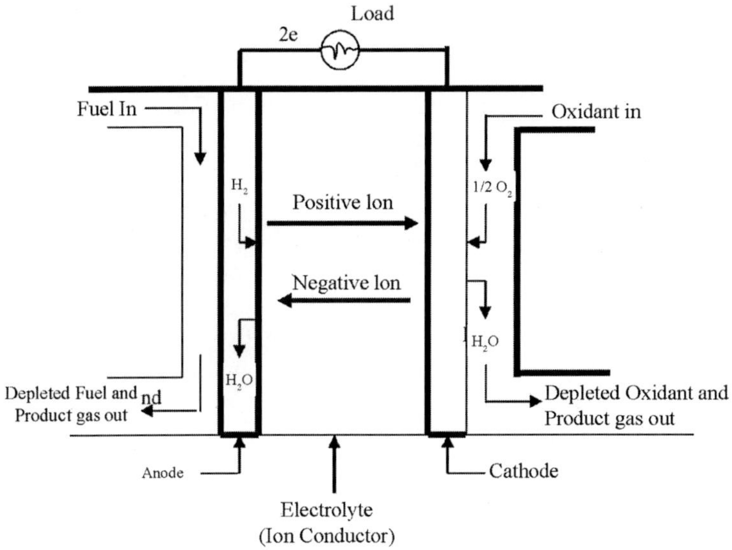

Fig 16.1: Schematic of a Fuel Cell

In a typical fuel cell, gaseous fuels are fed continuously to the anode (negative electrode) compartment and an oxidant i.e., oxygen compartment, the electrochemical reactions take place at the electrodes to produce an electric current. A fuel cell, although having components and characteristics similar to those of a typical battery, differs in several respects. The battery is an energy storage device. The maximum energy available is determined by the amount of chemical reactant stored within the battery itself. The battery will cease to produce electrical energy when the chemical reactants are consumed (i.e., discharged). In a secondary battery, the reactants are regenerated by recharging, which involves putting energy into the battery from an external source. The fuel cell, on the other hand is an energy conversion device that theoretically has the capability of producing electrical energy for a period till the fuel and oxidant are supplied to the electrodes. In reality degradation primarily corrosion or malfunction of components limits the practical operating life of fuel cells.

The ion can be either a positive or a negative charge carrier (surplus or deficit of electrons). The fuel or oxidant gases flow past the surface of the anode or cathode opposite the electrolyte and generate electrical energy by the electro chemical oxidation of fuel, usually hydrogen, and the electrochemical reduction of the oxidant, usually oxygen.

Gaseous hydrogen is the fuel of choice for most applications in fuel cell, because of following reasons:-

(1) High reactivity when suitable catalysts are used.

(2) Its ability to be produced from hydrocarbons for terrestrial applications.

(3) Its high energy density when stored cryogenically for closed environment applications, such as in space.

Similarly, the most common oxidant is gaseous oxygen, which is readily and economically available from air for terrestrial applications and again easily stored in a closed environment. A three-phase interface is established among the reactants, electrolyte and catalyst in the region of the porous electrode. Thc nature of this interface play a critical role in the electrochemical performance of a fuel cell, particularly in those fuel cells with liquid electrolytes. In such fuel cells, the reactant gases diffuse through a thin electrolyte film that wets portions of the porous electrode and react electro chemically on their respective electrode surface.

If the porous electrode contains an excessive amount of electrolyte, the electrode may “flood” and restrict the transport of gaseous species in the electrolyte phase to the reaction sites. The consequence is a reduction in the electro chemical performance of the porous electrode. Thus, a delicate balance must be maintained among the electrode, electrolyte and gaseous phases in the porous electrode structure. Much of the recent effort in the development of fuel cell technology has been devoted to reducing the thickness of cell components while refining and improving the electrode structure and the electrolyte phase, with the aim of obtaining a higher and more stable electrochemical performance while lowering cost.

The electrolyte not only transports dissolved reactants to the electrode, but also conducts ionic charge between the electrodes and thereby completes the cell electric circuit, as illustrated in Fig. 16.1. It also provides a physical barrier to prevent the fuel and oxidant gas streams from directly mixing.

The functions of porous electrodes in fuel cells are:

1. To provide a surface site where gas/liquid ionization or de-ionization reactions can take place.
2. To conduct ions away from or into the three-phase interface once they are formed (so an electrode must be made of materials having good electrical conductance)
3. To provide a physical barrier that separates the bulk gas phase and the electrolyte.

In order to increase the rates of reactions, the electrode material should be catalytic as well as conductive, porous rather than solid. The catalytic function of electrodes is more important in lower temperature fuel cells and less so in high temperature fuel cells because ionization reaction rates increase with temperature. It is also a corollary that the porous electrodes must be permeable to both electrolyte and gases, but not such that the media can be easily "flooded" by the electrolyte or "dried" by the gases in a one-sided manner.

A variety of fuel cells are in different stages of development. They can be classified depending on the combination of type of fuel and oxidant whether the fuel is processed outside (external reforming) or inside (internal reforming) the fuel cell, the type of electrolyte, the temperature of operation, whether the reactants are fed to the cell by internal or external manifolds, etc.

16.2 Classification of Fuel Cells

The most common classification of fuel cells is based on the type of electrolyte used and are classified as.

(A) Polymer Electrolyte Fuel Cell (PEFC)

(B) Alkaline Fuel Cell (AFC)

(C) Phosphoric Acid Fuel Cell (PAFC)

(D) Molten Carbonate Fuel Cell (MCFC)

(E) Intermediate Temperature Solid Oxide Fuel Cell (ITSOFC) and

(F) Tubular Solid Oxide Fuel Cell (TSOFC).

The operating temperature of these fuel cells are ranging from ~80°C for PEFC, ~100°C for AFC, ~200°C for PAFC, ~650°C for MCFC, ~800°C for ITSOFC, and 1000°C for TSOFC. The operating temperature and useful life of a fuel cell depend on the physico chemical and thermo mechanical properties of materials used in the cell components (i.e., electrodes, electrolyte, interconnect, current collector, etc.). Aqueous electrolytes are limited to temperatures of about 200°C or lower because of their high water vapor pressure and/or rapid degradation at higher temperatures. The operating temperature also depends on the type of fuel that can be used in a fuel cell. The low-temperature fuel cells with aqueous electrolytes are in most practical application, restricted to hydrogen as a fuel. In high-temperature fuel cells CO and even CH_4 can be used because of the inherently rapid electrode kinetics and the lesser need for high catalytic activity at high temperature. The higher temperature cells can favor the conversion of CO and CH_4 to hydrogen, then use the equivalent hydrogen as the actual fuel.

(a) **Polymer Electrolyte Fuel Cell (PEFC):** The electrolyte in this fuel cell is an ion exchange membrane (fluorinated sulfonic acid polymer or other similar polymer) that is an excellent proton conductor. The only liquid in this fuel cell is water thus corrosion problems are minimal. Water management in the membrane is critical for efficient performance of the fuel cell. It must operate under conditions where the byproduct water does not evaporate faster than it is produced because the membrane must be hydrated. Because of the limitation on the operating temperature imposed by the polymer, usually less than 120 °C and because of problems with water balance, a H_2-rich gas with minimal or no CO is used. Higher catalyst loading (Pt in most cases) than that used in PAFCs is required for both the anode and cathode.

(b) **Alkaline Fuel Cell (AFC):** The electrolyte in this fuel cell is concentrated (85 wt%) KOH in fuel cells operated at high temperature (~250°C) or less concentrated (35-50 wt%) KOH for lower temperature (<120°C) operation. The electrolyte is retained in a matrix (usually asbestos) and a wide range of electrocatalysts can be used (e.g., Ni, Ag, Metal Oxides, Spinels, and Noble metals). The fuel supply is limited to non-reactive constituents except for hydrogen. CO is a poison and CO_2 will react with the KOH to form K_2CO_3, thus altering the electrolyte. Even the small amount of CO_2 in air must be considered with the alkaline cell.

(c) **Phosphoric Acid Fuel Cell (PAFC):**Phosphoric acid concentrated to 100% is used for the electrolyte in this fuel cell which operates at 150 to 220°C. At lower temperatures phosphoric acid is a poor ionic conductor and CO poisoning of the Pt electrocatalyst in the anode becomes severe. The relative stability of concentrated phosphoric acid is high compared to other common acids, consequently the PAFC is capable of operating at the high end of the acid temperature range (100 to 220°C). In addition, the use of concentrated acid (100%) minimizes the water vapor pressure so water management in the cell is not difficult.

(d) **Molten Carbonate Fuel Cell (MCFC):**The electrolyte in this fuel cell is usually a combination of alkali carbonates which is retained in a ceramic matrix of $LiAlO_2$. The fuel cell operates at 600 to 700°C, where the alkali carbonates form a highly conductive molten salt with carbonate ions providing ionic conduction. At the high operating temperatures in MCFC, Ni (anode) and nickel oxide (cathode) are adequate to promote reaction. Noble metals are not required.

(e) **Intermediate Temperature Solid Oxide Fuel Cell (ITSOFC):**The electrolyte and electrode materials in this fuel cell are basically the same as used in the TSOFC. The ITSOFC operates at a lower temperature, typically between 600 to 800°C. For this reason, thin film technology is being developed to promote ionic conduction, alternative electrolyte materials are also being developed.

(f) **Tubular Solid Oxide Fuel Cell (TSOFC):**The electrolyte in this fuel cell is a solid, nonporous metal oxide, usually Y_2O_3-stabilized ZrO_2. The cell operates at 1000°C where ionic conduction by oxygen ions takes place. Typically, the anode is Co-ZrO_2 or Ni-ZrO_2 cermet, and the cathode is Sr-doped $LaMnO_3$.

In low-temperature fuel cells (PEFC, AFC, PAFC) protons or hydroxyl ions are the major charge carriers in the electrolyte, whereas in the high-temperature fuel cells, MCFC, ITSOFC, and TSOFC carbonate ions and oxygen ions are the charge carriers, respectively.

Even though the electrolyte has become the predominant means of characterizing a cell, another important distinction is the method used to produce hydrogen for the cell reaction. Hydrogen can be reformed from natural gas and steam in the presence of a catalyst starting at a temperature of ~760°C. The reaction is endothermic. MCFC, ITSOFC, and TSOFC operating temperatures are high enough that reforming reactions can occur within the cell, a process referred to as internal reforming. The reforming reaction is driven by the decrease in hydrogen as the cell produces power. This internal reforming can be beneficial to system efficiency because there is an effective transfer of heat from the exothermic cell reaction to satisfy the endothermic reforming reaction. A reforming catalyst is needed adjacent to the anode gas chamber for the reaction to occur. The cost of an external reformer is eliminated and system efficiency is improved, but at the expense of a more complex cell configuration and increased maintenance issues of the reaction sites. Porous electrodes used in fuel cells, achieve much higher current densities. These high current densities are possible because the electrode has a high surface area, relative to the geometric plate area that significantly increases the number of reaction sites and the optimized electrode structure has favorable mass transport properties. In an idealized porous gas fuel cell electrode, high current densities at reasonable polarization are obtained when the liquid (electrolyte) layer on the electrode surface is sufficiently thin so that it does not significantly impede the transport of reactants to the electro active sites, and a stable three-phase (gas/electrolyte/electrode surface) interface is established. When an excessive amount of electrolyte is present in the porous electrode structure, the electrode is considered to be "flooded" and the concentration polarization increases to a large value.

16.3 Advantages and Disadvantages

The fuel cell differs in their materials of construction, fabrication techniques and system requirements etc. These distinctions result in individual advantages and disadvantages that decide the potential of the various cells for different applications.

(i) ***PEFC:*** The PEFC, like the SOFC, has a solid electrolyte. As a result, this cell exhibits excellent resistance to gas crossover. In contrast to the SOFC, the cell operates at a low temperature (80 °C). This result in a capability to bring the cell to its operating temperature quickly, but the rejected heat cannot be used for cogeneration or additional power. This type of cell can operate at very high current densities compared to the other cells. However, heat and water management issues may limit the operating power density of a practical system. The PEFC tolerance for CO is in the low ppm level.

(ii) ***AFC:*** The AFC was one of the first modern fuel cells developed in 1960. The application at that time was to provide on-board electric power for the Apollo space vehicle. Desirable attributes of the AFC include its excellent performance on hydrogen (H_2) and oxygen (O_2) compared to other fuel cells due to its active O_2 electrode kinetics and its flexibility to use a wide range of electro catalysts, an attribute that provides development flexibility. When development was in progress for space application, terrestrial applications were also investigated for this type of cell. It was investigated that pure hydrogen would be required in the fuel stream because CO_2 in any reformed fuel reacts with the KOH electrolyte to form a carbonate reducing the electrolyte's ion mobility. Pure H_2 could be supplied to the anode by passing a reformed, H_2-rich fuel stream by a precious metal (palladium/silver) membrane. The H_2 molecule is able to pass through the membrane by absorption and mass transfer and into the fuel cell anode. However, a significant pressure differential is required across the membrane and the membrane is prohibitive in cost. Even the small amount of CO_2 in ambient air the source of O_2 for the reaction, would have to be scrubbed.

(iii) ***PAFC:*** The CO_2 in the reformed fuel gas stream and the air does not react with the electrolyte in a phosphoric acid electrolyte cell, but is a diluent. This attribute and the relatively low temperature of the PAFC made it a prime, early choice for terrestrial application. Although its performance is some what lower than the alkaline cell because of the cathode's slow oxygen reaction rate and although the cell still requires hydrocarbon fuels to be reformed into an H_2-rich gas, the PAFC system efficiency improved because of its higher temperature environment and less complex fuel conversion (no membrane and attendant pressure drop). There is no need to scrub CO_2 from the process air. The rejected heat from the cell is high enough in temperature to heat water or air in a system operating at atmospheric pressure. Some steam is available in PAFCs, a key point in expanding cogeneration applications.

PAFC systems achieve about 37 to 42% electrical efficiency (based on the LHV of natural gas). This is at the low end of the efficiency goal for fuel cell power plants. PAFCs use high cost precious metal catalysts such as platinum. The fuel has to be reformed external to the cell and CO has to be shifted by a water gas reaction to below 3 to 5 vol % at the inlet to the fuel cell anode or it will affect the catalyst. These limitations have prompted development of the alternate, higher temperature cells, MCFC, and SOFC.

(iv) ***MCFC:*** Many of the disadvantages of the lower temperature as well as higher temperature cells can be alleviated with the higher operating temperature MCFC (approximately 650°C). This temperature level results in several benefits like

1 The cell can be made of commonly available sheet metals that can be stamped for less costly fabrication.

2 The cell reactions occur with nickel catalysts rather than with expensive precious metal catalysts.

3 Reforming can take place within the cell provided a reforming catalyst is added (results in a large efficiency gain).

4 CO is a directly usable fuel and the rejected heat is of sufficiently high temperature to drive a gas turbine and produce a high pressure steam for use in a steam turbine or for cogeneration.

5 It operates efficiently with CO_2-containing fuels such as bio-fuel derived gases. This benefit is derived from the cathode performance enhancement resulting from CO_2 enrichment.

The MCFC has some disadvantages, like the electrolyte is very corrosive and mobile, and a source of CO_2 is required at the cathode (usually recycled from anode exhaust) to form the carbonate ion. Sulfur tolerance is controlled by the reforming catalyst and is low which is the same for the reforming catalyst in all cells. Operation requires use of stainless steel as the cell hardware material. The higher temperatures promote material problems, particularly mechanical stability that impacts life.

(v) ***ITSOFC:*** The intermediate temperature solid oxide fuel cell combines the best available attributes of fuel cell technology development with intermediate temperature (600-800°C) operation. Ceramic components are used for electrodes and electrolytes, carbon doesnot deposit on these ceramic materials therefore this fuel cell may accept hydrocarbons and carbon monoxide in the fuel. Internal reforming is practically at temperatures above 650°C. Moreover, use of solid state components avoids design issues such as corrosion and handling, inherent in liquid electrolyte fuel cells. The reduced temperature from the TSOFC allows stainless steel construction, which represents reduced manufacturing costs over more exotic metals. The disadvantages of ITSOFCs are that electrolyte conductivity and electrode kinetics drop significantly with lowered temperature.

(vi) ***TSOFC:*** The TSOFC is the fuel cell with the longest continuous development period, starting in the late 1950s, several years before the AFC. The solid ceramic construction of the cell alleviates cell hardware corrosion problems characterized by the liquid electrolyte cells and has the advantage of being impervious to gas cross-over from one electrode to the other. The absence of liquid also eliminates the problem of electrolyte movement or flooding in the electrodes. The kinetics of the cell are fast and CO is a directly useable fuel as it is in the MCFC and ITSOFC. There is no requirement for CO_2 at the cathode as with the MCFC. At the temperature of presently operating TSOFCs (~1000°C), fuel can be reformed within the cell. The temperature of a TSOFC is significantly higher than that of the MCFC and ITSOFC. However, some of the rejected heat from a TSOFC is needed to preheat the incoming process air. The high temperature of the TSOFC has its drawbacks. There are thermal expansion mismatches among materials, and sealing between cells is difficult in the flat plate configurations. The high operating temperature places severe constraints on materials selection and results in difficult fabrication processes.

16.4 Cell Energy Balance

The energy balance around the fuel cell is based on the energy absorbing and releasing processes (e.g., power produced, reactions, heat loss) that occur in the cell. As a result, the energy balance varies for the different types of cells because of the differences in reactions that occur according to cell type. In general, the cell energy balance states that the enthalpy flow of the reactants entering the cell will equal the enthalpy flow of the products leaving the cell plus the sum of following terms:

(1) The net heat generated by physical and chemical processes within the cell,

(2) The DC power output from the cell, and

(3) The rate of heat loss from the cell to its surroundings.

Component enthalpies are readily available on per mass basis from data. A typical energy balance calculation is the determination of the cell exit temperature knowing the reactant composition, the temperature, H_2 and O_2 utilization, the expected power produced, and a percent heat loss.

16.5 Cell Efficiency

The thermal efficiency of an energy conversion device is defined as the amount of useful energy produced relative to the change in stored chemical energy (commonly referred to as thermal energy) that is released when a fuel is reacted with an oxidant.

$$\eta = \frac{\text{Useful Energy Production}}{\Delta(\text{Change in store Chemical Energy})}$$

Hydrogen (a fuel) and oxygen (an oxidant) can exist in each other's presence at room temperature, but if heated to 580°C, they explode violently. The combustion reaction can be forced for gases lower than 580 °C by providing a flame, such as in a heat engine. A catalyst and an electrolyte in the fuel cell, can increase the rate of reaction of H_2 and O_2 at temperature lower than 580 °C. It is also observed that a non-combustible reaction can occur in fuel cells at temperature over 580 °C because of controlled separation of the fuel and oxidant. The heat engine process is thermal whereas the fuel cell process is electrochemical.

16.6 Magneto Hydro Dynamics: Principles and Applications

Magneto Hydrodynamic Generator (MHD) is a power generation device which converts kinetic energy fluid into electrical power by interaction of the fluid with a magnetic field without using a conventional electric generator.

This system eliminates all the intermediate linking conversion processes thereby improving the efficiency. It can be used with any high-temperature heat source like nuclear, solar energy, chemical etc. Its work principle is based on Faraday's Law. Which state that when a conductor is moved in a magnetic field an EMF is induced in the conductor.

In concept it is not different from conventional electrical generators where the conductor is a solid matter, usually copper but in detail it is very different because the conductor is a fluid & usually it is a compressible fluid. Sometimes it is known as magneto plasma dynamics (MPD) or magneto gas dynamics (MGD) or magneto fluid dynamics (MFD) etc.

The compressible flow MHD generator consists essential of a device in which a high temperature gas is exploded through a nozzle to high velocity and then passed through a magnetic field region, which contains a series of pairs of electrodes at either side of the magnetic channel. The electrode pairs fare each other and lie in the plane parallel to the plane described by the flow and magnetic field directions, which are at right angles to each other the temperature of gases may be produced by a gaseous fission or fusion reactor, or by a highenergy combustor and to increase the electrical conductivity of the gases it is necessary to "seed" then with an easily invisible substance such as potassium or calcium when the conducting fluid enters the magnetic field, interaction will occur and an induced electric field is established normal to the flow and magnetic field direction and between the electrode pairs. This field can then be used to drive a current via the electrodes to an external load where it is allowed to perform work. In doing so the kinetic or pressure energy of the gas is converted into electrical energy.

16.7 Essential Requirements for MHD Generator

Essentially MHD system has three important inputs, which are as follows:

(a). Magnetic fluid Simple magnetic field (having 2 T strength) or super conductive magnetic field (having 5 T strength)

(b). Fluid A ionized gaseous conductor, having temperature in the range of 2500 - 3000 K.

(c). The fluid must be moved into the magnetic field at a quite high speed i.e. 900 - 1000 m/s.

For MHD interaction to occur & for electricity to be generated from a moving gas it is necessary for the gas to be electrically conducting. Gases become electrically conducting when the neutral atoms or molecular are ionized into electron & positive ions. Both these types of particles are capable of carrying current being carried by the electrons and the gas is thus electronically conducting. The ionization of atoms and molecules requires energy, which basically originates from electrons, positive ion or neutral particle impact on an atom or molecule or from the absorption of quanta of radiation.

One way of increasing the conductivity of gas, without the need of exceeding high temperature, is to introduce into the gas a material whose ionization potential is lower than that of the gas atoms themselves. This means that the additions of seeding agent will ionize more readily than the gas itself, thus enhance the electrical conductivity.

16.8 Basic MHD Generator Configurations

It is possible to specify four basic MHD generator configurations:

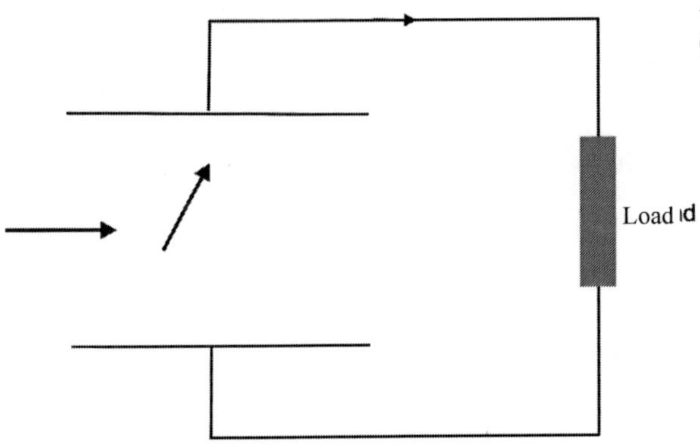

Fig. 16.2: Faraday Generator (Single Load)

(1) Continuous electrode Faraday generator

(2) Segmented electrode Faraday generator

(3) Hall generator

(4) Series or Cross Connected segmented electrode generator.

(A) Continuous Electrode Faraday Generator

Diagrammatically the continuous electrode Faraday generator is shown in Fig 16.2. This configuration is the simplest of all MHD generators and operates with a single load. The continuous equi potential electrodes force the electric field to be perpendicular to the flow.

(B) Segmented Electrode Faraday Generator

It is possible to eliminate the ohmic losses of the continuous electrode Faraday generator and restore the full conductivity of the gas by segmenting the electrodes and connecting each pair through its own load. This arrangement is shown in Fig 16.3.

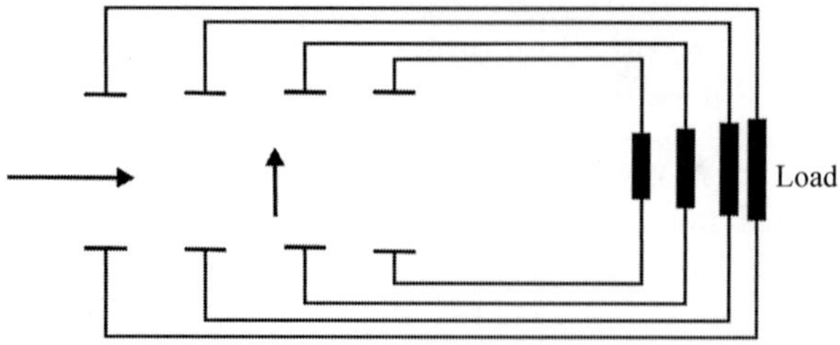

Fig. 16.3: Segmented Faraday Generator (Multiple Load)

(C) Hall Generator

The hall generator is one in which the segmented electrode pairs are short circuited and the external load is connected between the initial and final electrode pair as shown in Fig 16.4. The power extraction form the generator is by axial electric field and current.

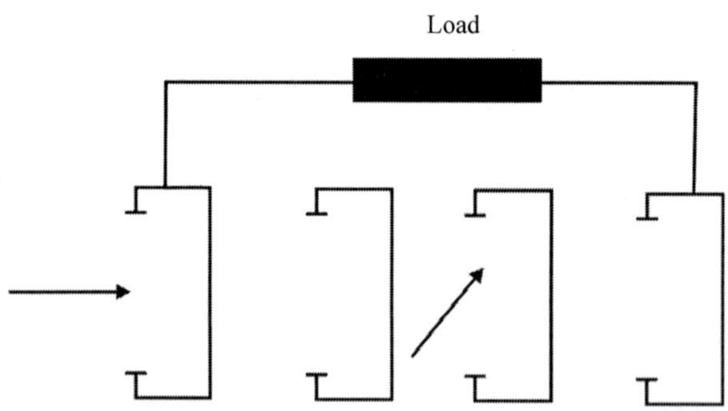

Fig. 16.4: Hall Generator (Single Load)

(D) The Series or Cross Connected Generator

The segmented electrode Faraday generator has the greatest specific power output, but suffers the impracticability of a multiplicity of subloads. This impracticability can be circumvented by loading the generator in the manner as shown in Fig 16.5, where one or few loads are connected.

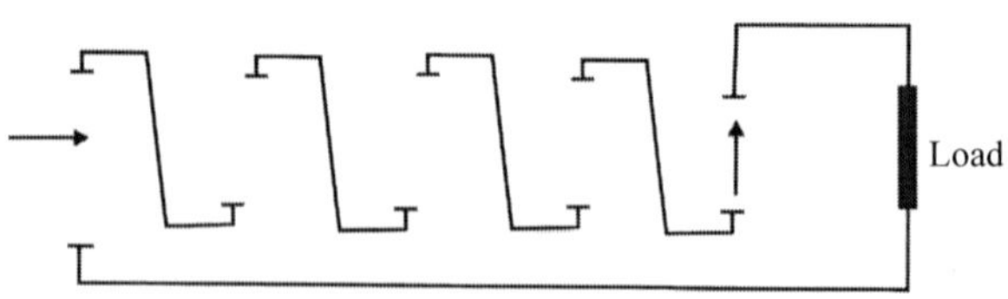

Fig. 16.5: Cross or Series Connected Generator

16.9 Advantages and Disadvantages of MHD Generators

Following advantages associated with MHD

(a) The conventional electricity generation require reduction in the cost of electricity and this can be achieved for fixed fuel costs, by increasing the plant energy conversion efficiency, reducing the installation capital cost and reducing the operating and maintenance costs. MHD offers the prospect of increasing the energy conversion efficiency above that found in the Rankine Power cycles used in conventional generators.

(b) It converts heat energy directly into electrical energy without mechanical moving part, hence conversion efficiency is high. Better fuel utilization reduce input cost and conserve energy resources.

(c) In a short period of time it can start and shut down.

(d) No harmful gas emission (pollution free)

(e) Most of conventional power plant working on 40 % efficiency whereas, MHD based power plant work on about 50% efficiency.

These are the disadvantages of MHD system

(a) In MHD system working fluid moving with very high velocity and having high fluid friction and heat transfer losses.

(b) A large voltage drop takes place across the gas film.

(c) The strong and large-sized magnets is required for smooth operations which increases the cost of an MHD system.

(d) The MHD system work at very high temperature which causes fast corrosion of components.

17

Techno – Economic Analysis

17.1 Introduction

It is well known fact that planners as well as end users have to consider any resources as technology from economic point of view. Therefore, it is important to have techno – economic analysis of each energy source option. In the feasibility study of particular energy option, three tier judgment approaches is required, which include technical feasibility, economical viability and social acceptability. Hence in the rural energy planning programme one has to considered techno socio economic analysis of an individual energy option. Since in integrated rural energy planning, locally available energy resources are used, presuming that it is socially recognized. Hence in the IREP project, the need and scope of techno economic analysis in terms of economic analysis of different technical feasible options, before between "economic" and "finical" analysis should be kept in mind. An economic analysis is needed to verify that a particular energy source will provide "net benefit" to society as a whole, before its final integration in the energy context, while a financial analysis integrated in the energy practices. In simple words, a financial analysis is concerned with goods and services, money inflow and outflows and the transfer of payments. An economic analysis also considers the social benefits and costs of goods and services which are not traded in a market.

17.2 Framework for Economic Evaluation of Energy Practice

On economic evaluation of a particular energy source/practice should include the following sequential tasks-

1. A specification of the objective of the particular energy source considered, that is, a reorganization task –

(a) What the source is trying to achieve?

(b) Who will be benefited from the source?

(c) What problem is the source attempting to overcome?

2. An identification of the alternatives for achieving the specified objectives of a particular energy source. For each of the alternatives identified, the following steps are carried out: -

(a) Identification and quantification of the physical inputs and outputs involved in the particular energy source.

(b) Determination of the unit value in terms of both market prices and economic values, for the input and out.

(c) Development of "value flow" tables, showing the total values of benefits and costs that are estimate to occur over the "useful life" of the particular energy source.

(d) Calculation of the measure of particular energy source worth and other indices required to answer the questions of decision makers and user's.

(e) Conduction of a sensitivity analysis in order to know the measures of energy source which might be altered with changed in the input – output relationship values.

17.2.1 Identification and Quantification of Physical Input – output Relationships

An evaluation of an energy source initially required the identification and quantification of the bio physical benefits resulting from the utilization of that source. Subsequently, economic evaluation attempt to "monetize" the quantitative and quantitative aspects of these benefits. Unfortunately, the task of quantifying these "input – output" linkage is difficult because.

(a) The bio physical relationship between a particular energy source and resulting benefit are not well known in many instances, since energy source in interrelation to ecology and environment.

(b) The information that links the source and benefit together in economic analysis often is not in a form that can be used readily.

One example to understand this approach is to consider different options for cooking energy requirement. Biogas is one of the options. It can easily identify and quantified physical input – output relationship for a particular size biogas plant. Say for 2 cum biogas plant, 50 kg dung plus 50-liter water as input and 2 cum biogas plus enriched manure as output. Further, one can think conventional utilization of 50 kg dung through cakes. However, an exact benefit derived from environmental point of view in quantification is a difficult task. The effects and benefit of a particular energy source are defined as changes or difference "with or without" the source being considered. In essence of biogas plant, when one says "increased energy availability" or "reduce environmental degradation", one should be referring to the difference in the energy availability and environmental degradation between conditions with and without the Biogas as energy source. Importantly, energy availability still can be declining or environmental degradation still cab increasing with a particular energy source, but at slower rates than without that source.

It is critical to keep the "with and without" concept in mind throughout an economical evaluation of a particular energy source. Ignoring this concept, result in significant error in measurement and analysis. The concept of "with and without is illustrating in Fig 17.1.

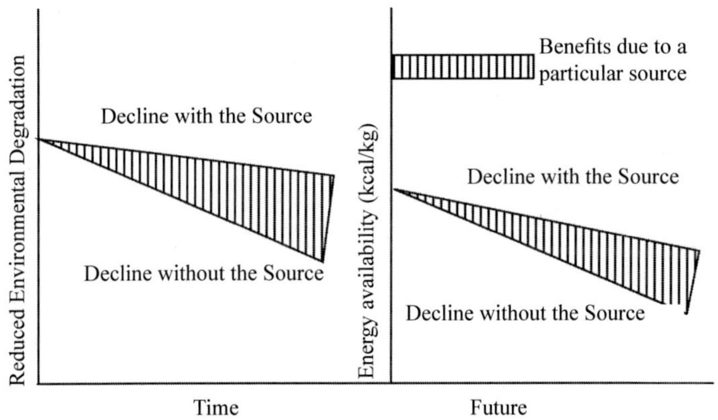

Fig. 17.1: With and Without Concept for Economic Evaluation of Energy Sources

(A) Identifying and Quantifying Input: - The inputs in a particular energy source are those derived from the additional use of the factors of production, that is, land, labour, capital, equipment, raw material, structures and civil works etc. Input are considered to be divided between "on-site" and "off-site" effects. For analysis, the inputs to a Biogas plant can be categorized as follows: -

1. Land: It depends on the size of plant.
2. Labour: It includes skilled and unskilled labour used during construction of a plant daily labour used.
3. Equipment: It includes the complete list of equipment required for construction, management and utilization of biogas and effluent slurry.
4. Raw Material: It is daily requirement plus water.
5. Structures and Civil Works: It consists of exact physical structure of biogas plant i.e. Input – outlet digestion assembly, pipe line lifting etc. in terms of main power and materials.

(B) Identifying and Quantifying Output: - In many cases, it is necessary to conduct studies to "transform" input into output. Using the "with or without" approach, technical relationship, such as those presented in Table 17.1, can be developed for a particular energy source. In those technical relationships, the dependent variables are the output and the independent variables are the inputs.

Table 17.1: Technical Relationship in an Energy Source (Example Biogas Plant)

S.No.	Dependent variables	Independent variable
1.	Annual Direct Energy Availability (cum per in case of a biogas plant)	Change in annual pollution rates.
2.	Annual Environmental Degradation	Energy sources (Conventional dung use practice characteristics in terms of environmental loss)
3.	Annual Indirect Energy Availability (Such as enrich manure tonne per year)	Agricultural lands, Agricultural infrastructure (proportion of area)

17.2.2 Determination of Unit Values

The approach employed in valuing the benefits and cost associate with a particular energy source is straight forward. Market prices are used to value those inputs and output traded in the market place and non-market benefits and costs are ignored. There are three approached to assign monetary values to the benefits and costs associated with an energy source.

1. **Market Price:** - the market price itself can be used in the economic evaluation of an energy source as a reflection of economic value of the margin.

2. **Surrogate Market Price:** - In the situation of benefits and costs not themselves valued in the market place, but for which "clear" substitute which are valued in the market exist, one can use appropriately adjusted market price for the substitute to develop surrogate values for benefits and cost being valued. Surrogate market approaches are often employed in deriving shadow prices, especially for unpriced "environmental benefits"

3. **Hypothetical Valuation:** In some cases, where there are no possibilities to derive acceptable market price measure of value, it is possible to drive value information through survey or through "expert judgment". However, because these approaches do not rely on market price, directly or indirectly, the results have to be interpreted carefully.

17.2.3 Development of "Value Flow" Chart

A "Value flow" chart is the economic analysis equivalent of a "financial cash flow" table, is that it lists all of the benefits and all of the costs of a particular energy source on a year-by-year basis over the period of the analysis or the life of the entire energy source.

Considering Value Tables in general, following points need to be made:

1. Economic efficiency prices are used throughout. Sometimes, market price will be adequate measures of these economic efficiency prices (Energy Prices), in other cases, shadow prices will be employed or price determined by the use of other valuation method such as cost of environmental benefits.
2. All of the prices are in constant monetary limits. Inflation is not included.
3. All the on-site and off-site benefits and cost of a particular energy sources should be included in a value flow chart.

17.2.4. Calculation of the measures of energy source worth

Once the value flow chart has been prepared, it is necessary to evaluate the "stream" of benefits and cost to compare alternative energy sources. A benefit cost analysis is a systematic approach to make this comparison. In practice, net present worth, integrate rate of return and benefit cost ratio are three principal value measures which is utilized to know the project worth. However, annualized cost method or life cycle costing (LCC) method has also been preferred for economic evaluation owing to its simplicity and yet performing the task of comparing economic viability effectively. All these measure are calculated by using the some data making identical assumptions, with the exception that no discount rate is employed in calculation of the internal rate of return.

(A) Net Present Worth: - This measure, also known as net present value (NPV), is based on the desire to determine the present value of net benefits from a particular energy source. If the goal of the economic analysis is to determine the total net benefits to society from a particular energy source, the net present worth (NPW) criteria provide a consistent ranking among the alternatives. The formula for the calculation of NPW is

$$NPW = \sum_{t=1}^{n} \frac{B_t - C_t}{(1+r)^t}$$

Where

B_t, C_t = Benefit or cost in year t

R = Discount rate

T = Time, from year 1 to n

(B) Internal Rate of Return: The internal rate of return (IRR) is also frequently used to evaluate an energy source. Unlike the NPW or the benefit cost ratio, the IRR does not use a predetermined discount rate. Rather, the IRR determines the discount rate that sets the present value of the benefits equal to the present value of the costs. That is, the IRR is the discount rate r, such that:

$$\sum_{t=1}^{n} \frac{B_t - C_t}{(1+r)^t} = 0$$

The IRR utilizes the same information that is used in NPW calculation. Although the discount is not prescribed but determined as a result of the IRR calculation. This does not eliminate the use o prescribed discount rate. In an evaluation involving the IRR measure of worth, the calculation IRR is compared to some prescribed discount rate to decide whether or not an energy source is worthwhile.

(C) Benefit Cost Ratio: a third measure of worth is the benefit cost ratio (B/C ratio). This ration simply compares the present value of the benefit to the present value of the cost, as shown below:

$$B/C\ \text{Ratio} = \frac{\sum_{i=1}^{n} \frac{B_t}{(1+r)^t}}{\sum_{i=1}^{n} \frac{C_t}{(1+r)^t}}$$

If the B/C ration is greater than 1, the present value of the benefits is greater than the present value of the costs. When this happens, the particular energy source meets the criteria of profitability. In fact, NPV, IRR and B/C ration (all these three measures) of an energy source worth are closely related. This fact is not surprising, since the same data on benefits and costs are used in calculating the three measures of worth. The values of these three measures of worth have the following general relations;

NPW	IRR	B/C Ratio
If >0	Then IRR≥ r	and >1
If <0	Then IRR<r	and <1
If =0	Then IRR=r	and =1

Although all of the three measure of worth use the same data in their calculations and have a “symmetry” in their results, it is possible that, when a set of alternative energy sources are examined, different evaluation criteria can giving different ranking to the alternatives. This raises the question of which measure of worth NPW, IRR or the B/C ratio, to use. It is the NPW that is the "economic objectives" as it is this measure that seeks to maximize for the investment of available, but scare, resources. Importantly, NPW always must be a part of any "choice criteria' or ranking scheme for accepting or rejecting an energy source. The IRR and the B/C ratio measures give no indication of the magnitude of the net benefits. Therefore, the reliance on either the IRR or the B/C ration can be led to a result in which the total net benefits from an energy source selected are smaller that those benefits would be by using the NPW criterion.

(D) Annualized cost method or life cycle costing (LCC): it takes into account all costs including that of the investment spread over the entire useful life of the machine. It works on the principle that any new system should have the annualized cost (worked out from following equation) at least equal to that of an existing system to justify its replacement.

$$AC = (AC)_i + AFC$$

$$(AC)_i = CRF + AOMR + AFC$$

$$CRF = \frac{r}{1-(1+r)^{-n}}$$

$$AMOR = F_1$$

Where

AC = Annualized cost of the system under consideration (Rupees)

(AC)i = Annualized cost component representable as a function of initial investment.

CRF = Capital Recovery Factor

AOMR = Annual cost of Operation, Maintenance & Repairs (Rs/yr)

AFC = Annual fuel cost (Rs/year)

R = annual discount rate reflecting on the cost of capital advance to the user as per norms of funding agency

n = Useful life in years deduced from the expected yearly usage and total number of operating hours, the system could be run with ordinary repair and maintenance before a major replacement is necessary.

F_1 = Annual operating, maintenance and repair (AMOR) factor.

17.2.5 Conduction of a Sensitivity Analysis

Any time one becomes concerned with the future and one becomes concerned with uncertainty. Despite what any one might say it is not possible to predict the future, more the probabilities of different future. In general, uncertainty of an energy sources is associated with:

(a) **Natural Factors:** Especially those associated with natural climatic processes and their irregularities in occurrence such as Solar energy and Wind energy etc.

(b) **Human Factors:** Particularly those associated with utilization practices, management capabilities and impact of social, cultural and economic factors.

(c) **Technological Factors:** Including those associated with "structural" element in energy source.

Therefore, in an economic evaluation of an energy sources especially for Integrated Rural Energy Planning Programme, it is necessary to deal with uncertainty in the planning process. There are two interrelated steps involved in the consideration of uncertainty in the planning of a particular energy source.

(a) Identification of the likely sources of uncertainty. For each likely source, estimation of a "reasonable" rage of values that the parameter involved might take. Here, one can rely on the past experience and available evidence.

(b) Suggestion of appropriate way to re – design an energy source to reduce the chances of unacceptable results.

From a sensitivity analysis a planner can determine the magnitude of change in the measures of worth that is NPW, IRR, or the B/C ratio resulting from input – output relationship, value, size and scale use of a particular energy source.

17.3 Techno – Economic Energy Planning

The new and appropriate options or decentralized energy planning in the integrated Rural Energy Planning Programme could be considered in four major sectors as Cooking energy planning (domestic energy), lighting energy planning (Electricity generation strategy), Transport energy planning (Strategy for Electricity Generation).

17.3.1 Cooking Energy Planning

There are different sources available which can be used for meeting cooking energy requirement. Theses sources could be conventional as well as non conventional. However, it is essential to use energy source according to grade in cooking energy planning. This is not only required from energy conservation point of view, but also better quality of cooking food. In this context, it is important to know that every energy sources must be associated with a grade or quality, which may determined by its temperature or the temperature it can producer.

Such an approach along with optimum utilization of available resources based on techno–economic analysis for meeting out cooking energy requirement in rural area in given in Table 17.2. For cooking any item, the temperature requirement is approximately 100 -125 °C for certain duration depending on the nature of the material to be cooked. This can be fulfilled through cow dung cake, biogas, kerosene and electricity.

Table 17.2 revels that biogas is the optimum choice for meeting cooking energy requirement in rural areas. Further, biogas is not only limited for recycle of dung but in fact now, it is recognized as waste recycling and resource recovery system as waste material like biomass, forest waste, excreta, industrial waste and other municipal waste may also be anaerobically digested and may produce combustible gas. In addition to this, this approach may also eliminate the problem of pollution, which otherwise is being created on account of these wastes.

In rural areas biomass, agriculture and forest wastes are available in abundance. As such these materials are not efficiently used and are creating pollution therefore safe disposal of these materials is essential. Presently, in rural area mainly cattle dung is used for biogas generation.

However, night soil produces three times biogas as compared to cattle dung. Mixing it with dung or with water hyacinth and algae could boost methane yield from 55% to 75%. Further, rather than burning agricultural and forest wastes as ordinary fuel, it is better to anaerobically digest it for biogas generation or briquettes. It has been observed that feeding agricultural waste to the animal and the use the dung in biogas plants, gives an overall efficiency of 24%, but charging these wastes directly into the biogas digester could give 60% efficiency.

17.3.2 Lighting Energy Planning

Electricity seems to be optimum choice for meeting out lighting energy requirement for small cottage industries, other production unit and other activities in rural areas. But centralized electricity generation is costly, besides high transmission and distribution losses. Decentralized electricity generation can be a cost-effective power generation system based on renewable energy sources. Some option being investigated such as Wind energy, Biomass, Biogas, Solar thermal, Solar Photovoltaic etc are being investigated. Among

Table 17.2: Different Option for Cooking Energy in Rural Areas

S. No.	Alternatives Sources / Heating value in Indian context	Conversion Efficiencies (Percent)	Total Energy utilized (kcal)	Cost of useful 1000 kcal available	Other Constraint
1.	Cow dung Cake 3000 kcal /kg cost Rs. 5 /kg	10 Traditional chulha	300	Rs. 16.60	Low conversion efficiency, More deposit on Utensil, Effect on house wife's health
2.	Firewood 3600 kcal/kg Cost Rs. 6/kg	-do-	360	Rs. 16.60	Low conversion efficiency, Effect on house wife's health, Deforestation ad ecological imbalance
3.	Biogas 4700 kcal/cum cost Rs.35/cum	60 (Biogas burner)	2820	Rs 12.50	Better fuel for cooking, No effect on user's health, Yield enrich manure, Convenient
4.	Kerosene 8900 kcal/lit Cost Rs. 25/lit	15 (Traditional Stove)	1339	Rs.18.60	Inconvenient, Costly, Not available easily
5.	Electricity 860 kcal /kwh cost Rs. 8/kwh	70 (Hot Plate)55.17	602	Rs. 13.50	Not suitable, Require more attention during cooking
6.	LPG (Button Stove) 12000 kcal/kg 800 per 14.5 kg	60 (LPG Burner)	8400	Rs. 4.59	Dangerous in use
7.	Solar cooker	Only for 250 days/year can be used effectively Only one meal can be prepared Completely in the hands of natures Only about 0.50 kg of product can be cooked at a time in conventional one box type cooker			

the various renewable options available today, wood-based power generation through energy plantation is receiving increasing attention in the developing countries in view of its cost effectiveness and inherent storage capacity. Table 17.3 gives the comparison of different centralized and decentralized energy sources for power generation i.e. electricity generation.

However, biogas could also be considered for electricity generation. It could be directly used through gas mantles or be converted into electricity by using duel fuel gensets.

17.3.3 Agriculture Energy Planning

As far as Agriculture Energy Planning is concede, it could be viewed through two angles. Firstly minimizing the dependence upon chemical fertilizer and other fossil fuel usage in the agriculture production. One of the best options for this purpose is to anaerobically digest waste material for biogas production. The gas production through this approach could be used for meeting out cooking energy requirement and for replacing petroleum for running engines for agricultural operations. In addition, this technology would yield enriched manure in the form of digested slurry, which could go a long way in reducing fertilizer consumption.

Table 17.3: Option for Power Generation

S.No.	Alternative Options	Economics and other constraints under Indian context.
Centralized Power Generation		
1.	Coal based Thermal Pants (fossil fuel)	• Rs 3.00 per kWh. This greatly depends upon the location and the transportation cost. • Low conversion efficiency (41%)
2.	Diesel Power	• Rs 4.00 Per kWh, high initial capital investment.
3.	Gas based thermal Power Plant	• Cost varying from Rs4.00 to Rs 4.50 per kWh
4.	Nuclear Energy based Power Plant	• Social constraints • Non availability of raw material
5.	Hydro Power Generation	• Capital intensive • Site specific
Decentralized Power Generation		
1.	Biogas	• Can replace about 100 percent petrol and 80 percent diesel in the operation. • Require large quantity of biogas, about 0.65 cum gas per hour per kW electricity. • Cost about Rs. 2.25 Per kWh
2.	Solar	• Can be generated through thermal and photovoltaic route. • High capital investment • The cost is at least 2 to 5 times the price of power generation through wind farm

3.	Wind Energy	• Aero-generators are site specific. • Require sufficient torque or economically harnessed at 5 kmph speed. • The cost of power from wing farms will be around Rs 2.50 to 3.00 in the first year depending on the machine and other factors. • It can compute with the conventional power generation provided sufficient quantity of wind energy is available at the site.
4.	Wood based power generation	• Technology is prefect and make use different type of biomass efficiently. • Cost of power is around Rs 3.00 Per kWh at the first year. • The economics of co-generation power plant based on biomass seems to be much better. However, the capital cost the plant with co – generation is nearly 3 times more than conventional system.
5.	Magneto Hydro Dynamics	• Conversion efficiency is more comparative to fossil fuel-based plant. • Technology is yet to commercialized refined. • High capital investment.
6.	Geothermal/Tidal/Wave energy	• Site specific. • Less potential is available. • Require high capital investment.

Second view of Agricultural Energy Planning is utilization of efficient Agricultural machinery. Here efficient indicate the energy efficiency of the machinery. The energy efficiency of Agricultural machinery could be increased through utilization of alternative fuels. These alternate fuels could be based on new and renewable energy sources. Such as Biogas, Solar energy, Biomass energy etc. However, biomass could provide best option for replacing petroleum. Special energy crops like tapioca, sugarbeet, sweet sorghum, sweet potato etc. do not use large energy input, but could provide us power alcohol, which is one of efficient replacement for petroleum. Hydrocarbon plants like black quince and copaiba could yield diesel substitute that need to no refining.

References

Abbot CG. The silver disk pyrheliometer (with one plate). Smithsonian Miscellaneous Collections. 1911; 56(19): 1-10

Akpinar EK. Mathematical modelling of thin layer drying process under open sun of some aromatic plants. J. Food Eng. 2006; 77; 864-870.

Akpinar EK. Drying of parsley leaves in a solar dryer and under open sun: modeling, energy and exergy aspects. J Food Process Eng 2011; 34: 27-48.

Angın D, Şensoz S. Effect of pyrolysis temperature on chemical and surface properties of biochar of rapeseed (Brassica napus L.). Int J Phytoremediation 2014; 16(7–8):684–693

Angrist SW. Direct Energy Conversion. 2nd ed. (Boston, AA: Allyn and Bacon, Inc. 1971).

Backus CE. Ed. Solar Cells, Institute of Electrical and Electronics Engineers (IEEE) Press, New York, 1976).

Ball RS. Natural sources of power. Constable, London, 1908.

Bathe G. Horizontal Windmills, Draft Mills and Similar Air flow Engines. Philadelphia, 1948.

Batten MI. English Windmills, Vol. I. Society for the protection of Ancient Buildings, Architectural press, London, 1930. Vol. II by Smith, D., 1932.

Beckman WA, et al. Design Considerations for a 50 Watt Photovoltaic Power System Using Concentrated Solar Energy", Solar Energy 1966; 10(3): 132-136.

Beedell S. Windmills. David & Charles, Newton Abbot, 1975.

Bejan A. Advanced Engineering Thermodynamics, John Wiley & Sons, New York 1997.

Bejan A. Entropy Generation through Heat and Fluid Flow. New York: Wiley-Interscience; 1982.

Bejan A. Fundamentals of exergy analysis, entropy generation minimization, and the generation of flow architecture. International Journal of Energy Research 2002; 26: 545-565.

Bejan A. Unification of three different theories concerning the ideal conversion of enclosed radiation. Solar Energy Engineering 1987; 109: 46-51.

Benko I. Possibilities of utilizing renewable energy sources. Procidia polytechnic, Mech Engg. 1981; 25; 67-86

Bergmann A, Colombo S, Hanley N. Rural versus urban preferences for renewable energy developments. Ecological Economics 2008; 65: 616–25.

Bickler DB, Costoque EN. Photovoltaic Cells and Arrays. in Record of the Photovoltaic Power Conditioning Workshop, Sandia Laboratories, Albuquerque, New Mexico (March 1977).

Bilgen S, Kaygusuz K, Sari A. Renewable energy for a clean and sustainable future. Energy Sources, Part A: Recovery, Utilization, and Environmental Effects 2004; 26: 1119–29.

Bradley WJ. Designing and Sitting Solar Power Plants," Consulting Eng. 1977; 48(3): 80 84.

British Wind Energy Association, Wind Energy for the Eighties. peter perigrinus, Stevenage, 1982.

Bruno R. A correlation procedure for separating direct and diffuse insolation on a horizontal surface. Solar Energy 1978; 20: 97 100.

Buch F, Fahrenbruch AL, Bube RH. Photovoltaic properties of five II□VI heterojunctions. Journal of Applied Physics, 1977; 48(4): 1596-1602.

Buckiust RO, King R. Diffuse solar radiation on a horizontal surface for a clear sky. Solar Energy 1978; 21: 503 509.

Buckland S. Lee's patent Windmill, 1744 1747. Wind & Watermill Section, Society for protection of Ancient Buildings, London, 1987.

Calvert NG. Wind power principles, Their Application on the Small Scale. C. Griffin & Co. Ltd., London, 1979.

Çengel YA. Cimbala JM. Fluid Mechanics: Fundamentals and Applications. New York: McGraw-Hill, 2006.

Choudhury NKD. Solar radiation at New Delhi. Solar Energy, 1963; 7(2): 44 52.

Collares-Pereira M, Rabl A. (1979). The average distribution of solar radiation-correlations between diffuse and hemispherical and between daily and hourly insolation values. Solar energy, 1979; 22(2): 155-164.

Cooper PI. Some factors affecting the absorption of solar radiation in solar stills. Solar Energy 1972; 13: 373 381.

Coulson KL. Solar and terrestrial radiation methods and measurements. Academic press, New York, 1975.

Cuomo JJ, Zieflar JF, Woodale JM. A new concept for solar energy thermal conversion. Appl. Phys. Lett., 1975; 26: 557.

Currin CG., et al. "Feasibility of Low Cost Silicon Solar Cells", Solar Cells (New York: IEEE Press, 1976).

Dave JV. Validity on the isotropic distribution approximation in solar energy estimation. Solar Energy 1977; 19: 331 333.

de Little RJ. The windmill Yesterday and Today. J. Baker London, 1972.

Demirbas A. Recent advances in biomass conversion technologies. Energy Educational Science and Technology 2000; 6: 19–40.

Dermatis SN, Faust, JW. Jr. Semiconductor Sheets for the Manufacture of Semiconductor Devices. IEEE Transactions Communications Electronics (New York: IEEE Press, 1963).

Devi BA, Kamalakkannan N, Prince PS. Supplementation of fenugreek leaves to diabetic rats. Effect on carbohydrate metabolic enzymes in diabetic liver and kidney. Phytotherapy Research 2003; 17: 1231-1233.

Dincer I, Al-Muslim H. Thermodynamic analysis of reheat cycle steam power plants. International Journal of Energy Research 2001; 25: 727-739.

Dincer I, Cengel YA. Energy, Entropy and exergy concepts and their roles in thermal engineering. Entropy 2001; 3: 116-149.

Dincer I, Rosen MA. Exergy- Energy, Environment and Sustainable Development. Elsevier, Science Publications, 2007

Dincer I. Energy and environmental impacts: present and future perspectives. Energy Sources, Part A: Recovery, Utilization, and Environmental Effects 1998;20(4):427–53.

Dincer I. Environmental Issues. I. Energy Utilization. Energy Sources, Part A: Recovery, Utilization, and Environmental Effects 2001;23:69–81.

Dincer I. Thermodynamic, exergy and environmental impact. Energy Sour 2000; 22: 723–32.

Doymaz I, Tugrul N, Pala M. Drying characteristics of dill and parsley leaves. J. Food Eng. 2006; 77: 559-565.

Doymaz I. Thin layer drying behaviour of mint leaves. J. Food Eng. 2006; 74: 370-375.

Drummeter LF, Jr. G. Hass, `Solar absorptance and thermal emittance of evaporated coatings', Physics of thin films, Academic Press. New York, 1964; 2: 305 361.

Drummond AJ. The extraterrestrial solar spectrum', Ed. A.J. Drummond and M.P. Thekaekara, Inst, Environ, Sci., Mt. Prospect. III, 1. 1973.

Duffie JA, Beckman WA. Solar energy thermal processes. Willey Interscience publication, 1974.

Duffie JA, Beckman WA. Solar Engineering of Thermal Processes. Wiley, New York, 1991.

Duffie JA, Beckman WA. Solar Radiation. Solar Engineering of Thermal Process, John Wiley and Sons, New York 1980:3-42.

Eckert ERG, Drake RM. Analysis of heat and mass transfer, Mc Graw Hill, Inc., 1972 : 586.

Ekechukwu OV, Norton B. 1997. Review of solar-energy drying systems II: an overview of solar drying technology. Energy Conversion & Management, 1999; 40: 615-655.

Ekechukwu OV. Review of solar-energy drying systems I: an overview of drying principles and theory Energy Conversion & Management, 1999; 40: 593-613

Fanger PO. Thermal Comfort Analysis and Applications in Environmental Engineering. McGraw Hill, New York 1977

Farhad S, Saffar-Avval M, Younessi-Sinaki. Efficient design of feedwater heaters network in steam power plants using pinch technology and exergy analysis. International Journal of Energy Research 2008; 32: 1–11.

Forson FK, Nazha MAA, Akuffo FO, Rajakaruna H. 2007. Design of mixed-mode natural convection solar crop dryers: Application of principles and rules of thumb. Renewable Energy 2007; 32: 2306–2319

Foster H, Mackenzie BA. Solar heat for grain drying: Selection, performance, management. West Lafayette, IN: Perdue University, Cooperative Extension Service 1980.

Fridleifsson IB. Geothermal energy for the benefit of the people. Renewable and Sustainable Energy Reviews 2001; 5: 299–312.

Gagge AP, Nishi Y, Nevins RG. The Role of Clothing in Meeting FEA Energy Conservation Guidelines. ASHRAE Trans. 1976; 82(2):234

Gaggoli RA, Wepfer WJ. Exergy Economics, Energy 1980; 5: 823-837.

Gallo WLR, Milanez LF. Choice of a Reference State for Exergetic Analysis, Energy 1990; 15: 113-121.

Garg HP, Chandra R, Rani U. Transient analysis of solar air heaters using finite difference techniques', Energy Research, 1981; 5: 243 252.

Garg HP, Gupta CL. Design data for direct solar utilization devices 1: System data', J. Inst, Engrs. (India). 1968; 47(9): Pt. GE3, 461 470.

Garg HP, Gupta CL. Design data for direct solar utilization devices, Part II Solar radiation data. Indian J., Met. Geop., 1980; 20(8): 221 226.

Gribik JA, Osterle JF. The second law efficiency of solar energy conversion. Solar energy engineering 1984; 106: 16-21.

Gupta MK, Kaushik SC. Exergetic performance evaluation and parametric studies of solar air heater. Energy, 2008; 33(11): 1691-1702.

Gupta MK, Kaushik SC. Exergetic utilization of solar energy for feed water preheating in a conventional thermal power plant. International Journal of Energy Research 2009; 33: 593-604.

Gupta MK, Kaushik SC. Performance evaluation of solar air heater for various artificial rougness geometry based on energy, effective and exergy efficiencies. Renewable Energy 2009; 34: 465-476.

Haines A, Kovats RS, Campbell-Lendrum D, Corvalan C. Climate change and human health: impacts, vulnerability and public health. Public Health 2006;120: 585-96.

Hall DO, Mynick HE, Williams RH. Cooling the greenhouse with bioenergy. Nature 1991; 353: 11–2.

Hammond AL. Photovolatics: The Semiconductor Revolution Comes to Solar. Science. 1977; 197(4302): 445 447.

Harverson M. Persian Windmills. The International Molinological Society, Reading, 1991

Haurwitz G. Insolation in relation to cloud type', J Meteor, 1948; 5: 110 113.

Hickok F. Handbook of Solar and Wind Energy (Boston, MA: Chaners Books, 1975).

Hollands KGT, UnnySE, Raithby GD, Konicek L. Free convective heat transfer across inclined air layers, J. Heat Transfer 1976; 98: 189.

Holman JP. Heat-transfer, 5th edn., McGraw-Hill, New York, 1981.

Horst GH, Hovorka AJ. Fuelwood: the "other" renewable energy source for Africa? Biomass and Bioenergy 2009; 33: 1605–16.

Hottel HC, Whillier A. Evaluation of flat plate solar collector performance. Transactions of the conference on the use of solar energy. Tucson. 1955; 3(2): 74 104.

Jain D, Tiwari GN. Modeling and optimal design of evaporative cooling system in controlled environment greenhouse. Energy Conversion and Management 2002; 43: 2235-2250.

Jain D. Modeling the performance of greenhouse with packed bed thermal storage on crop drying application. J Food Eng 2005;71(2):170–8.

Janjai , Bala BK, Solar Drying Technology. Food Eng Rev. 2012; 4: 16-54.

Janjai S, Intawee P, Kaewkiew J, Sritus C, Khamvongsa V. A large-scale solar greenhouse dryer using polycarbonate cover: Modeling and testing in a tropical environment of Lao People's Democratic Republic. Renew Energy 2011; 36: 1053-1062.

Janjai S. A greenhouse type solar dryer for small-scale dried food industries: Development and dissemination. Int. J. Energy and environment. 2012; 3: 383-398.

Kaushik SC, Gupta MK. Energy and exergy efficiency comparison of community-size and domestic-size paraboloidal solar cooker performance. Energy for Sustainable Development 2008; 12: 60-64.

Kaushik SC, Misra RD, Singh N. Second law analysis of a solar thermal power system. International Journal of Solar Energy 2000: 20; 239-253.

Kays WM, Crawford ME. Convective heat and mass transfer. New York: McGraw Hill; 1980.

Klein SA, Duffie JA, Backman WA. Transient considerations of flat plate solar collectors. J. Engineering for power, Trans ASME, 1974; 96A: 109 113.

Klein SA. Calculation of flat plate collector loss coefficients', Solar Energy, 1975; 17: 79 80.

Klein SA. Calculation of Monthly Average Insolation of Tilted Surface. Solar Energy 1977; 325

Klein WH, Goldberg B, Shropslire, W Jr. Instrumentation for the measurement of the variation, quantity and quality of sun and sky radiation. Solar Energy 1975; 19(2): 115 122.

Kondratyev K.Ya. Radiation processes in the atmosphere' World Meteorological Organisation, Geneva, WMO No. 309.

Kondratyev KYa, Nikolsky GA. Solar radiation and solar activity. Quart, J. Roy, Meteorol Soc., 1970; 96: 509 522.

Kotas TJ. The Exergy method of Thermal Plant Analysis, John Wiley & Sons Publication: 1984.

Kothari S, Panwar NL. Experimental Studies on Suitability of Greenhouse for Nursery Raising. Agricultural Engineering International: the CIGR Ejournal. Manuscript 2008; X: EE 08 003.

Kralova I, Sjöblom J. Biofuels-renewable energy sources: a review. Journal of Dispersion Science and Technology 2010; 31: 409–25.

Kreith F, Kreider JF. Principles of solar engineering. Hemisphere Publishing Corporation, Mc Graw Hill Book Company, 1978.

Lafforgue G, Magné B, Moreaux M. Energy substitutions, climate change and carbon sinks. Ecological Economics 2008;67:589–97.

Lampert CM. The use of coating for enhanced solar thermal energy collection', Lawrence Berkeley Laboratory, Report No. LBL. 8072, April, 1979.

Lausten J. Energy Efficiency Requirements in Building Codes, Energy Efficiency Policies for New Buildings. Paris, France: International Energy Agency, 2008

Lin BYH, Jordan RC. (1960). The interrelationship and characteristic distribution of direct, diffuse, and total solar energy. Solar Energy, 1960;4(3): 1-19.

Liu BYH, Jordan RC. Daily insolation on surfaces tilted towards the equator', ASRAE Journal, 1961; 3 (18): 53 59.

Majumdar NC, Mathur BL, Kaushik SB. Prediction of direct solar radiation for low atmospheric turbidity', Solar Energy, 1972; 13(4): 383 394.

Masek O, Brownsort P, Cross A, Sohi S. Influence of production conditions on the yield and environmental stability of biochar. Fuel 2013; 103:151–155

Mathur AN, Rathore NS, New and Renewable Energy Sources. Bohra Ganesh Publications, Udaipur 1996.

Mattox DM. Solar energy materials preparation techniques', J. Vac. Sci. Technol. 1975; 12(5): 1023 1031.

McGuigan D. Small Scale Wind power. Prism Press, Dorchester, 1978.

Mitchell D. Convective Heat transfer in Man and Other Animals, Heat loss from Animal and Man. Butterworth Publishing Inc. London 1974

Moran MJ. Availability Analysis, ASME Press, New York; 1989.

Muneer T, Hawas MM, Sahili K. Correlation between hourly diffuse and global radiation for New Delhi. Energy conversion and management,1984; 24(4): 265-267.

Muneer T, Hawas MM. Correlation between daily diffuse and global radiation for India. Energy conversion and management 1984; 24(2): 151-154.

National Aeronautics and Space Administration (NASA). Space Technology. 2019

Nielsen JBH, Seadi TA, Popiel PO. The future of anaerobic digestion and biogas utilization. Bioresource Technology 2009;100: 5478–84.

Orgill JF, Hollands KGT. Correlation equation for hourly diffuse radiation on a horizontal surface', Solar Energy, 1977; 19: 357 359.

Ozbek B, Dadali G. Thin-layer drying characteristics and modelling of mint leaves undergoing microwave treatment. J. Food Eng. 2007; 83: 541-549.

Panwar NL, Kaushik SC, Kothari S. Role of renewable energy sources in environmental protection: A review. Renewable and Sustainable Energy Reviews 2011; 15 : 1513–1524.

Panwar NL, Kaushik SC, Kothari S. State of the art of solar cooking: An overview. Renewable and Sustainable Energy Reviews 16 : 3776-3785

Panwar NL, Kothari R, Tyagi VV. Thermo chemical conversion of biomass - Eco friendly energy routes. Renewable and Sustainable Energy Reviews 2012; 16 : 1801-1816

Panwar NL, Rathore NS. Potential of surplus biomass gasifier based power generation: A case study of an Indian state Rajasthan. Mitigation and Adaptation Strategies for Global Change 2009; 14:711-720.

Panwar NL, Shrirame HY, Rathore NS, Sudhakar J, Kurchania AK. Performance evaluation of a diesel engine fueled with methyl ester of castor seed oil. Applied Thermal Engineering 2010; 30(2-3): 245-249.

Panwar NL. Energetic and Exergetic Analysis of Walk-in Type Solar Tunnel Dryer for Kasuri Methi (Fenugreek) Leaves Drying. Int. Journal of Exergy 2014; 4: 519-531.

Panwar NL. Thermal modeling, energy and exergy analysis of animal feed solar cooker. Journal of Renewable Sustainable Energy 5, 043105.

Petela R. Exergy analysis of the solar cylindrical-parabolic cooker. Solar energy 2005;79:221-33.

Petela R. Exergy of heat radiation. Journal of Heat Transfer 1964; 86: 187-192.

Petela R. Exergy of undiluted thermal radiation. Solar Energy 2003; 74: 469–488.

Peterson RE, Ramsay JW. Thin film coating in solar thermal power systems', J. Vac. Sci. Technol., 1975; 12: 471.

Peterson RE, Ramsay JW. Thin film coating in solar thermal power systems', J. Vac. Sci. Technol., 1975; 12: 471.

Rao KR, Seshadri TN. Solar insolation curves. Ind. J. Met. & Geop., 1961; 12(2): 267 272.

Rathore NS, Panwar NL, Kothari S. Biomass production and Utilization Technology Himanshu Publication, Udaipur (Rajasthan) 2007.

Rathore NS, Panwar NL. (2007). Renewable energy sources for sustainable development. New India Publishing Agency, New Delhi, India. 2007

Rathore NS, Panwar NL. Design and development of energy efficient solar tunnel dryer for industrial drying. Clean Techn Environ Policy 2011; 13: 125-132.

Rathore NS, Panwar NL. Experimental studies on hemi cylindrical walk-in type solar tunnel dryer for grape drying. Applied Energy 2010;87: 2764-2764.

Rathore NS, Panwar NL. Experimental studies on hemi cylindrical walk in type solar tunnel dryer for grape drying. Applied Energy 2010; 87: 2764-2767.

Ravindranath NH, Hall DO. Biomass, energy, and environment: a developing country perspective from India. Oxford, United Kingdom: Oxford University Press; 1995.

Robinson AL. Amorphous Silicon: A New Direction for Semiconductors. Science, 1977; 197(4306): 851-853.

Rosen MA, Dincer I. A study of industrial steam process heating through exergy analysis. International Journal of Energy Research 2004; 28: 917-930.

Rosen MA, Dincer I. On exergy and environmental impact. International Journal of Energy Research 1997; 21: 643-654.

Rosen MA, Hooper FC. Second law analysis of thermal energy storage systems. Proc., 1st Trabzon Int. Energy and Environment Symp., Trabzon, Turkey, 1996: 361–371.

Rosen MA. Exergy and Government Policy: Is There a Link, Exergy An International Journal 2002; 2: 224-226.

Rosen MA. Second law analysis: approaches and implications. International Journal of Energy Research 1999; 23: 415-429.

Ruth DW, Chant RE. The relationship of diffuse radiation to total radiation in Canada. Solar Energy 1976; 18: 153 154.

Sarasavadia PN, Sawhney RL, Pangavhane DR, Singh SP. Drying behaviour of brined onion slices. Journal of Food Engineering 1999; 40: 219–226.

Sharma RD. Effect of fenugreek seeds and leaves on blood glucose and serum insulin responses in human subjects. Nutrition Research 1986; 6: 1353-1364.

Shukla A, Tiwari GN, Sodha MS, Experimental study of effect of an inner thermal curtain in evaporative cooling system of cascade greenhouse, Solar Energy 2008; 82: 61-72.

Shukla BD, Singh G. Drying and Dryers (Food and Agricultural Crops), Jain Brothers, New Delhi 2003.

Sims REH. Bioenergy to mitigate for climate change and meet the needs of society, the economy and the environment. Mitigation and Adaptation Strategies for Global Change 2003;8:349–70.

Sims REH. Renewable energy: a response to climate change. Solar Energy 2004;76:9–17.

Singh OK, Kaushik SC. Variables influencing the exergy based performance of a Steam Power Plant. International Journal of Green Energy 2013; 10:257-284.

Singh OK, Panwar NL. (Effects of thermal conductivity and geometry of materials on the temperature variation in packed bed solar air heater. Journal of Thermal Analysis and Calorimetry 2013;111: 839-847.

Singh OK, Panwar NL. Effects of thermal conductivity and geometry of materials on the temperature variation in packed bed solar air heater. J Therm Anal Calorim 2013; 111: 839–847.

Smith EVP, Gottlieb DM. Space Science Rev., 1974; 16: 771.

Sodha MS, Dang A, Bansal PK, Sharma SB (1985) An analytical and experimental study of open sun drying and a cabinet type drier. Energy Conversion & Management 1985; 25(3):263-71.

Spokas KA. Review of the stability of biochar in soil: predictability of O:C molar ratios. Carbon management 2010;1(2):289–303

Sreekumar A. Techno-economic analysis of a roof-integrated solar air heating system for drying fruit and vegetables. Energy Conversion and Management 2010; 51: 2230-2238.

Stair R, Ellis HT. The solar constant based on new spectral irradiance data from 3100 to 5300 Augstroms.J.Ap. Meteorol, 1968; 7: 635 644

Tabor H. Selective surfaces for solar collectors. Low temperature Engineering Application of Solar Energy. ASHRAE, New York, 1967, Chapter 4,41 52.

Thekackara MP. Data on incident solar energy. Supplement to the Proceedings of the 20th Annual Meeting of the Institute for Environmental Science, 1974, 21.

Thekaekara MP. Solar radiation measurement: techniques and instrumentation. Solar Energy, 1976; 18: 309 325.

Threlkeld JL, Jordan RC. Direct solar radiation available on clear days. Am. J. Heating Piping air Conditioning, 1957; 29: 135 145.

Threlkeld JL. Thermal environmental engineering (Vol. 11). Prentice Hall. 1970

Tingem M, Rivington M. Adaptation for crop agriculture to climate change in Cameroon: turning on the heat. Mitigation and Adaptation Strategies for Global Change 2009;14:153–68.

Van Dijk HJ, Goedhart PD, Windpumps for Irrigation, C.W.D., Amsterdam, 1990.

Van Wylen GJ, Sonntag RE, Borgnakke C. Fundamentals of Classical Thermodynamics, John Wiley and Sons; 1994.

Verma LR, Bucklin RA, Endan JB, Wratten FT. Effects of drying air parameters on rice drying models. T. ASAE 1985; 28: 296-301.

Watmuff JH, Charters WWS, Proctor D. Solar and wind induced external coefficients for solar collectors. Comples 1977; 2:56

Weinberger H. The physics of the solar pond', Solar Energy 1964; 8(2): 45.

Weiss A. Algorithms for the calculation of moist air properties on a hand calculator. Trans ASAE, 1977; 20: 1133-1136.

Whillier A. Design factors influencing solar collectors in Low temperature engineering applications of solar energy. ASHRAE, New York, 1967.

Whillier A. Solar energy collection and its utilization for house heating, ScD. Thesis, MIT, 1953.

Wiebelt JA, Henderson JB. Selected ordinates for total solar radiation property evaluation from spectral data. Trans. Am. Soc. Mech. Engg., J. Heat Transfer, 101, 1979.

Wolff AB. The Windmill as Prime Mover, 2nd edn. Wiley, New York, 1885.

Wood BD. Solar energy measuring instruments. Solar Energy Engineering, Academic Press, London, 1977.

Worrell E, Bernstein L, Roy J, Price L, Harnisch J. Industrial energy efficiency and climate change mitigation. Energy Efficiency 2009;2: 109–23.

Zakhidov RA. Central Asian countries energy system and role of renewable energy sources. Applied Solar Energy 2008;44:218–23.

Zhao Z, Houchati M, Beitelmal, A. An Energy Efficiency Assessment of the Thermal Comfort in an Office building. Energy Procedia, 2017;134: 885-893.

Kothari S, Panwar NL. Experimental Studies on Suitability of Greenhouse for Nursery Raising. Agricultural Engineering International: the CIGR Ejournal. Manuscript 2008; X: EE 08 003.

Kralova I, Sjöblom J. Biofuels-renewable energy sources: a review. Journal of Dispersion Science and Technology 2010; 31: 409–25.

Kreith F, Kreider JF. Principles of solar engineering. Hemisphere Publishing Corporation, Mc Graw Hill Book Company, 1978.

Lafforgue G, Magné B, Moreaux M. Energy substitutions, climate change and carbon sinks. Ecological Economics 2008;67:589–97.

Lampert CM. The use of coating for enhanced solar thermal energy collection', Lawrence Berkeley Laboratory, Report No. LBL. 8072, April, 1979.

Lausten J. Energy Efficiency Requirements in Building Codes, Energy Efficiency Policies for New Buildings. Paris, France: International Energy Agency, 2008

Lin BYH, Jordan RC. (1960). The interrelationship and characteristic distribution of direct, diffuse, and total solar energy. Solar Energy, 1960;4(3): 1-19.

Liu BYH, Jordan RC. Daily insolation on surfaces tilted towards the equator', ASRAE Journal, 1961; 3 (18): 53 59.

Majumdar NC, Mathur BL, Kaushik SB. Prediction of direct solar radiation for low atmospheric turbidity', Solar Energy, 1972; 13(4): 383 394.

Masek O, Brownsort P, Cross A, Sohi S. Influence of production conditions on the yield and environmental stability of biochar. Fuel 2013; 103:151–155

Mathur AN, Rathore NS, New and Renewable Energy Sources. Bohra Ganesh Publications, Udaipur 1996.

Mattox DM. Solar energy materials preparation techniques', J. Vac. Sci. Technol. 1975; 12(5): 1023 1031.

McGuigan D. Small Scale Wind power. Prism Press, Dorchester, 1978.

Mitchell D. Convective Heat transfer in Man and Other Animals, Heat loss from Animal and Man. Butterworth Publishing Inc. London 1974

Moran MJ. Availability Analysis, ASME Press, New York; 1989.

Muneer T, Hawas MM, Sahili K. Correlation between hourly diffuse and global radiation for New Delhi. Energy conversion and management,1984; 24(4): 265-267.

Muneer T, Hawas MM. Correlation between daily diffuse and global radiation for India. Energy conversion and management 1984; 24(2): 151-154.

National Aeronautics and Space Administration (NASA). Space Technology. 2019

Nielsen JBH, Seadi TA, Popiel PO. The future of anaerobic digestion and biogas utilization. Bioresource Technology 2009;100: 5478–84.

Orgill JF, Hollands KGT. Correlation equation for hourly diffuse radiation on a horizontal surface', Solar Energy, 1977; 19: 357 359.

Ozbek B, Dadali G. Thin-layer drying characteristics and modelling of mint leaves undergoing microwave treatment. J. Food Eng. 2007; 83: 541-549.

Panwar NL, Kaushik SC, Kothari S. Role of renewable energy sources in environmental protection: A review. Renewable and Sustainable Energy Reviews 2011; 15 : 1513–1524.

Panwar NL, Kaushik SC, Kothari S. State of the art of solar cooking: An overview. Renewable and Sustainable Energy Reviews 16 : 3776-3785

Panwar NL, Kothari R, Tyagi VV. Thermo chemical conversion of biomass - Eco friendly energy routes. Renewable and Sustainable Energy Reviews 2012; 16 : 1801-1816

Panwar NL, Rathore NS. Potential of surplus biomass gasifier based power generation: A case study of an Indian state Rajasthan. Mitigation and Adaptation Strategies for Global Change 2009; 14:711-720.

Panwar NL, Shrirame HY, Rathore NS, Sudhakar J, Kurchania AK. Performance evaluation of a diesel engine fueled with methyl ester of castor seed oil. Applied Thermal Engineering 2010; 30(2-3): 245-249.

Panwar NL. Energetic and Exergetic Analysis of Walk-in Type Solar Tunnel Dryer for Kasuri Methi (Fenugreek) Leaves Drying. Int. Journal of Exergy 2014; 4: 519-531.

Panwar NL. Thermal modeling, energy and exergy analysis of animal feed solar cooker. Journal of Renewable Sustainable Energy 5, 043105.

Petela R. Exergy analysis of the solar cylindrical-parabolic cooker. Solar energy 2005;79:221-33.

Petela R. Exergy of heat radiation. Journal of Heat Transfer 1964; 86: 187-192.

Petela R. Exergy of undiluted thermal radiation. Solar Energy 2003; 74: 469–488.

Peterson RE, Ramsay JW. Thin film coating in solar thermal power systems', J. Vac. Sci. Technol., 1975; 12: 471.

Peterson RE, Ramsay JW. Thin film coating in solar thermal power systems', J. Vac. Sci. Technol., 1975; 12: 471.

Rao KR, Seshadri TN. Solar insolation curves. Ind. J. Met. & Geop., 1961; 12(2): 267 272.

Rathore NS, Panwar NL, Kothari S. Biomass production and Utilization Technology Himanshu Publication, Udaipur (Rajasthan) 2007.

Rathore NS, Panwar NL. (2007). Renewable energy sources for sustainable development. New India Publishing Agency, New Delhi, India. 2007

Rathore NS, Panwar NL. Design and development of energy efficient solar tunnel dryer for industrial drying. Clean Techn Environ Policy 2011; 13: 125-132.

Rathore NS, Panwar NL. Experimental studies on hemi cylindrical walk-in type solar tunnel dryer for grape drying. Applied Energy 2010;87: 2764-2764.

Rathore NS, Panwar NL. Experimental studies on hemi cylindrical walk in type solar tunnel dryer for grape drying. Applied Energy 2010; 87: 2764-2767.

Ravindranath NH, Hall DO. Biomass, energy, and environment: a developing country perspective from India. Oxford, United Kingdom: Oxford University Press; 1995.

Robinson AL. Amorphous Silicon: A New Direction for Semiconductors. Science, 1977; 197(4306): 851-853.

Rosen MA, Dincer I. A study of industrial steam process heating through exergy analysis. International Journal of Energy Research 2004; 28: 917-930.

Rosen MA, Dincer I. On exergy and environmental impact. International Journal of Energy Research 1997; 21: 643-654.

Rosen MA, Hooper FC. Second law analysis of thermal energy storage systems. Proc., 1st Trabzon Int. Energy and Environment Symp., Trabzon, Turkey, 1996: 361–371.

Rosen MA. Exergy and Government Policy: Is There a Link, Exergy An International Journal 2002; 2: 224-226.

Rosen MA. Second law analysis: approaches and implications. International Journal of Energy Research 1999; 23: 415-429.

Ruth DW, Chant RE. The relationship of diffuse radiation to total radiation in Canada. Solar Energy 1976; 18: 153 154.

Sarasavadia PN, Sawhney RL, Pangavhane DR, Singh SP. Drying behaviour of brined onion slices. Journal of Food Engineering 1999; 40: 219–226.